Functional Nanomaterials for Optoelectronics and Photocatalysis

Functional Nanomaterials for Optoelectronics and Photocatalysis

Editors

Protima Rauwel
Erwan Rauwel

Basel • Beijing • Wuhan • Barcelona • Belgrade • Novi Sad • Cluj • Manchester

Editors
Protima Rauwel
Estonian University of Life Sciences
Tartu, Estonia

Erwan Rauwel
Estonian University of Life Sciences
Tartu, Estonia

Editorial Office
MDPI
St. Alban-Anlage 66
4052 Basel, Switzerland

This is a reprint of articles from the Special Issue published online in the open access journal *Nanomaterials* (ISSN 2079-4991) (available at: https://www.mdpi.com/journal/nanomaterials/special_issues/optoelect_photocataly).

For citation purposes, cite each article independently as indicated on the article page online and as indicated below:

Lastname, A.A.; Lastname, B.B. Article Title. *Journal Name* **Year**, *Volume Number*, Page Range.

ISBN 978-3-0365-9238-1 (Hbk)
ISBN 978-3-0365-9239-8 (PDF)
doi.org/10.3390/books978-3-0365-9239-8

Cover image courtesy of Patricio Paredes

© 2023 by the authors. Articles in this book are Open Access and distributed under the Creative Commons Attribution (CC BY) license. The book as a whole is distributed by MDPI under the terms and conditions of the Creative Commons Attribution-NonCommercial-NoDerivs (CC BY-NC-ND) license.

Contents

About the Editors vii

Protima Rauwel and Erwan Rauwel
Functional Nanomaterials for Optoelectronics and Photocatalysis
Reprinted from: *Nanomaterials* 2023, 13, 2694, doi:10.3390/nano13192694 1

Guy L. Kabongo, Gugu H. Mhlongo and Mokhotjwa S. Dhlamini
Unveiling Semiconductor Nanostructured Based Holmium-Doped ZnO: Structural, Luminescent and Room Temperature Ferromagnetic Properties
Reprinted from: *Nanomaterials* 2021, 11, 2611, doi:10.3390/nano11102611 3

Protima Rauwel, Augustinas Galeckas and Erwan Rauwel
Enhancing the UV Emission in ZnO–CNT Hybrid Nanostructures via the Surface Plasmon Resonance of Ag Nanoparticles
Reprinted from: *Nanomaterials* 2021, 11, 452, doi:10.3390/nano11020452 19

Shivangani Shivangani, Maged F. Alotaibi, Yas Al-Hadeethi, Pooja Lohia, Sachin Singh, D. K. Dwivedi, et al.
Numerical Study to Enhance the Sensitivity of a Surface Plasmon Resonance Sensor with BlueP/WS$_2$-Covered Al$_2$O$_3$-Nickel Nanofilms
Reprinted from: *Nanomaterials* 2022, 12, 2205, doi:10.3390/nano12132205 31

Keshav Nagpal, Erwan Rauwel, Elias Estephan, Maria Rosario Soares and Protima Rauwel
Significance of Hydroxyl Groups on the Optical Properties of ZnO Nanoparticles Combined with CNT and PEDOT:PSS
Reprinted from: *Nanomaterials* 2022, 12, 3546, doi:10.3390/nano12193546 47

Mariya Zvaigzne, Alexei Alexandrov, Anastasia Tkach, Dmitriy Lypenko, Igor Nabiev and Pavel Samokhvalov
Optimizing the PMMA Electron-Blocking Layer of Quantum Dot Light-Emitting Diodes
Reprinted from: *Nanomaterials* 2021, 11, 2014, doi:10.3390/nano11082014 61

Mingyue Hou, Zhaohua Zhou, Ao Xu, Kening Xiao, Jiakun Li, Donghuan Qin et al.
Synthesis of Group II-VI Semiconductor Nanocrystals via Phosphine Free Method and Their Application in Solution Processed Photovoltaic Devices
Reprinted from: *Nanomaterials* 2021, 11, 2071, doi:10.3390/nano11082071 71

Yuzhou Xia, Ruowen Liang, Min-Quan Yang, Shuying Zhu and Guiyang Yan
Construction of Chemically Bonded Interface of Organic/Inorganic g-C$_3$N$_4$/LDH Heterojunction for Z-Schematic Photocatalytic H$_2$ Generation
Reprinted from: *Nanomaterials* 2021, 11, 2762, doi:10.3390/nano11102762 85

Patricio Paredes, Erwan Rauwel and Protima Rauwel
Surveying the Synthesis, Optical Properties and Photocatalytic Activity of Cu$_3$N Nanomaterials
Reprinted from: *Nanomaterials* 2022, 12, 2218, doi:10.3390/nano12132218 97

Patricio Paredes, Erwan Rauwel, David S. Wragg, Laetitia Rapenne, Elias Estephan, Olga Volobujeva and Protima Rauwel
Sunlight-Driven Photocatalytic Degradation of Methylene Blue with Facile One-Step Synthesized Cu-Cu$_2$O-Cu$_3$N Nanoparticle Mixtures
Reprinted from: *Nanomaterials* 2023, 13, 1311, doi:10.3390/nano13081311 121

Yuri Hendrix, Erwan Rauwel, Keshav Nagpal, Ryma Haddad, Elias Estephan,
Cédric Boissière and Protima Rauwel
Revealing the Dependency of Dye Adsorption and Photocatalytic Activity of ZnO
Nanoparticles on Their Morphology and Defect States
Reprinted from: *Nanomaterials* **2023**, *13*, 1998, doi:10.3390/nano13131998 . . . . **139**

About the Editors

Protima Rauwel

Protima Rauwel is the Chair Professor of Energy Applications at the Institute of Technology of the Estonian University of Life Sciences. She obtained her PhD in 2005 from the University of Caen, France, in condensed matter and materials science with a special focus on nitride thin films with quantum dots and quantum wells. She has post-doctoral experience from the University of Aveiro, Portugal, and the University of Oslo, Norway, where she continued her research on thin films and freestanding nanomaterials. Her research focus is currently on the synthesis and characterization of nanomaterials for water remediation, optoelectronic, electrochemical, and anti-microbial applications. She has co-authored more than 90 publications, 3 books, and 5 book chapters. She is a co-inventor of two patents and has a Scopus H-index of 30.

Erwan Rauwel

Erwan Rauwel received his PhD in Materials Science from the University of Caen in 2003. He then began his postdoctoral studies in collaboration with STMicroelectronics at Minatec, Grenoble, France, and at the University of Aveiro (Marie Curie IE Fellowship), Portugal. He worked as a senior researcher at the University of Aveiro, in collaboration with Statoil, and then joined Taltech, Estonia, as a professor for five years. He is currently a professor at the Estonian University of Life Science, Estonia, where his group investigates the properties of nanoparticles best used for targeted applications that include water purification (heavy metal ions extraction, organic pollutant removal), the development of biocidal coating to fight AMR and COVID-19, and the development of nanocomposites for photocurrent generation and energy harvesting (with Minatec, Grenoble). His group is also investigating new possible treatments for cancer (with the LBN, Montpellier,). He has more than 83 peer-reviewed publications with an H-index of 25, 5 book chapters, and 6 patents. He is the chief scientist of his start-up company, which specializes in nanomaterials (PRO-1 NANOSolutions).

Editorial

Functional Nanomaterials for Optoelectronics and Photocatalysis

Protima Rauwel [1,*] and Erwan Rauwel [1,2]

[1] Institute of Forestry and Engineering Sciences, Estonian University of Life Sciences, 51006 Tartu, Estonia; erwan.rauwel@emu.ee
[2] Institute of Veterinary Medicine and Animal Sciences, Estonian University of Life Sciences, 51006 Tartu, Estonia
* Correspondence: protima.rauwel@emu.ee

The present energy crisis has encouraged the use of energy-efficient devices and green energy sources. In addition to their energy-efficient operation, it is now essential that the production of these devices is cost-effective. Devices requiring energy-efficient operation and production include light emitting diodes (LEDs) applied to general lighting systems or for specific applications in electronic devices. With regard to the production of energy, cost-effective and new materials are being intensively investigated. The new generation of devices consists of hybrid materials and nanomaterials, involving polymers coupled with inorganic counterparts. The advantages of these hybrid materials include lower production costs, an overall weight reduction in the device and easier recyclability. In this regard, functional nanomaterials appear to be the most suitable choice of materials for these applications. In electronic devices, they allow miniaturization, while in energy-harvesting applications, i.e., photovoltaics and photocatalysis, they allow for more efficient energy conversion owing to the higher surface-to-volume ratio. Since the active sites for energy conversion in these nanomaterials are localized on the surface, the volume of the device is therefore reduced. Hence, the energy produced per unit mass is higher, as a lower amount of material is required in the device. Along with cost-effective production techniques, the overall device costs are therefore lowered.

The present Special Issue focuses on functional nanomaterials applied to optoelectronics and photocatalysis with several common nanomaterials to both fields, in particular ZnO. The compilation is clearly divided into three categories: (i) optoelectronics, (ii) photovoltaics and (iii) photocatalysis. In the optoelectronics section, the publication of Kabongo et al. describes the synthesis of ZnO doped with Ho, exhibiting ferromagnetic properties under microwave excitation [1]. The second publication, by Rauwel et al., reports on the combination of ZnO nanoparticles with CNT and Ag nanoparticles, and emphasizes the plasmonic effect of Ag nanoparticles in the enhancement of UV emission owing to the Burstein–Moss effect [2]. The plasmonic effect of metal nanoparticles has also been theoretically studied by Shivangi et al. in the enhancement of an SPR-based sensor device of BlueP/WS_2-covered Al_2O_3-nickel nanofilms [3]. Another article, by Nagpal et al., describes the enhancement and suppression of the visible light emission of ZnO nanostructures with the addition of carbon nanotubes (CNTs) and poly(3,4-ethylenedioxythiophene) polystyrene sulfonate (PEDOT:PSS), respectively [4]. Polymers, such as Poly(methyl methacrylate) (PMMA), are also used as electron-blocking layers in a II–VI semiconductor QLED, as described by Zvaigzne et al. [5]. Similar II-VI semiconductor materials were also grown by Hou et al. via a phosphine-free method, and their photovoltaic properties were evaluated, which marks the second topic of this Special Issue [6]. The third topic, i.e., photocatalysis, is composed of four publications. This topic can be divided into two subgroups: (i) H_2 production and (ii) dye degradation. For H_2 production, Xia et al. report on a heterostructure of an organic/inorganic interface of g-C_3N_4/LDH that can be activated under visible light

Citation: Rauwel, P.; Rauwel, E. Functional Nanomaterials for Optoelectronics and Photocatalysis. *Nanomaterials* **2023**, *13*, 2694. https://doi.org/10.3390/nano13192694

Received: 22 September 2023
Accepted: 26 September 2023
Published: 3 October 2023

Copyright: © 2023 by the authors. Licensee MDPI, Basel, Switzerland. This article is an open access article distributed under the terms and conditions of the Creative Commons Attribution (CC BY) license (https://creativecommons.org/licenses/by/4.0/).

radiation [7]. This Special Issue contains three publications on the study of dye degradation using nanomaterials: The first publication is a review article by Paredes et al. that surveys the Cu$_3$N nanomaterials used to date [8]. The second publication by Paredes et al. describes the one-step synthesis of nanoparticle mixtures of Cu-Cu$_3$N-Cu$_2$O and their potential in the sunlight-driven photocatalytic degradation of azo dyes [9]. The last publication, by Hendrix et al., reports on the degradation of azo dyes using ZnO nanomaterials, and investigates for the first time, the influence of their morphology and defect states under both UV and sunlight [10].

We wish you a pleasant read and hope that this Special Issue on "Functional Nanomaterials for Optoelectronics and Photocatalysis" will serve as a valuable resource for researchers and PhD students in the field.

Acknowledgments: The Guest Editors would like to thank the Editor-in-Chief and the Editorial Assistants for their contribution in making the guest editing process smooth and efficient. We also acknowledge the authors for submitting their valuable work to this Special Issue, as without it, the successful completion of this Special Issue would not have been possible. Finally, a special thank you goes to all of the reviewers who participated in the peer review process of the submitted manuscripts.

Conflicts of Interest: The authors declare no conflict of interest.

References

1. Kabongo, G.L.; Mhlongo, G.H.; Dhlamini, M.S. Unveiling Semiconductor Nanostructured Based Holmium-Doped ZnO: Structural Luminescent and Room Temperature Ferromagnetic Properties. *Nanomaterials* **2021**, *11*, 2611. [CrossRef] [PubMed]
2. Rauwel, P.; Galeckas, A.; Rauwel, E. Enhancing the UV Emission in ZnO–CNT Hybrid Nanostructures via the Surface Plasmon Resonance of Ag Nanoparticles. *Nanomaterials* **2021**, *11*, 452. [CrossRef] [PubMed]
3. Shivangani; Alotaibi, M.F.; Al-Hadeethi, Y.; Lohia, P.; Singh, S.; Dwivedi, D.K.; Umar, A.; Alzayed, H.M.; Algadi, H.; Baskoutas, S. Numerical Study to Enhance the Sensitivity of a Surface Plasmon Resonance Sensor with BlueP/WS2-Covered Al2O3-Nickel Nanofilms. *Nanomaterials* **2022**, *12*, 2205. [CrossRef] [PubMed]
4. Nagpal, K.; Rauwel, E.; Estephan, E.; Soares, M.R.; Rauwel, P. Significance of Hydroxyl Groups on the Optical Properties of ZnO Nanoparticles Combined with CNT and PEDOT:PSS. *Nanomaterials* **2022**, *12*, 3546. [CrossRef] [PubMed]
5. Zvaigzne, M.; Alexandrov, A.; Tkach, A.; Lypenko, D.; Nabiev, I.; Samokhvalov, P. Optimizing the PMMA Electron-Blocking Layer of Quantum Dot Light-Emitting Diodes. *Nanomaterials* **2021**, *11*, 2014. [CrossRef] [PubMed]
6. Hou, M.; Zhou, Z.; Xu, A.; Xiao, K.; Li, J.; Qin, D.; Xu, W.; Hou, L. Synthesis of Group II-VI Semiconductor Nanocrystals via Phosphine Free Method and Their Application in Solution Processed Photovoltaic Devices. *Nanomaterials* **2021**, *11*, 2071. [CrossRef] [PubMed]
7. Xia, Y.; Liang, R.; Yang, M.-Q.; Zhu, S.; Yan, G. Construction of Chemically Bonded Interface of Organic/Inorganic g-C3N4/LDH Heterojunction for Z-Schematic Photocatalytic H2 Generation. *Nanomaterials* **2021**, *11*, 2762. [CrossRef] [PubMed]
8. Paredes, P.; Rauwel, E.; Rauwel, P. Surveying the Synthesis, Optical Properties and Photocatalytic Activity of Cu3N Nanomaterials. *Nanomaterials* **2022**, *12*, 2218. [CrossRef] [PubMed]
9. Paredes, P.; Rauwel, E.; Wragg, D.S.; Rapenne, L.; Estephan, E.; Volobujeva, O.; Rauwel, P. Sunlight-Driven Photocatalytic Degradation of Methylene Blue with Facile One-Step Synthesized Cu-Cu2O-Cu3N Nanoparticle Mixtures. *Nanomaterials* **2023**, *13*, 1311. [CrossRef] [PubMed]
10. Hendrix, Y.; Rauwel, E.; Nagpal, K.; Haddad, R.; Estephan, E.; Boissière, C.; Rauwel, P. Revealing the Dependency of Dye Adsorption and Photocatalytic Activity of ZnO Nanoparticles on Their Morphology and Defect States. *Nanomaterials* **2023**, *13*, 1998. [CrossRef] [PubMed]

Disclaimer/Publisher's Note: The statements, opinions and data contained in all publications are solely those of the individual author(s) and contributor(s) and not of MDPI and/or the editor(s). MDPI and/or the editor(s) disclaim responsibility for any injury to people or property resulting from any ideas, methods, instructions or products referred to in the content.

Article

Unveiling Semiconductor Nanostructured Based Holmium-Doped ZnO: Structural, Luminescent and Room Temperature Ferromagnetic Properties

Guy L. Kabongo [1,*], Gugu H. Mhlongo [2] and Mokhotjwa S. Dhlamini [1]

1. Department of Physics, University of South Africa, P.O. Box 392, Pretoria 0003, South Africa; dhlamms@unisa.ac.za
2. CSIR-National Centre for Nano-Structured Materials, P.O. Box 395, Pretoria 0001, South Africa; gmhlongo@csir.co.za
* Correspondence: geekale@gmail.com

Abstract: This research work describes the synthesis of ZnO nanostructures doped with Ho^{3+} ions using a conventional sol–gel synthesis method. The nanostructured produced exhibited a wurtzite hexagonal structure in both ZnO and ZnO:Ho^{3+} (0.25, 0.5, 0.75 mol%) samples. The change in morphology with addition of Ho^{3+} dopants was observed, which was assigned to Ostwald ripening effect occurring during the nanoparticles' growth. The photoluminescence emission properties of the doped samples revealed that Ho^{3+} was emitting through its electronic transitions. Moreover, reduced surface defects were observed in the Holmium doped samples whose analysis was undertaken using an X-ray Photoelectron Spectroscopy (XPS) technique. Finally, enhanced room temperature ferromagnetism (RT-FM) for Ho^{3+}-doped ZnO (0.5 mol%) samples with a peak-to-peak line width of 452 G was detected and found to be highly correlated to the UV–VIS transmittance results.

Keywords: ZnO; holmium; nanostructures; ferromagnetism; XPS

1. Introduction

The surge of interest observed on ZnO nanostructures research in the past three decades among the scientific community was due to the versatile optical, magnetic and surface properties that they display under room temperature conditions [1]. The ZnO wide direct tunable band-gap, large exciton binding energy at room temperature makes it to be a convenient material for device fabrication in a wide range of applications such as optoelectronic, photonics and dilute magnetic semiconductors (DMS) [2–5]. It has potentialities to be applied in other devices such as gas sensors [6], biolabels [7], solar cells [8] and piezoelectric nanogenerators [9]. Recently, several research groups have investigated the optical properties of ZnO nanostructures exhibiting different morphologies such as flower-like [10], nanorods [11], comb-like [12] and nanowire arrays [13]. The need to engineer ZnO defects has driven scientists to dope this semiconductor material with either transition metal ions or light emitting elements such as the promising rare earth ions which in most cases resulted in tailoring the band-gap of ZnO [14]. Despite its large magnetic moment, Ho^{3+} has attracted less interest within the scientific community. Recently, Popa et al. [15] investigated the impact of structural on optical properties of sol-gel derived holmium doped ZnO thin films using relatively higher doping concentrations (1, 3, 5 at%), and others have demonstrated elsewhere that holmium concentration variation is very effective in tuning ZnO properties [16]. Earlier, Khataee and co-workers [17] reported on holmium doped zinc oxide nanoparticles synthesized via sonochemistry for the evaluation of the Reactive Orange 29 degradation in catalysis. Moreover, Singh and colleagues have successfully measured the DC magnetization and resistivity in Ho doped ZnO nanoparticles produced via wet chemical synthesis [18]. Furthermore, in their investigation dedicated to unveiling the optical and dielectric properties of Ho-doped ZnO, a research group have successfully established the correlation of the abovementioned properties to ZnO related defects [19]. Interestingly, sol–gel remains to date one of the most efficient synthesis procedure to produce ultrasmallZnO nanoparticles [20,21]. In the current study, the observed change in morphology in Ho^{3+}-doped ZnO nanocrystals is reported and it was assigned to Ostwald ripening effect which occurred during the growth process of ZnO nanostructures. Moreover, a comparison of the defect state XPS core levels in un-doped and Ho^{3+}-doped ZnO (0.5 mol%) samples is presented and discussed in detail. Furthermore, in spite of the low concentration of Ho^{3+} dopant, XPS Ho 4d core levels were detected. Finally, the current research work is a novel experimental study on room temperature ferromagnetism based on microwave absorption of Ho^{3+} ions doped in ZnO nanostructures owing to the large magnetic moment of Ho^{3+}. The ferromagnetism enhancement observed which is related to the number of spins participating to the ferromagnetic resonance is found to be in accordance with the UV–VIS transmittance and band-gap results (Scheme 1).

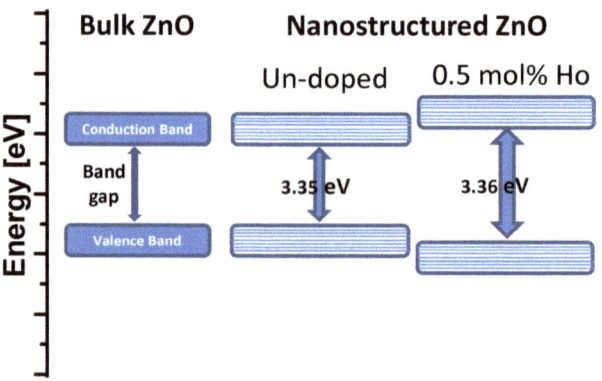

Scheme 1. ZnO Band-gap engineering.

2. Experimental Section

The samples used in the present study were all synthesized in the same conditions except the molar concentration of Ho dopant which varied. The synthesis was conducted using zinc acetate ($Zn(CH_3COO)_2 \cdot 2H_2O$), sodium hydroxide (NaOH) and holmium nitrate pentahydrate ($Ho(NO_3)_3 \cdot 5H_2O$), which were all purchased from Sigma-Aldrich (Kempton Park, Gauteng, South Africa) and used as received. ZnO:Ho^{3+} nanocrystals were synthesized by a sol–gel method following the same procedure as reported in our previous works [22–24]. The solution of sodium hydroxide dissolved in ethanol was prepared separately, then cooled in ice water and added dropwise judiciously to the ethanol solution of Zn^{2+} ions. For preparation of Ho^{3+}-doped ZnO samples with different concentrations of Ho^{3+} (0.25, 0.5 and 0.75 mol%), the ethanol solution of Holmium nitrate pentahydrate was added into the hydrolyzed Zn^{2+} solution prepared following the above route. The obtained clear solution was kept at room temperature for 24 h and then washed several times in a mixture of ethanol and heptane (1:2 molar ratio) to eliminate unreacted Na^+ and CH_3COO^- ions. The resulting precipitates were then re-dispersed in ethanol and dried at 200 °C for 2 h in an electric oven (ambient atmosphere).

The structure of the obtained samples was characterized on an X'Pert PRO PANalytical diffractometer with CuKα at λ = 0.15405 nm. Transmission electron microscopy (TEM) images were taken by using a JEOL-Jem2100 microscope. Morphology and chemical composition of the samples (Cu-grid deeped in ethanolic solution of the sample) were analyzed using a JEOL JSM-7500F field-emission scanning electron microscope (FE-SEM JEOL Ltd.,Tokyo, Japan) equipped with energy dispersive X-ray spectrometer (EDX) (an Oxford Instruments, High Wycombe, UK). Photoluminescence (PL) and Time Resolved PL properties for the un-doped and doped samples were analyzed using a JobinYvonFluorolog

3 spectrofluorometer equipped with a Xenon lamp and nanoLED for excitation at room temperature. The transmittance measurements were carried out using a Perkin-Elmer Lambda 1050 UV/Vis/NIR spectrophotometer. The X-ray photoelectron spectroscopy (XPS) core levels were carried out using a PHI 5000 Versaprobe-Scanning ESCA Microprobe (ULVAC-PHI, Inc. Kanagawa, Japan). Finally, the room temperature magnetization measurements were examined through the microwave absorption measurements collected using a JEOL X-band electron spin resonance (ESR; JEOL Ltd.,Tokyo, Japan) spectrometer (JES FA 200; JEOL Ltd.,Tokyo, Japan) operating at 9.4 GHz equipped with an Oxford ESR900 gas-flow cryostat and a temperature controller (Scientific instruments 9700; Oxford Instruments plc, Abingdon, UK). In order to evade saturation during measurements, the microwave power was maintained at 5 mW. The DC static field HDC was slowly swept between 0 and 8360 Gauss. The DC field was modulated with an AC field whose amplitude was kept constant at 100 kHz frequency. The microwave absorption output was measured as a derivative signal.

3. Results and Discussion

3.1. EDS, SEM, TEM, XRD Analysis

Considering that during the course of experiment the photoluminescence optimization which revealed that optimum luminescence was obtained from 0.5 mol% Ho^{3+}-doped ZnO, particular attention was devoted to elucidating the microstructural behavior of 0.5 mol% Ho^{3+}-doped ZnO sample relative to the un-doped ZnO. The elemental composition of the sample in Figure 1a was surveyed by EDS to demonstrate qualitatively the presence of holmium dopant in the investigated area of the sample in addition to zinc and oxygen. However, the observed copper peak originates from SEM sample preparation (i.e., Cu-grid) [14]. It is worth mentioning that the expected chemical species were all found to be well distributed throughout the analyzed area (see Figure S1) of the as-synthesized sample as shown in Figure 1b. Figure 1c,d shows classical SEM micrographs for the un-doped and 0.5 mol% Ho^{3+}-doped samples, respectively. The images show that spherical-like ZnO particles in the case of un-doped sample and rod-like particles in the case of 0.5 mol% Ho^{3+}-doped samples were formed during the wet chemical synthesis route. Interestingly, the observed mutation of particle morphology from un-doped to 0.5 mol% Ho^{3+}-doped ZnO samples is attributed to the Ostwald ripening effect [25,26].

TEM measurements were undertaken carefully on both un-doped and 0.5 mol% Ho^{3+}-doped ZnO samples (see Figure 2). The measurements revealed that the nanostructures were evenly distributed and were highly crystalline. However, a peculiar phenomenon based on the mutation of morphology was observed and was attributed to doping with holmium ions, more precisely due to the growth of nanoparticles in solution through a diffusion limited Ostwald ripening process known to be the most predominantly growth mechanism so far [27]. However, further investigations are required in order to effectively elucidate on the observed mechanism of morphology mutation. It is however suggested that studies on colloidal nanocrystals growth under TEM using the so-called liquid cell electron microscopy [28,29] should be performed. Such study has been undertaken previously but not quite intensively, especially on the ZnO nanoparticles growth [30]. TEM analysis revealed that the particles diameter and width were ~10 nm for un-doped and 0.5 mol% Ho^{3+}-doped ZnO, respectively, the former contained a mixture of spherical like and rod-like particles. The TEM result is found to be consistent with former SEM particles morphology characterization.

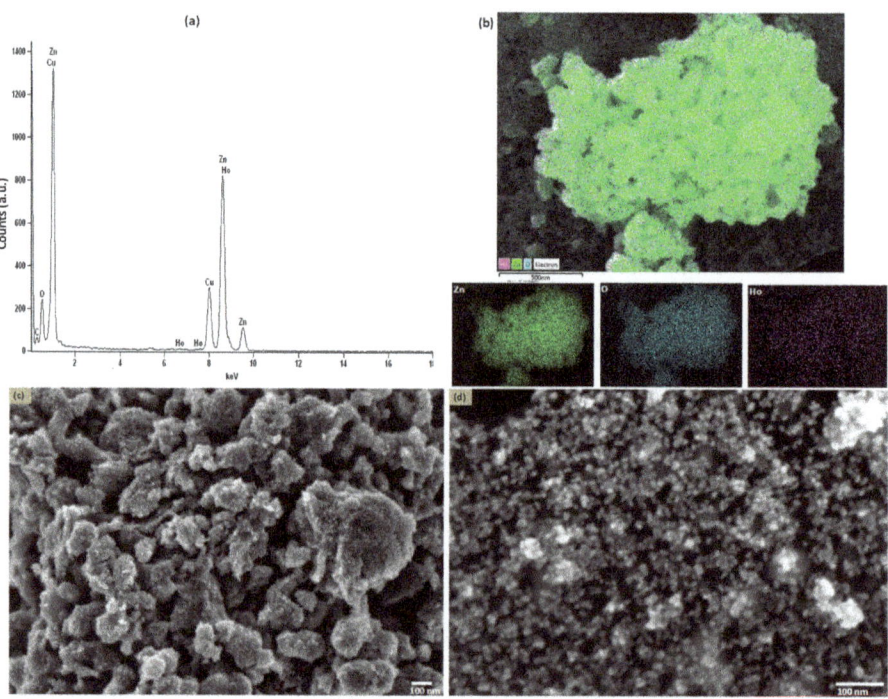

Figure 1. EDS (**a**) spectrum and (**b**) elemental mapping for 0.5 mol% Ho^{3+}-doped ZnO nanocrystals. SEM micrographs for (**c**) un-doped ZnO and (**d**) (0.5 mol%) Ho^{3+}-doped ZnO nanocrystals.

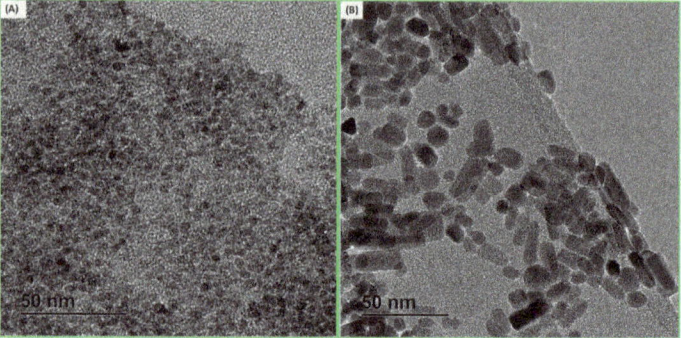

Figure 2. TEM images for (**A**) un-doped and (**B**) (0.5 mol%) Ho^{3+}-doped ZnO nanocrystals.

Figure 3 illustrates the X-ray diffraction (XRD) profiles of the ZnO and ZnO:Ho^{3+} nanocrystals. The analysis revealed that the as-synthesized nanocrystals ranging between 4–8 nm in diameter (See Table 1) were highly crystalline and exhibited the hexagonal wurtzite structure (space group $P6_3mc$) indexed to JCPDS card # 36-1451. Furthermore, no second phase originating from Ho$_2$O$_3$ was observed in the Ho^{3+}-doped ZnO samples [31] which confirms that the dopants successfully substituted the Zn^{2+} ions within the ZnO lattice structure. The Scherrer equation was employed to estimate the crystallite size [32]:

$$D = \frac{k\lambda}{\beta \cos O} \quad (1)$$

where D, λ, β, Θ and k are the crystallite size, the wavelength of the incident X-ray CuK radiation (0.15405 nm), the full width at half maximum (FWHM), the diffracting angle and a numerical constant (0.89), respectively. The results obtained are consistent with the TEM results, where the particles were in the nanometer range (see Figure S2).

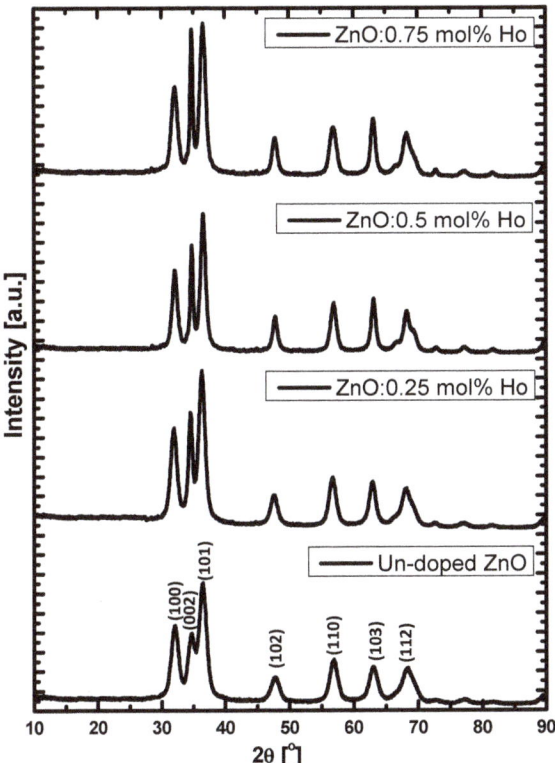

Figure 3. XRD patterns for undopedZnO and Ho^{3+}-doped ZnO nanocrystals dried at 200 °C.

Table 1. The XRD parameters and the average estimated crystallite size (D) from Scherrer equation.

Sample	2θ(101) [Degrees]	Δ(2θ) [Rad] FWHM = β	D [nm]
Un-doped ZnO	36.52	0.02913	5.0
0.25 mol% Ho	36.34	0.02234	6.5
0.5 mol% Ho	36.55	0.01768	8.2
0.75 mol% Ho	36.55	0.02042	7.1

3.2. UV–VIS Transmittance and Photoluminescence for Undoped and Ho^{3+}-Doped ZnO Nanostructures

The transmittance spectra in the range of 300–700 nm are depicted on Figure 4a. A sharp UV cut off band at approximately 360 nm was observed and assigned to the band-to-band transition of ZnO. The bandedge peak is found to be highly blue shifted as compared to the bulk ZnO (~386 nm) [33] owing to quantum confinement effect [34]. The un-doped ZnO sample exhibited a transmittance of the order of 40% far below the 0.5 mol% Ho^{3+} tramittance which is in the order of 60%. This implies that the incorporation of Ho^{3+} dopant state within the band-gap of the ZnO matrix has enhanced the transparency of the material. However, the transmittance was dropped after incorporation of 0.75 mol% Ho^{3+} defects owing to the segregation of the dopant. The optical band-gap (Figure 4b) was extrapolated using the Tauc's formula [35]:

$$\alpha(x)h\nu = A(h\nu - E_g)^{\frac{1}{2}} \quad (2)$$

where A is the constant, $h\nu$ is the photon energy, E_g the optical band-gap, and $\alpha(x)$ is the absorption coefficient of ZnO nanoparticle. The absorption coefficient can be calculated from the Beer–Lambert law:

$$\alpha(x) = -\frac{1}{d}\ln(T) \quad (3)$$

where T is the normalized transmittance. The band-gaps were obtained from the Tauc plot depicted in Figure 4b. The extrapolation revealed optical band-gaps of ~3.35, 3.34, 3.36 and 3.33 eV corresponding to un-doped ZnO, 0.25, 0.5 and 0.75 mol% Ho^{3+}-doped ZnO, respectively. These obtained band-gaps values are in good agreement with previous reports [36].

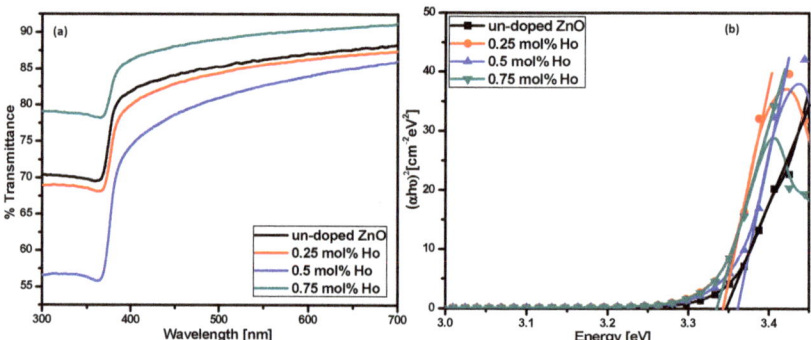

Figure 4. (a) UV–VIS transmission spectra for un-doped and Ho^{3+}-doped ZnO nanostructures (b) optical band–gap extrapolation from Tauc plot.

Based on the previous characterization, the emission spectrum of the as-synthesized 0.5 mol% Ho^{3+}-doped ZnOnano-phosphor sample was first examined by exciting with 325 nm electromagnetic radiations using a xenon lamp. It was found that Ho^{3+} (0.5 mol%) doped sample exhibited the optimum emission intensity. Figure 5a compares the PL spectra for both un-doped and 0.5 mol% Ho^{3+}-doped ZnO samples using 325 nm (3.82 eV) UV light radiations for excitation. The most common reported ZnO defects emission trend was observed, and the exciton emission owing to the recombination of electron-hole pairs was also detected at 381 nm (3.25 eV) for the un-doped ZnO sample, while 379 nm (3.27 eV) was observed in the case of 0.5 mol% Ho^{3+}-doped ZnO [37]. It is worth noting that after doping with Ho^{3+} ions, the free-exciton emission in the ZnO matrix undergoes two different kinds of modifications. The major alteration observed in the free-exciton emission was the slight blue shift (2 nm) observed to be derived from Ho^{3+} doping in the ZnO matrix as compared to un-doped ZnO, and this shift implies a change in the band structure [38–40] (see Figure 5b). Elilarassi et al. [41] elaborated on such phenomenon, which was assigned to a decrease of the transition probability taking place within the band-gap of the doped ZnO matrix among the oxygen vacancy and the Zn vacancy after the substitution of Zn^{2+} ions by the dopant in the host lattice [39,40]. The second modification observed was the severe increase in intensity of the free-exciton emission observed with holmium doping, which was attributed to energy transfer from the Ho^{3+} ions to ZnO host and/or to the reduction in concentration of oxygen defects as reported in our previous study [14]. The observed enhancements of the exciton emission confirmed the ability of spontaneous lasing action in the ultraviolet spectral range, thus, positioning 0.5 mol% Ho^{3+}-doped ZnO nanostructure as a candidate material for spintronic applications (see Figure 5a) [42,43].

Considering the current extensive controversies about the origin of the visible emission in ZnO, one can speculate on the defects that caused this emission [44]. A change in surface defect occurring on the surface of ZnO owing to Ho^{3+} doping could be the main factor causing the blue shift observed in the visible emission. Furthermore, the doping with holmium has occasioned band bending at the surface of ZnO resulting from the change of the Fermi energy level [45–47]. It is however important to undertake further investigation to elucidate this hypothesis.

Moreover, the ZnO:Ho^{3+} (0.5 mol%) emission and excitation spectra are depicted in Figure 5c,d. Five prominent excitation peaks at 383, 397, 439, 467 and 492 nm were observed from the spectrum. From these excitation wavelengths, we have selected the 492 nm wavelength to measure the emission spectrum of 0.5 mol% Ho^{3+}-doped ZnO. A number of emission bands are detected spreading from the visible to the near infrared (NIR) regions at 540 ($^5S_2,^5F_4 \rightarrow ^5I_8$), 606 ($^5F_5 \rightarrow ^5I_8$), 670 ($^5F_5 \rightarrow ^4I_8$), 757 ($^5S_2,^5F_4 \rightarrow ^5I_7$) and 808 ($^5I_5 \rightarrow ^5I_8$) nm (Figure 5d) [48–52]. The most prominent emission is observed in the green region centered at 540 nm with a less intense peak at about 574 nm; both peaks are due to the hypersensitive ($^5S_2,^5F_4 \rightarrow ^5I_8$) transition.

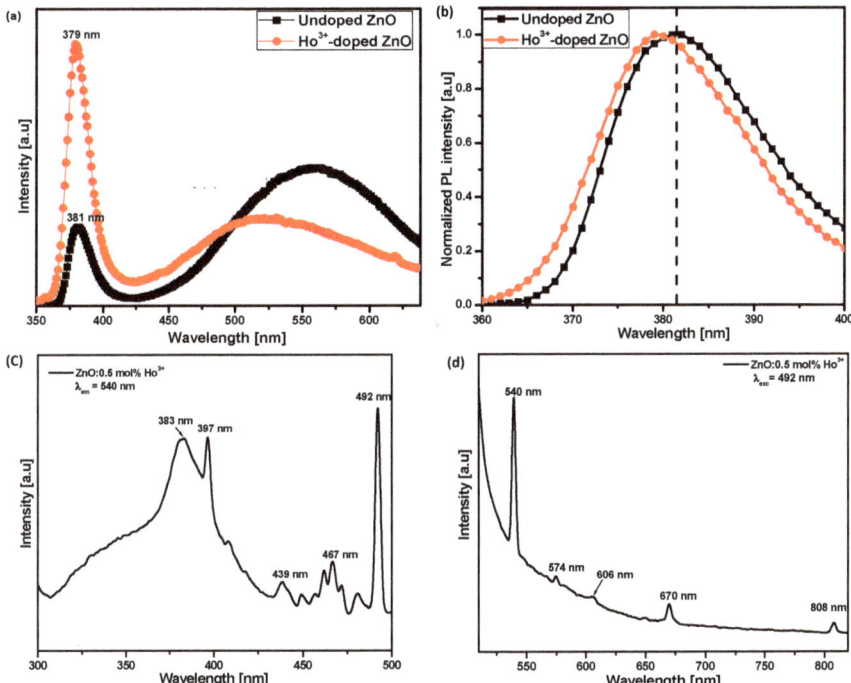

Figure 5. (**a**) Emission spectra for as prepared un-doped and 0.5 mol% Ho^{3+}-doped ZnO samples (λ_{exc} = 325 nm), (**b**) shows the magnification of the normalized exciton emission peak. (**c**) Excitation and (**d**) emission spectra for 0.5 mol% Ho^{3+}-doped ZnO nanocrystals.

3.3. Surface State XPS Characterization

Figure 6 shows the XPS spectra for both un-doped and 0.5 mol% Ho-doped samples, which allow the elucidation of the oxidation state of different species present in the as-prepared nanostructures, namely, Zinc, Oxygen and Holmium. The survey scans (not presented here) for both samples were identical at the sole exclusion of Ho4d core level present in the Ho^{3+}-doped ZnO samples. Besides Zn 2p, O 1s, and Ho 4d core levels, a C 1s core level arising from extrinsic surface impurities was also detected. The binding energy correction was applied using the C 1s (284.8 eV) core level. For both samples, the Zn 2p core level which appears as a doublet due to the interaction of the spin and orbital magnetic moments did not encounter any peak shift (not shown here). This observation implies that zinc is not sparingly sensitive to a change in the oxidation state [53]. The de-convoluted O 1s core level for (0.5 mol%) Ho^{3+}-doped sample (see Figure 6) exhibited three oxygen components. The peaks appeared at 529.21 ± 0.05, 530.32 ± 0.05 and 531.27 ± 0.05 eV ascribed to O^{2-} ions on the hexagonal wurtzite structure of ZnO; O^{2-} ions in oxygen deficient regions within the ZnO matrix and chemisorbed species on the surface of ZnO respectively, were observed [10]. As compared to the un-doped sample, the intensity of the O$_2$ peak was found to decrease owing to a decrease in concentration of oxygen defects in the 0.5 mol% Ho^{3+}-doped ZnO sample. This finding has a direct implication in the luminescence quenching of the green emission observed in the PL spectra, which was also demonstrated in our previous report [14] and similarly by Kumar et al. [54].However, the surface cleaning with Ar$^+$ sputtering has an influence on the intensity of O$_3$ peak which is slightly decreased as compared to the peak before Ar$^+$ cleaning. Furthermore, it is important to focus on the detection of Ho 4d core level peak at about 160.09 ± 0.05 eV binding energy (Figure 6e) [55]. The Ho 4d core level was de-convoluted and displayed two bands at about 161.04 ± 0.05 eV and 163.93 ± 0.05 eV before Ar$^+$ sputtering. It is worth noting that the small shoulder appearing at about 175 eV could be assigned to multiplets splitting.

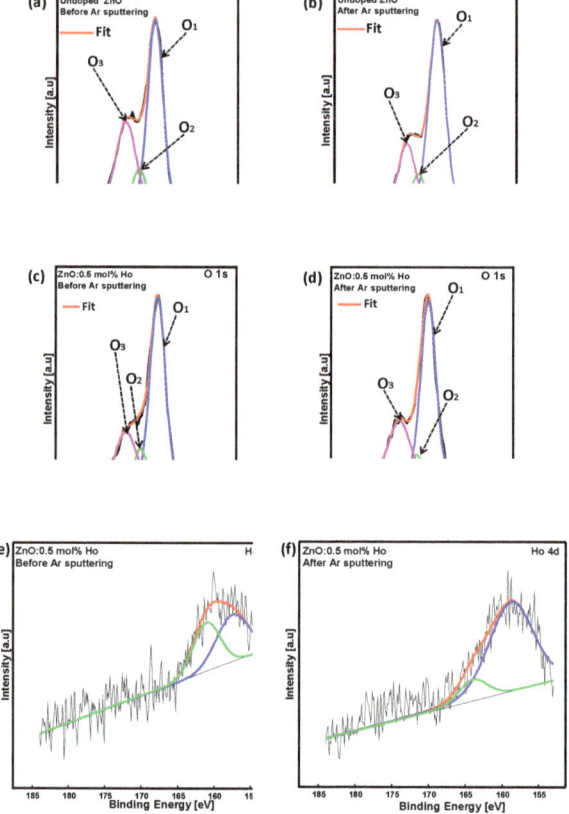

Figure 6. XPS O 1s core levels spectra for un-doped ZnO (**a**) before and (**b**) after Ar$^+$ sputtering and 0.5 mol% Ho^{3+}-doped ZnO (**c**) before and (**d**) after Ar$^+$ sputtering. XPS Ho 4d core levels (**e**) before and (**f**) after Ar$^+$ sputtering.

3.4. Room Temperature Ferromagnetic (RT FM) Properties

Being one of the most reliable spectroscopic techniques to probe defects at the surface and interface of materials, the ESR technique was used in the current study. ZnO is naturally diamagnetic even though there is still a controversial debate on the matter among scientists [56,57]. However, most scientific reports admitted the probability to observe two types of paramagnetic behaviour in ZnO. The low-field signal often assigned to unpaired electron trapped at an oxygen vacancy site (g = 2.0023) and the high-field signal owing to shallow donor centers (g = 1.96) [58]. The spectroscopic g-factor of the free electron can be calculated from the following equation:

$$g = \frac{h\nu}{\beta B} \quad (4)$$

where ν is the microwave frequency, h is Planck's constant, β is the Bohr magneton, and B is the magnetic field.

The investigation of the magnetic behavior of Ho^{3+}-doped ZnO nanostructures was successfully undertaken in this study. Due to the high sensitivity of the technique to defects, we conducted measurements on various concentrations in order to completely understand the effect of the doping concentration on the number of spins. It is a well-known fact that a number of non-magnetic materials have been observed to exhibit ferromagnetic properties at the nanoscale [10,59–63]. Scientists worldwide have found interest in investigating the RT-FM in un-doped and Ho^{3+}-doped ZnO nanostructures [64,65]. It is worth mentioning that the 4f rare earth elements exhibit exceptional magnetic properties as compared to the 4d transition metal (TM) elements [66,67]. However, few studies reported the RT-FM of Ho^{3+} into ZnO[68–71]. At first sight, the ESR results obtained in the current investigation seem not to correlate well with the PL due to the fact that the Ho^{3+} doping into ZnO have induced a change in the nature of defects within the matrix. A similar trend was previously reported by Garcia et al. [72] who investigated the d^0 ferromagnetism in ZnO capped with organic molecules. He hypothesized that the organic species present in ZnO samples could induce ferromagnetic-like behaviour (see Figure 7).

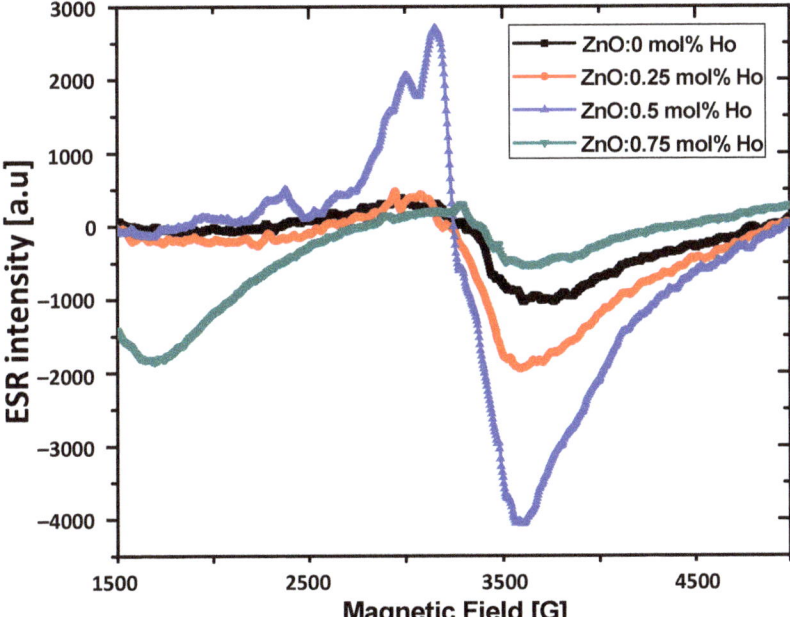

Figure 7. X-band (9.4 GHz) ESR spectra for un-doped and Ho^{3+}-doped ZnO samples at room temperature.

In spite of an abundant literature on the study of the RT-FM mechanism in ZnO nanostructures, the debate on its origin is still open and promising. Xiaoyong Xu et al. [73] have reported on the size dependence RT-FM in ZnO quantum dots; in their study they established a correlation between the RT-PL and the RT-FM in un-doped ZnO nanoparticles denoting that the d^0 ferromagnetism is largely related to the concentration of native defect such as oxygen vacancy in the samples. However, contrary to previous studies on un-doped ZnO, the doping with Ho^{3+} could be responsible for the magnetic intensity enhancement observed in this study which is not in accordance with the correlated PL/FM result reported elsewhere [73,74]. This observation is probably due to the important magnetic moment of the Ho^{3+} ions dopants, which effectively dominated on the RT-FM as compared to the concentration of oxygen defects. From the current observation and based on previous studies, we can speculate on the strong RT-FM observed in the Ho^{3+}-doped samples which could be ascribed to the s-f coupling between the ZnO host and the Ho^{3+} dopant due to its large magnetic moment and the incorporation of more ferromagnetic defects within the ZnO matrix [63,64,66,75]. The collected ESR spectra revealed the occurrence of strong microwave absorption at about 3248 Gauss which may be assigned to unpaired electron trapped at an oxygen vacancy site. The obtained g-factor values corresponding to the ferromagnetic resonance of the un-doped and 0.5 mol% Ho^{3+}-doped ZnO were found to be 2.032 and 2.067, respectively. Moreover, the 0.5 mol% Ho^{3+}-doped ZnO sample exhibited an additional peak at about 3004 G (g = 2.24) and 2372 G (g = 2.83) whose features may originate from Ho^{3+}. On the other hand, the peak-to-peak line widths (ΔH) obtained were found to be 633, 648, 417 and 347 G for un-doped 0.25, 0.5 and 0.75 mol% Ho^{3+}-doped samples, respectively. Furthermore, the number of spins (N_s) contributing to the ferromagnetic resonance was found to be 1.576×10^8, 2.906×10^8, 3.35×10^8 and 0.284×10^8 for the 0, 0.25, 0.5 and 0.75 mol% Ho^{3+} samples, respectively (see Figure 8). It is worth mentioning that the observed enhanced ferromagnetism is related to the number of spins participating in the ferromagnetism and further related to the concentration of Ho^{3+} [64]. In fact, the maximum microwave absorption intensity was attained at the critical concentration of 0.5 mol%; above this concentration, a severe decrease was observed (see Figure 7). This observation could be attributed to the ferromagnetism saturation due to excess of holmium ions being segregated on the surface of the ZnO host matrix. ESR spectra of Ho^{3+}-doped ZnO samples showed a similar intensity trend as compared to the UV–VIS transmittance spectra.

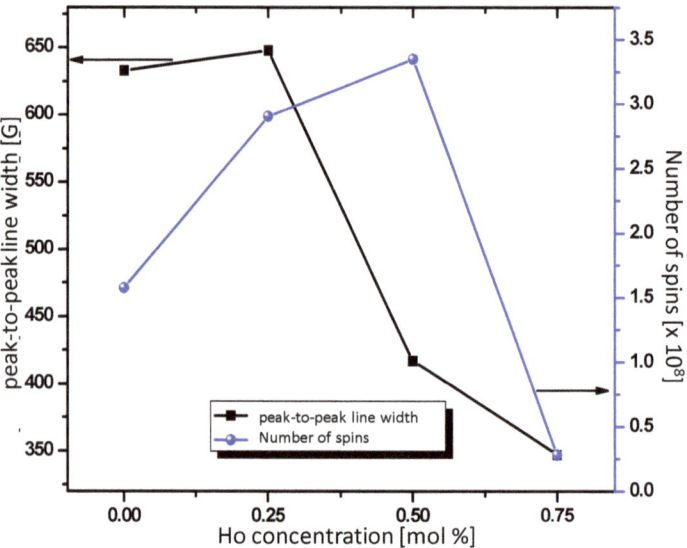

Figure 8. Peak-to-peak line width and number of spins as a function of Ho^{3+} concentration.

3.5. Time-Resolved Photoluminescence Lifetime Analysis

Time-resolved photoluminescence (TRPL) decay analysis of un-doped and 0.5 mol% Ho^{3+}-doped ZnO samples, which focused on the exciton (380 nm) and defect (520 nm) emission is depicted in Figure 9a–d, and the related fitted parameters are presented in Table 2. The TRPL data collected were analyzedusing multiple exponential functions in DAS-6 software platform. The obtained TRPL decays were found to be non-exponential, meaning that several emissive states were present in all TRPL data with the exception of ZnO:Ho^{3+} (0.5 mol%) data which exhibited exponential behaviour at 380 nm (Figure 9c). The decay profile of ZnO exciton (380 nm) has been previously reported to contain two emissive states, a fast and slow component [76–79]. The fast decay component (τ_1) has been assigned to non-radiative de-activation, and the slowdecaytrace (τ_2) resulted from radiative lifetime of free-excitons [79]. However, in the case of ZnO defect emission (520 nm), the faster decay trace (τ_1) has been assigned to arise from (i) radiative recombination of shallowly trapped electrons and deep trapped holes and (ii) the recombination of donor acceptor pair [77].

Moreover, due to the non-exponential nature of the TRPL decays under investigation, a bi- and tri-exponential function were the most reasonable approach to fit the data using the model presented equation below:

$$I(t) = I_0 \sum_{i=1}^{n} A_i e^{-\left(\frac{t}{\tau_i}\right)} \qquad (5)$$

where $I(t)$ is the fluorescence intensity at time t, I_0 is the initial fluorescence intensity, τ_i are lifetimes, and A_i are pre-exponential factors.

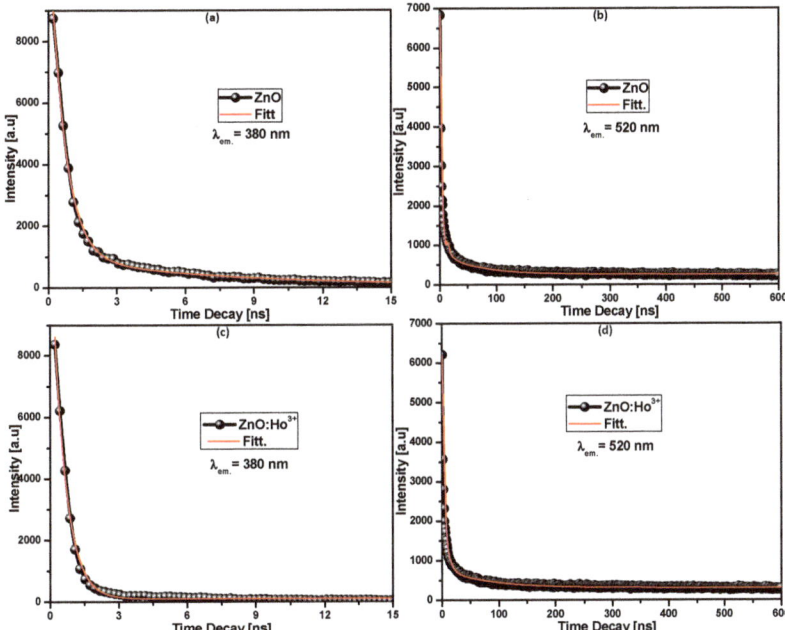

Figure 9. Exciton and defects TRPL lifetime decay of (**a**,**b**) ZnO and (**c**,**d**) ZnO:Ho^{3+} (0.5 mol%) nanostructures.

By monitoring the emission at 380 nm (3.26 eV), it has been observed that the complete de-activation was achieved in about 15 ns, indicating the fast decay of the free-exciton transition. On the other hand, the 520 nm (2.38 eV) emission revealed the occurrence of a much longer decay due to the dominance of deep traps kind of defects. However, for both un-doped and 0.5 mol% Ho^{3+}-doped samples, the best fit revealed the existence of three emissive states at 520 nm Table 2a,b. It is worth noting that the averaged slow decay observed in un-doped ZnO defect emission results from more efficient trapping of the charge carriers due to high density of defects [77].

Interestingly, the decay profile of the ZnO:Ho^{3+} (0.5 mol%) at 380 nm exhibited a single emissive state much faster than that of un-doped ZnO which contained two emissive states. This could be the result of the distortion in the band-structure induced by Ho^{3+} ions doping. Moreover, we can speculate on the behaviour of the slow decay observed in the TRPL at 380 nm, which completely vanishes, this phenomenon is indicative of an energy transfer from the ZnO exciton to Ho^{3+} [79]. This result is in accordance with the photoluminescence results reported earlier in this study. Finally, the longer decay of (τ_1) observed in the doped sample indicates the good optical quality of ZnO:Ho^{3+} (0.5 mol%) as compared to the un-doped ZnO sample.

Table 2. (a) PL Decay of ZnO nanocrystals dispersed in ethanol under $\lambda_{ex.}$ = 320 nm using NanoLED (b) PL Decay of ZnO:Ho^{3+} (0.5 mol%) nanostructures dispersed in ethanol under $\lambda_{ex.}$ = 320 nm using NanoLED.

		(a)		
$\lambda_{em.}$ (nm)	τ_1 (ns)	τ_2 (ns)	τ_3 (ns)	χ^2
380	0.651	6.769	-	0.99
520	0.828	7.085	65.257	0.99
		(b)		
$\lambda_{em.}$ (nm)	τ_1 (ns)	τ_2 (ns)	τ_3 (ns)	χ^2
380	0.566	-	-	0.99
520	0.803	7.452	79.292	0.99

4. Conclusions

In summary, the use of the sol–gel method allows facile successful synthesis of Ho^{3+} doped ZnO (0.5 mol%) nanocrystals in which the Ho^{3+} ions were found to be emissive through the 4f-4f electronic transitions. Holmium was actively emissive at about 540 ($^5S_2, ^5F_4 \rightarrow ^5I_8$), 574, 606, 670 ($^5F_5 \rightarrow ^4I_8$) and 808 nm allied with the respective intra-ionic transitions. Moreover, optimal 0.5 mol% Ho^{3+} incorporation in ZnO relatively increased the crystallite size from 4 to 8 nm, tuned the band-gap and modified the initial morphology of un-doped ZnO spherical-like to rods-like which was attributed to Ostwald ripening effect. Furthermore, enhanced room temperature ferromagnestism related to the number of spins participating to the ferromagnetic resonance was reported. Finally, Time-resolved photoluminescence analysis revealed the change induced in the relaxation charge carriers in defects states as a result of 0.5 mol% Ho^{3+} doping in ZnO.

Supplementary Materials: The following are available online at https://www.mdpi.com/article/10.3390/nano11102611/s1, Figure S1: SEM image used to conduct EDS mapping, Figure S2: HRTEM images for (A) un-doped and (B) (0.5 mol%) Ho^{3+}-doped ZnO nanocrystals.

Author Contributions: G.L.K. was responsible for the concept, methodology and experiments of the study as well as for writing the manuscript. G.H.M. supported the methodology development and contributed to the writing of the manuscript. M.S.D. supported the concept development and design of the experiment. M.S.D. provided funding and contributed to the writing of the manuscript. All authors have read and agreed to the published version of the manuscript.

Funding: G.L.K. is thankful to the University of South Africa for its generous Post-Doctoral Fellowship. This work was supported by the National Research Foundation of South Africa under grant # 88028, and by the CSIR.

Data Availability Statement: Data is contained within the article or Supplementary Material.

Acknowledgments: The authors are grateful to the CSIR and National Research Foundation (NRF) for their financial grant (# 88028). The National Nano Surface Characterization Facility of the University of the Free State is acknowledged for XPS measurements. The funding support from the University of South Africa (UNISA) is gratefully acknowledged. S.S. Nkosi is acknowledged for ESR measurements.

Conflicts of Interest: The authors declare no conflict of interest.

References

1. Wojnarowicz, J.; Chudoba, T.; Lojkowski, W. A Review of Microwave Synthesis of Zinc Oxide Nanomaterials: Reactants, Process Parameters and Morphologies. *Nanomaterials* **2020**, *10*, 1086. [CrossRef] [PubMed]
2. Özgür, Ü.; Alivov, Y.I.; Liu, C.; Teke, A.; Reshchikov, M.A.; Doğan, S.; Avrutin, V.; Cho, S.J.; Morkoç, H. A comprehensive review of ZnO materials and devices. *J. Appl. Phys.* **2005**, *98*, 041301. [CrossRef]
3. Ohno, H. Making nonmagnetic semiconductors ferromagnetic. *Science* **1998**, *281*, 951. [CrossRef] [PubMed]
4. Venkatesan, M.; Fitzgerald, C.B.; Lunney, J.G.; Coey, J.M.D. Anisotropic ferromagnetism in substituted zinc oxide. *Phys. Rev. Lett.* **2004**, *93*, 177206. [CrossRef] [PubMed]
5. Norberg, N.S.; Kittilstved, K.R.; Amonette, J.E.; Kukkadapu, R.K.; Schwartz, D.A.; Gamelin, D.R. Synthesis of colloidal Mn^{2+}:ZnO quantum dots and high-TC ferromagnetic nanocrystalline thin films. *J. Am. Chem. Soc.* **2004**, *126*, 9387. [CrossRef]
6. Waclawik, E.R.; Chang, J.; Ponzoni, A.; Concina, I.; Zappa, D.; Comini, E.; Motta, N.; Faglia, G.; Sberveglieri, G. Functionalised zinc oxide nanowire gas sensors. Beilstein. *J. Nanotechnol.* **2012**, *3*, 368.
7. Gulia, S.; Kakkar, R. ZnO Quantum Dots for Biomedical Applications. *Adv. Mater. Lett.* **2013**, *4*, 876. [CrossRef]
8. Kuwabara, T.; Omura, Y.; Yamaguchi, T.; Taima, T.; Kohshin, T.; Higashimine, K.; Vohra, V.; Murata, H. Factors affecting the performance of bifacial inverted polymer solar cells with a thick photoactive layer. *J. Phys. Chem. C* **2014**, *118*, 4050. [CrossRef]
9. Yang, R.S.; Qin, Y.; Dai, L.M.; Wang, Z.L. Power generation with laterally packaged piezoelectric fine wires. *Nat. Nanotechnol.* **2009**, *4*, 34. [CrossRef]
10. Mhlongo, G.H.; Motaung, D.E.; Nkosi, S.S.; Swart, H.C.; Malgas, G.F.; Hillie, K.T.; Mwakikunga, B.W. Temperature-dependence on the structural, optical, and paramagnetic properties of ZnO nanostructures. *Appl. Surf. Sci.* **2014**, *293*, 62. [CrossRef]
11. Motaung, D.E.; Mhlongo, G.H.; Nkosi, S.S.; Malgas, G.F.; Mwakikunga, B.W.; Coetsee, E.; Swart, H.C.; Abdallah, H.M.I.; Moyo, T.; Ray, S.S. Shape-selective dependence of room temperature ferromagnetism induced by hierarchical ZnO nanostructures. *ACS Appl. Mater. Interfaces* **2014**, *6*, 8981. [CrossRef]
12. Ahmad, U. Growth of Comb-like ZnO Nanostructures for Dye-sensitized Solar Cells Applications. *Nanoscale Res.Lett.* **2009**, *4*, 1004.
13. Wang, Z.L.; Song, J.H. Piezoelectric nanogenerators based on zinc oxide nanowire arrays. *Science* **2006**, *312*, 242. [CrossRef]
14. Kabongo, G.L.; Mhlongo, G.H.; Malwela, T.; Mothudi, B.M.; Hillie, K.T.; Dhlamini, M.S. Microstructural and photoluminescence properties of sol–gel derived Tb^{3+} doped ZnO nanocrystals. *J. Alloys Compd.* **2014**, *591*, 156–163. [CrossRef]
15. Popa, M.; Pop, L.C.; Schmerber, G.; Bouillet, C.; Ersen, O. Impact of the structural properties of holmium doped ZnO thin films grown by sol–gel method on their optical properties. *Appl. Surf. Sci.* **2021**, *562*, 150159. [CrossRef]
16. Aydin, S.; Turgut, G. Synthesis and investigation of some physical properties of pure and Ho-loaded ZnO nano-rods. *Appl. Phys. A* **2019**, *125*, 622. [CrossRef]
17. Khataee, A.; Saadi, S.; Vahid, B.; Joo, S.W. Sonochemical synthesis of holmium doped zinc oxide nanoparticles: Characterization, sonocatalysis of reactive orange 29 and kinetic study. *J. Ind. Eng. Chem.* **2016**, *35*, 167–176. [CrossRef]
18. Shubra, S.; Divya, D.J.N.; Ramachandran, B.; Ramachandra, R.M.S. Synthesis and comparative study of Ho and Y doped ZnO nanoparticles. *Mat. Lett.* **2011**, *65*, 2930–2933.
19. Franco, A., Jr.; Pessoni, H.V.S. Optical band-gap and dielectric behavior in Ho-doped ZnO nanoparticles. *Mat. Lett.* **2016**, *180*, 305–308.
20. Kashif, M.; Ali, S.M.U.; Ali, M.E.; Abdoulgafour, H.I.; Hashim, H.; Wilander, M.; Hassan, Z. Morphological, optical, and Raman characteristics of ZnO nanoflakes prepared via a sol–gel method. *Phys. Status Solidi A* **2012**, *209*, 143–147. [CrossRef]
21. Kashif, M.; Ali, M.E.; Ali, S.M.U.; Hashim, U. Sol–gel synthesis of Pd doped ZnO nanorods for room temperature hydrogen sensing applications. *Ceram. Int.* **2013**, *34*, 6461–6466. [CrossRef]
22. Kabongo, G.L.; Mhlongo, G.H.; Mothudi, B.M.; Mbule, P.S.; Hillie, K.T.; Dhlamini, M.S. Structural, photoluminescence and XPS properties of Tm^{3+} ions in ZnO nanostructures. *J. Lum.* **2017**, *187*, 141–153. [CrossRef]
23. Kabongo, G.L.; Mbule, P.S.; Mhlongo, G.H.; Mothudi, B.M.; Hillie, K.T.; Dhlamini, M.S. Photoluminescence quenching and enhanced optical conductivity of P3HT-derived Ho^{3+}-doped ZnO nanostructures. *Nanoscale Res. Lett.* **2016**, *11*, 1–11. [CrossRef]
24. Kabongo, G.L.; Mbule, P.S.; Mhlongo, G.H.; Mothudi, B.M.; Dhlamini, M.S. Time-resolved fluorescence decay and Gaussian analysis of P3HT-derived Ho^{3+}- and Tm^{3+}-doped ZnO nanostructures. *Bull. Mater. Sci.* **2020**, *43*, 48. [CrossRef]
25. Lifshitz, I.M.; Slyozov, V.V. The kinetics of precipitation from supersaturated solid solutions. *J. Phys.Chem. Solids* **1961**, *19*, 35. [CrossRef]
26. Wagner, C. Theorie der Alterung von Niderschlagen durch Umlösen (Ostwald Reifung). *Z. Elektrochem.* **1961**, *65*, 581.

27. Viswanatha, R.; Sarma, D.D. Growth of Nanocrystals in Solution. In *Nanomaterials Chemistry: Recent Developments and New Directions*; Rao, C.N.R., Müller, A., Cheetham, A.K., Eds.; Wiley-VCH Verlag GmbH & Co. KgaA: Weinheim, Germany, 200 [CrossRef]
28. Liao, H.-G.; Cui, L.; Whitela, S.; Zheng, H. Real-time imaging of Pt_3Fe nanorod growth in solution. *Science* **2012**, *336*, 101 [CrossRef]
29. Liu, Y.; Lin, X.-M.; Sun, Y.; Rajh, T. Cu-Catalyzed Asymmetric Borylative Cyclization of Cyclohexadienone-Containing 1,6-Enyne *J. Am. Chem. Soc.* **2013**, *135*, 3764. [CrossRef]
30. Layek, A.; Mishra, G.; Sharma, A.; Spasova, M.; Dhar, S.; Chowdhury, A.; Bandyopadhyaya, R. A Generalized Three-Stage Mechanism of ZnO Nanoparticle Formation in Homogeneous Liquid Medium. *J. Phys. Chem. C* **2012**, *116*, 24757. [CrossRef]
31. De la Rosa, L.S.; Chavez Portillo, M.; Mora-Ramirez, M.A.; Carranza Tellez, V.; Pacio Castillo, M.; Juarez Santiesteban, H.; Cortes Santiago, A.; Portillo Moreno, O. Synthesis of holmium oxide (Ho_2O_3) nanocrystal by chemical bath deposition. *Optik* **2020**, *21*, 164875. [CrossRef]
32. Patterson, A.L. The Scherrer Formula for X-Ray Paticle size. *Phys. Rev.* **1939**, *56*, 978. [CrossRef]
33. Li, X.-H.; Xu, J.-H.; Jin, M.; Shen, H.; Li, X.M. Electrical and optical properties of bulk ZnO single crystal grown by flux Bridgman method. *Chin. Phys. Lett.* **2006**, *23*, 3356.
34. Bang, J.; Yang, H.; Holloway, P.H. Enhanced luminescence of SiO_2:Eu^{3+} by energy transfer from ZnO nanoparticles. *J. Chem. Phys.* **2005**, *123*, 084709. [CrossRef] [PubMed]
35. Tauc, J. *Optical Properties of Solids*; Abeles, F., Ed.; North-Holland Pub. Co.: Amsterdam, The Netherlands, 1972; ISBN 10:0720402042/13:9780720402049.
36. Morkoç, H.; Hadis, Ü.Ö. *Zinc Oxide: Fundamentals, Materials and Device Technology*; Wiley-VCH Verlag GmbH & Co., KgaA Weinheim, Germany, 2009; ISBN 978-3-527-40813-9.
37. Wahab, R.; Hwang, I.H.; Kim, Y.-S.; Musarrat, J.; Siddiqui, M.A.; Seo, H.-K.; Tripathye, S.K.; Shin, H.-S. Non-hydrolytic synthesis and photo-catalytic studies of ZnO nanoparticles. *Chem. Eng. J.* **2011**, *175*, 450. [CrossRef]
38. Fox, M.A.; Dulay, M.T. Heterogeneous photocatalysis. *Chem. Rev.* **1993**, *93*, 341. [CrossRef]
39. Liu, Y.M.; Fang, Q.Q.; Wu, M.Z.; Li, Y.; Lv, Q.R.; Zhou, J.; Wang, B.M. Structure and photoluminescence of arrayed $Zn_{1-x}Co_xO$ nanorods grown via hydrothermal method. *J. Phys. D Appl. Phys.* **2007**, *40*, 4592.
40. Baiqi, W.; Xudong, S.; Qiang, F.; Iqbal, J.; Yan, L.; Honggang, F.; Dapeng, Y. Photoluminescence properties of Co-doped ZnO nanorods array fabricated by the solution method. *Phys. E* **2009**, *41*, 413. [CrossRef]
41. Elilarassi, R.; Chandrasekaran, G. Synthesis and optical properties of Ni-doped zinc oxide nanoparticles for optoelectronic applications. *Optoelectron. Lett.* **2010**, *6*, 6. [CrossRef]
42. Mustaqima, M.; Liu, C. ZnO-based nanostructures for dilute magnetic semiconductor. *Turk. J. Phys.* **2014**, *38*, 429. [CrossRef]
43. Xian, F.L.; Li, X.Y. Effect of nd doping level on optical and structural properties of ZnO:Nd thin films synthesized by the sol-gel route. *Opt. Laser Technol.* **2013**, *45*, 508. [CrossRef]
44. Zhang, W.; Zhao, J.; Liu, Z.; Liu, Z. Structural, optical and magnetic properties of $Zn_{1-x}Fe_xO$ powders by sol-gel method. *Appl. Surf. Sci.* **2013**, *284*, 49. [CrossRef]
45. Ramani, M.; Ponnusamy, S.; Muthamizhchelvan, C. Zinc oxide nanoparticles: A study of defect level blue-green emission. *Opt. Mater.* **2012**, *34*, 817. [CrossRef]
46. Spanhel, L.; Haase, M.; Weller, H.; Henglein, A. Photochemistry of colloidal semiconductors. 20. Surface modification and stability of strong luminescing CdS particles. *J. Am.Chem. Soc.* **1987**, *109*, 5649. [CrossRef]
47. Zhang, D.H. Properties of ZnO Films Prepared by Ionbeam Assisted Reactive Deposition and by rf Bias Sputtering. Ph.D. Thesis, University of Waterloo, Waterloo, ON, Canada, 1993.
48. Jean-Claude, G. Bünzli and Anne-Sophie Chauvin. Lanthanides in Solar Energy Conversion. In *Handbook on the Physics and Chemistry of Rare Earths*; Jean-Claude, G.B., Vitalij, K.P., Eds.; Elsevier: Amsterdam, The Netherlands, 2014; Volume 44, pp. 169–28 ISBN 978-0-444-62711.
49. Bai, Y.; Wang, Y.; Peng, G.; Yang, K.; Zhang, X.; Song, Y. Enhanced upconversion photoluminescence intensity by doping Li^+ in Ho^{3+} and Yb^{3+} codoped Y_2O_3 nanocrystals. *J. Alloys Compd.* **2009**, *478*, 676. [CrossRef]
50. Boyer, J.C.; Vetrone, F.; Capobianco, J.A.; Speghini, A.; Bettinelli, M. Optical transitions and upconversion properties of Ho^{3+} doped ZnO-TeO_2 glass. *J. Appl. Phys.* **2003**, *93*, 9460. [CrossRef]
51. Dhlamini, M.S.; Mhlongo, G.H.; Swart, H.C.; Hillie, K.T. Energy transfer between doubly doped Er^{3+}, Tm^{3+} and Ho^{3+} rare earth ions in SiO_2 nanoparticles. *J. Lum.* **2011**, *131*, 790. [CrossRef]
52. Xiushan, Z.; Peyghambarian, N. High-power ZBLAN glass fiber lasers: Review and prospect. *Adv. Optoelectron.* **2010**, *2010*, 501956. [CrossRef]
53. Wöll, C. The chemistry and physics of zinc oxide surfaces. *Prog. Surf. Sci.* **2007**, *82*, 55. [CrossRef]
54. Kumar, V.; Swart, H.C.; Ntwaeaborwa, O.M.; Kroon, R.E.; Terblans, J.J.; Shaat, S.K.K.; Yousif, A.; Duvenhage, M.M. Origin of the red emission in zin oxide nanophosphors. *Mater. Lett.* **2013**, *101*, 57. [CrossRef]
55. Moulder, J.F.; Stickle, W.F.; Sobol, P.E.; Bomben, K.D. *Handbook of X-ray Photoelectron Spectroscopy*; Perkin-Elmer Corporation: Eden-Prairie, MN, USA, 1992.
56. Gehlhoff, W.; Hoffmann, A. Acceptors in ZnO nanocrystals: A reinterpretation. *Appl. Phys. Lett.* **2012**, *101*, 262106. [CrossRef]

37. Teklemichael, S.T.; Hlaing, O.W.M.; Mc Cluskey, M.D.; Walter, E.D.; Hoyt, D.W. Acceptors in ZnO nanocrystals. *Appl. Phys. Lett.* **2011**, *98*, 232112. [CrossRef]
38. Zeng, H.; Duan, G.; Li, Y.; Yang, S.; Xu, X.; Cai, W. Blue luminescence of ZnO nanoparticles based on Non-Equilibrium Processes: Defect Origins and Emission controls. *Adv. Funct. Mater.* **2010**, *20*, 561. [CrossRef]
39. Dietl, T.; Ohno, H.; Matsukura, F.; Cibert, J.; Ferrand, D. Zener model description of ferromagnetism in zinc-blende magnetic semiconsuctors. *Science* **2000**, *287*, 1019. [CrossRef] [PubMed]
40. Radovanovic, P.V.; Gamelin, D.R. High-temperature ferromagnetism in Ni^{2+}-doped ZnO aggregates prepared from colloidal dilute magnetic semiconductor quantum dots. *Phys. Rev. Lett.* **2003**, *91*, 157202/1. [CrossRef] [PubMed]
41. Schwartz, D.A.; Gamelin, D.R. Reversible 300 K ferromagnetic ordering in a dilute magnetica semiconductor. *Adv. Mater.* **2004**, *16*, 2115. [CrossRef]
42. Baik, J.M.; Lee, J.L. Fabrication of vertically well-alligned (Zn,Mn)O nanorods with room temperature ferromagnetism. *Adv.Mater.* **2005**, *17*, 2745. [CrossRef]
43. Bishnoi, S.; Khichar, N.; Das, R.; Kumar, V.; Kotnala, R.K.; Chawla, S. Triple excitation with dual emission in paramagnetic ZnO:Er^{3+} nanocrystals. *RSC Adv.* **2014**, *4*, 32726. [CrossRef]
44. Rai, G.M.; Iqbal, M.A.; Xu, Y.; Will, I.G.; Zhang, W. Influence of rare earth Ho^{3+} doping on structural, microstructure and magnetic properties of ZnO bulk and thin film systems. *Chin. J. Chem. Phys.* **2011**, *24*, 353.
45. Popa, M.; Schmerber, G.; Toloman, D.; Gabor, M.S.; Mesaros, A.; Petrisor, T. Magnetic and electrical properties of undoped and holmium doped ZnO thin films grown by sol-gel method. In *Advanced Engineering Forum*; Trans Tech Publications Ltd.: Stafa-Zurich, Switzerland, 2013; Volume 8, pp. 301–308. [CrossRef]
46. Dalpian, G.M.; Wei, S.-H. Electron-induced stabilization of ferromagnetism in $Ga_{1-x}Gd_xN$. *Phys. Rev. B* **2005**, *72*, 115201. [CrossRef]
47. Shi, H.; Zhang, P.; Li, S.-S.; Xia, J.-B. Magnetic coupling properties of rare-earth metals (Gd, Nd) doped ZnO: First-principles calculations. *arXiv* **2010**, arXiv:1005.1115v1. [CrossRef]
48. Chen, Q.; Wang, J. ferromagnetism in Nd-doped ZnO nanowires and the influence of oxygen vacancies: Ab initio calculations. *Phys. Chem. Chem. Phys.* **2013**, *15*, 17793. [CrossRef]
49. Wang, D.D.; Chen, Q.; Xing, G.Z.; Yi, J.B.; Bakaul, S.R.; Ding, J.; Wang, J.L.; Wu, T. Robust room temperature ferromagnetism with Giant anisotropy in Nd-doped ZnO nanowire arrays. *Nano Lett.* **2012**, *12*, 3994. [CrossRef] [PubMed]
50. Ungureanu, M.; Schmidt, H.; Xu, Q.Y.; Wenckstern, H.V.; Spemann, D.; Hochmuth, H.; Lorenz, M.; Grundmann, M. Electrical and magnetic properties of RE-doped ZnO thin films (RE = Gd,Nd). *Superlattice Microst.* **2007**, *42*, 231. [CrossRef]
51. Potzger, K.; Zhou, S.Q.; Eichhorn, F.; Helm, M.; Skorupa, W.; Mucklich, A.; Fassbender, J.; Herrmannsdorfer, T.; Bianchi, A. ferromagnetic Gd-implanted ZnO single crystals. *J. Appl.Phys.* **2006**, *99*, 063906. [CrossRef]
52. Garcia, M.A.; Merino, J.M.; Pinel, E.F.; Quesada, A.; De la Venta, J.; Ruíz González, M.L.; Castro, G.R.; Crespo, P.; Llopis, J.; González-Calbet, J.M.; et al. Magnetic properties of ZnO nanoparticles. *Nano Lett.* **2007**, *7*, 1489. [CrossRef] [PubMed]
53. Xu, X.; Xu, C.; Dai, J.; Hu, J.; Li, F.; Zhang, S. Size dependence Defect-induced room temperature ferromagnetism in undoped ZnO nanoparticles. *J. Phys. Chem. C* **2012**, *116*, 8813. [CrossRef]
54. Xing, G.; Wang, D.; Yi, J.; Yang, L.; Gao, M.; He, M.; Yang, J.; Ding, J.; Sum, T.C.; Wu, T. correlated d0 ferromagnetism and photoluminescence in undoped ZnO nanowires. *Appl. Phys. Lett.* **2010**, *96*, 112511. [CrossRef]
55. Dhar, S.; Brandt, O.; Ramsteiner, M.; Sapega, V.F.; Ploog, K.H. colossal magnetic moment of Gd in GaN. *Phys. Rev. Lett.* **2005**, *94*, 037205. [CrossRef] [PubMed]
56. Zhong, Y.; Djurisic, A.B.; Hsu, Y.F.; Wong, K.S.; Brauer, G.; Ling, C.C.; Chan, W.K. Exceptionally long exciton photoluminescence lifetime in ZnO tetrapods. *J. Phys.Chem. C* **2008**, *112*, 16286. [CrossRef]
57. Layek, A.; Manna, B.; Chowdhury, A. Carrier recombination dynamics through defect states of ZnO nanocrystals: From nanoparticles to nanorods. *Chem. Phys. Lett.* **2012**, *539–540*, 133. [CrossRef]
58. Lee, S.-K.; Chen, S.L.; Hongxing, D.; Sun, L.; Chen, Z.; Chen, W.M.; Buyanova, I.A. Long lifetime of free excitons in ZnO tetrapod structures. *Appl. Phys. Lett.* **2010**, *96*, 083104. [CrossRef]
59. Guidelli, E.J.; Baffa, O.; Clarke, D.R. Enhanced UV emission from Silver/ZnO and Gold/ZnO Core-Shell nanoparticles: Photoluminescence, radioluminescence, and optically stimulated luminescence. *Sci. Rep.* **2015**, *5*, 14004. [CrossRef]

Enhancing the UV Emission in ZnO–CNT Hybrid Nanostructures via the Surface Plasmon Resonance of Ag Nanoparticles

Protima Rauwel [1,*], Augustinas Galeckas [2] and Erwan Rauwel [1]

1. Institute of Technology, Estonian University of Life Sciences, 51014 Tartu, Estonia; erwan.rauwel@emu.ee
2. Department of Physics, University of Oslo, P.O. Box 1048 Blindern, 0316 Oslo, Norway; augustinas.galeckas@fys.uio.no
* Correspondence: protima.rauwel@emu.ee; Tel.: +372-7-313-322

Abstract: The crystal quality and surface states are two major factors that determine optical properties of ZnO nanoparticles (NPs) synthesized through nonaqueous sol–gel routes, and both are strongly dependent on the growth conditions. In this work, we investigate the influence of the different growth temperatures (240 and 300 °C) on the morphology, structural and crystal properties of ZnO NP. The effects of conjoining ZnO NP with carbon nanotubes (CNT) and the role of surface states in such a hybrid nanostructure are studied by optical emission and absorption spectroscopy. We demonstrate that depending on the synthesis conditions, activation or passivation of certain surface states may occur. Next, silver nanoparticles are incorporated into ZnO–CNT nanostructures to explore the plasmon–exciton coupling effect. The observed enhanced excitonic and suppressed defect-related emissions along with blue-shifted optical band gap suggest an intricate interaction of Burstein–Moss, surface plasmon resonance and surface band-bending effects behind the optical phenomena in hybrid ZnO–CNT–Ag nanocomposites.

Keywords: ZnO; CNT; nanohybrid; photoluminescence; surface states; Ag nanoparticles; surface plasmon resonance; Burstein–Moss effect

1. Introduction

ZnO is a well-acknowledged efficient semiconductor phosphor with a wide direct band gap (~3.3 eV), high electron mobility (155 cm^{-2} V^{-1} s^{-1}) and large exciton binding energy (60 meV) that ensures effective luminescence at room temperature (RT) [1]. These attractive optoelectronic properties make ZnO a promising multifunctional material for a variety of short-wavelength, light-emitting and detection applications, such as ultraviolet (UV) light emitting diodes (LED), UV lasers and tunable UV photodetectors [2]. In general, ZnO is a highly luminous material exhibiting two distinctive emission signatures in the UV and visible regions of the spectrum, respectively. The former is associated with the near-band-edge (NBE) emission that involves excitonic and shallow-state optical transitions. The other characteristic signature, a broad band in the visible region, originates from a variety of defect states within the band gap and is identified as deep-level emission (DLE) [3]. The luminescent deep states in ZnO are mostly of intrinsic origin, i.e., involving oxygen and zinc vacancies (V_O, V_{Zn}), interstitials (Zn_i, O_i) and related complexes. Such point defects may be represented differently at the surface and in the bulk of the crystal; they manifest different charge states and act as donors or acceptors contributing to the visible emission in ZnO [4]. For example, the green emission at 2.3 eV is a surface-related Vo acting as an acceptor state, while the green emission at 2.5 eV is the volume-related Vo acting as a donor state. It is important to note that a visible-light response is of particular interest for various photocatalytic activities [5]. In an attempt to extend the absorption into the visible range, several techniques, such as ion implantation and metal or nonmetal doping, have

been investigated [6–8]. However, these methods do not appear to be industrially viable in terms of complexity and fabrication costs.

A similar but more cost-effective band gap modification is achievable by introducing Zn and O interstitials and vacancies during the synthesis process itself. In our previous work, we have demonstrated that by using the appropriate precursor it is possible to promote the generation of native defects and consequently modify the absorption and emission properties of ZnO [9]. In that study, nonaqueous sol–gel routes were used for the synthesis of ZnO nanoparticles (NP), and the effects of hydroxyl groups on the physical properties of ZnO were highlighted. Furthermore, the effect of hybridizing ZnO NPs with carbon nanotubes (CNT) further indicated that both the NBE and DLE emissions were enhanced [10]. In all cases, the emission spectra indicated that ZnO had undergone surface modification when linked to CNT, with the newly observed luminescence features linked to the band bending at the ZnO–CNT interface.

Furthermore, significant photoluminescence (PL) enhancement occurs upon hybridizing ZnO NP with CNT. However, in order to maximize photocatalytic activity, these radiative recombination pathways need to be suppressed [11]. Indeed, in photocatalysis, the electron–hole (e-h) pairs migrate to the surface of the nanoparticles, where the electrolyte undergoes redox reactions producing O and OH radicals [12] potent enough to degrade organic pollutants and exterminate microbes [13]. In antimicrobial applications in dry media, reactive oxygen species are produced by e-h pair separation on the surfaces of the nanoparticles. For both of these activities, the efficient e-h pair separation would require restrained radiative recombination pathways manifesting as a quenched PL emission. In addition, UV light irradiation is a complex and expensive method for degrading organic and neutralizing microbes. Therefore, visible-light-activated nanocatalysts, such as ZnO and its hybrids, are being actively pursued as cost-effective alternatives.

An optical band gap increase due to the Burstein–Moss effect has been observed in heavily doped ZnO with a variety of dopants such as Ga [14], Al [15], Gd [16], In [17], Bi [18], Fe [19], Ti [20] and Mg [21]. However, such heavy doping modifies the atomic arrangement and, in turn, the intrinsic properties of ZnO, as it tends to generate secondary phases. Surface plasmon resonance (SPR) induced by noble metal nanoparticles linked to ZnO offers similar advantages in terms of band gap increase and enhanced UV emission without modifying the intrinsic properties of ZnO. In addition, several studies report enhanced photocatalytic and antimicrobial properties along with enhanced UV emission after adding Ag NP to ZnO nanostructures [22]. Prolonging the lifetime of charge carriers by adding other noble metal nanoparticles, such as Au, has also been demonstrated, owing to their SPR effect [23]. However, in terms of cost-effectiveness, Ag appears to be the preferred material when compared to Au. In addition, it is less prone to oxidation as compared to Cu nanoparticles, which also exhibit SPR [24]. In the present study, we address the effects of synthesis conditions and the plasmonic enhancement on the UV emission in ZnO–CNT–Ag nanohybrids. We demonstrate that depending upon the synthesis temperature and the nature of the surface defects in ZnO, the enhancement of PL emission is possible with the addition of Ag NP [25].

2. Materials and Methods
2.1. Synthesis

The procedure for synthesizing ZnO NP was carried out under air; zinc acetate (3.41 mmol) (99.99%, Aldrich) was added to 20 mL (183 mmol) of benzylamine ($\geq$99.0%, Aldrich). The mixture was poured into a stainless steel autoclave and firmly closed. Thereafter, the autoclave was placed in a furnace at temperatures of 240 or 300 °C for 2 days. The milky suspensions were centrifuged; the precipitates were thoroughly washed with ethanol and dichloromethane and subsequently dried in air at 60 °C. For the ZnO–CNT–Ag nanohybrid synthesis, NANOCYL NC7000 multi-walled carbon nanotubes (MWCNT) with an average diameter of 10 nm and length of 1.5 µm were homogeneously dispersed into the solution of zinc acetate and benzylamine before the solution was transferred into an autoclave for a similar reaction synthesis. Surfactant-free metallic silver nanoparticles of 3 nm diameter were provided by PRO-1 NANOSolutions with >99.9% Ag purity. The supplier guarantees a shelf life of at least 1 year and oxidative stability up to 800 °C indicated by TGA measurements.

2.2. Characterization

X-ray diffraction (XRD) patterns were collected using a Panalytical Empyrean diffractometer (Malvern-Panalytical, Netherlands) with a Cu Kα1 radiation source (λ = 0.15406 nm). CHN measurement was conducted using a Leco TruSpec Micro CHNS Analyzer model 630-200-200 (Verder Scientific, Germany) at a temperature of 1075 °C. Carbon, hydrogen and sulfide were measured by infrared absorption, and nitrogen was measured by thermal conductivity. Scanning electron microscopy (SEM) images were recorded on an FEI Quanta 200FEG (FEI, Netherlands). High-resolution transmission electron microscopy (HRTEM) was carried out on a probe-corrected Titan G2 80–200 kV (FEI, Netherlands)and Jeol 200cx (JEOL, Japan), both operating at 200 kV in TEM mode and providing a point-to-point resolution of 2.4 Å. Before the study, the powder was crushed and dissolved in ethanol, and the solution was spread on a carbon-coated grid. Optical absorption properties were derived from the diffuse reflectance measurements performed at room temperature using a ThermoScientific EVO-600 UV–Vis spectrophotometer. Photoluminescence (PL) was investigated at 300 K by employing the 325 nm wavelength of a He–Cd CW laser with an output power of 10 mW as an excitation source. The emission was collected by a microscope and directed to a fiber-optic spectrometer (Ocean Optics, USA) USB4000, spectral resolution 2 nm. The density of the compact powders with and without CNTs was estimated to be approximately the same in all measurements.

3. Results and Discussion

Table 1 provides the list of ZnO based samples that were studied in this work. ZnO1 and ZnO2 correspond to ZnO samples synthesized at 240 and 300 °C, respectively. ZnO1–CNT and ZnO2–CNT refer to the two ZnO samples ZnO1 and ZnO2 after conjoining with CNT. ZnO1–CNT–Ag and ZnO2–CNT–Ag are samples containing ZnO1 (synthesized at 240 °C) and ZnO2 (synthesized at 300 °C), combined with CNT and Ag NP. In this study, the terms ZnO–CNT and ZnO–CNT–Ag refer to both ZnO1 and ZnO2 together.

Table 1. List of samples, their composition and synthesis temperature of ZnO nanoparticles.

Sample	Composition	ZnO Synthesis Temperature (°C)
ZnO1	ZnO	240
ZnO2	ZnO	300
ZnO1–CNT	ZnO and CNT	240
ZnO2–CNT	ZnO and CNT	300
ZnO1–CNT–Ag	ZnO, CNT and Ag	240
ZnO2–CNT–Ag	ZnO, CNT and Ag	300

3.1. Structure and Morphology

XRD patterns of the ZnO–CNT synthesized at 240 and 300 °C are provided in Figure 1a. The diffraction patterns correspond to the hexagonal wurtzite-analogous (P6$_3$mc) crystal structure (a = 3.25 Å and c = 5.20 Å). The XRD patterns show that ZnO NP are well crystallized without the presence of any secondary phases, with sizes ranging from 80 to 115 nm depending upon the temperature of synthesis [10]. Figure 1b is a typical XRD pattern of the Ag NPs used in this study. It can be clearly seen from the XRD pattern that Ag NP are single-phase and highly crystalline and only exhibit the characteristic diffraction peaks of the Ag face-centered cubic (Fm-3m) structure (JCPDS File No. 87-0720). No secondary phase structure was detected by XRD. The morphology and size distribution of Ag NP were also studied by SEM and TEM. The SEM images in Figure 2a,b illustrates that Ag NP are agglomerated. Nanoparticles smaller than 100 nm can be identified in Figure 2b, which could also be agglomerates of smaller Ag NP. TEM study was subsequently performed to confirm the size of Ag NP. The TEM micrograph presented in Figure 2c illustrates that Ag NP are spherical with an average diameter of 3 nm. The size distribution histogram in Figure 2d provides a mean nanoparticle size of 2.78 nm.

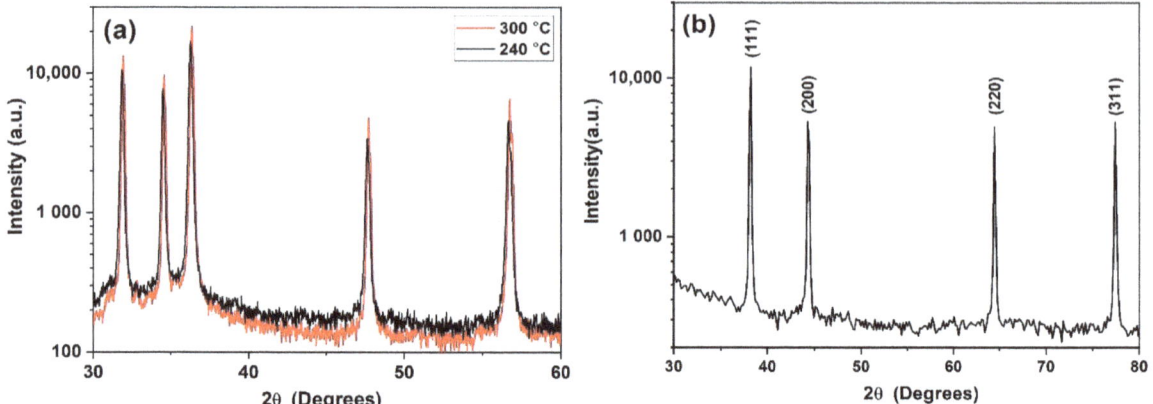

Figure 1. XRD pattern of (**a**) ZnO NP synthesized at 240 and 300 °C and (**b**) free-standing Ag NP.

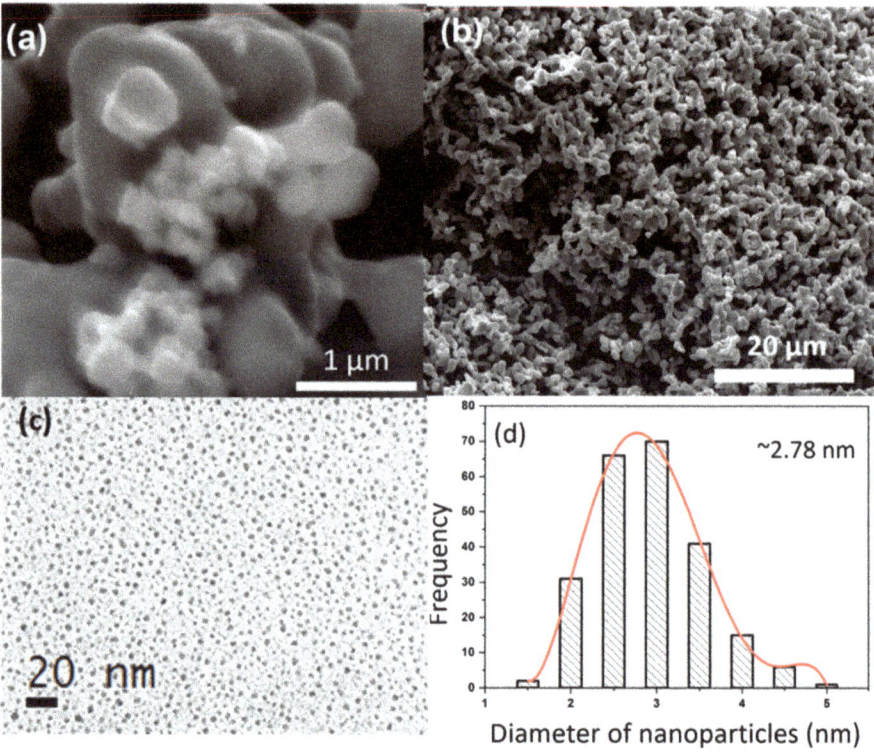

Figure 2. SEM images of Ag NP (**a**,**b**); TEM overview of Ag NP spread on carbon grid (**c**) and their size distribution histogram (**d**).

CHN measurements were performed on the Ag nanopowders to measure the quantities of organic elements, if any, on their surfaces. The analysis of 2.0 mg of Ag NP showed that the sample only contains 0.035 wt % carbon, which is most likely due to air contamination.

TEM micrographs in Figure 3a,b provide an overview of the ZnO–CNT nanohybrids. The interlaced CNT and ZnO NP form a matrix, thereby ensuring good contact between ZnO and CNTs. Figure 3c presents the hybrid ZnO–CNT–Ag nanocomposite and demonstrates that Ag NP are in direct contact with the ZnO NP. The Ag NPs are circled for clarity in Figure 3c, as at that scale, the smaller Ag NP (2.8 nm in diameter) are barely noticeable against much larger ZnO NP measuring 100–200 nm. The low concentration of only 1 wt % Ag of ZnO–CNT makes their detection challenging.

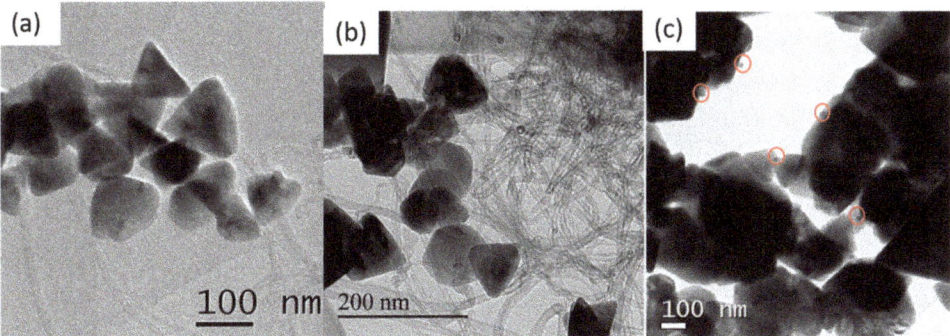

Figure 3. TEM micrographs of (**a**) ZnO1–CNT, (**b**) ZnO2–CNT and (**c**) ZnO1–CNT–Ag. The red circles highlight the Ag NP.

3.2. Optical Properties

The optical absorption properties of the samples are summarized in Figure 4 with Tauc plots derived from the diffuse reflectance measurements. For pure Ag NP with an average size of 2.8 nm, as follows from the TEM size distribution histogram in Figure 2d, the longitudinal component of the surface plasmon resonance is observed at ~3.88 eV (Figure 4a). This value appears close to the value typically reported in the literature, where an absorption peak at around 4 eV is generally observed. The size reduction and the surfactant or organic ligands on the Ag NP surfaces lead to the redshift of absorption [26]. The Ag NP employed in the present study were surfactant-free (i.e., 0.035 wt % C), and thus the observed minor redshift could only be due to the reduced size of the Ag NP measuring on an average 2.8 nm in diameter. Moreover, once the Ag NP were combined with CNT, the SPR absorption peak slightly redshifted further to 3.7 eV (Figure 4d) due to the realignment of the Fermi levels at the Ag NP and CNT interface, which facilitated electron transfer between Ag NP and CNT [27]. In our previous study, a detailed optical absorption study of the NANOCYL 700 MWCNT revealed their metallic nature [28].

The absorption properties of the ZnO–CNT nanocomposites labelled ZnO1–CNT and ZnO2–CNT are displayed in Figure 4b,c, both exhibiting similar absorption edges at 3.26 eV that match the band gap of ZnO. There are no shoulders extending from the absorption edges into the visible region for either of the two samples. By contrast, for the hybrid nanocomposite containing Ag NP labelled ZnO1–CNT–Ag, the absorption edge at 3.57 eV is observed along with a minor absorption in the range 2.3–3.5 eV (Figure 4e). The band edge energy exceeding the fundamental band gap of ZnO in such a ZnO–CNT–Ag nanohybrid system is indicative of plasmon coupling between the semiconducting ZnO NPs and metallic Ag NP. Indeed, the increased optical band gap of ZnO NP that follows from the blueshifted absorption edge can be explained by the Burstein–Moss (BM) effect typically observed in highly excited or heavily doped semiconductors [29,30]. For a degenerate n-type semiconductor or semimetal, the Fermi level is positioned within the conduction band due to free electrons partially filling this band; i.e., only states above the Fermi level are unoccupied and available for photon capture in a direct optical transition. Consequently, the onset of absorption occurs for photon energies higher than the fundamental band gap and the magnitude of this BM shift is proportional to the carrier concentration as $n^{2/3}$ [14]. In the present nanohybrids, the carrier concentration increase in ZnO NPs arises from the excess electrons provided by the Ag NP under photoexcitation. As a rough estimation, referring to standard BM effect calculation procedures, the difference between the band gap of ZnO ($E_{g(ZnO)}$ = 3.26 eV) and ZnO–CNT–Ag ($E_{g(ZnO–CNT–Ag)}$ = 3.57 eV) (cf. Figure 4b,e) creates a BM shift (ΔBM) of ~0.31 eV. This suggests an electron concentration of the order $n_e \approx 4 \times 10^{20}$ cm^{-3}, and the Fermi level is therefore positioned within the conduction band (CB) at around $E_c + 0.3$ eV. In turn, these parameters provide the necessary clues for further assessment of Schottky or Ohmic barriers and ultimately for building a feasible energy band model behind the optical phenomena of SPR, BM and surface-band-bending (SBB) which is discussed in what follows.

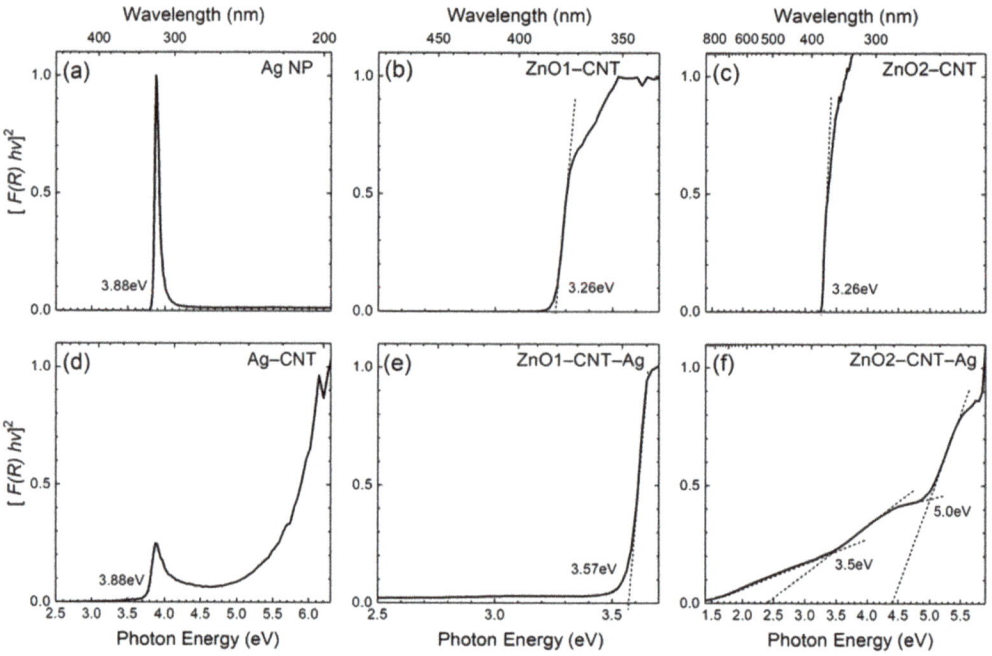

Figure 4. Tauc plots of the (**a**) free-standing Ag NP, (**b**) ZnO1–CNT, (**c**) ZnO2–CNT, (**d**) Ag–CNT, (**e**) ZnO1–CNT–Ag and (**f**) ZnO2–CNT–Ag nanohybrids.

For the hybrid ZnO2–CNT–Ag nanocomposite, two absorption shoulders at ~3.5 and 5 eV can be observed on the Tauc plot in Figure 4f. The absorption edge at 3.5 eV is associated with the BM-shifted optical band gap of ZnO2–CNT–Ag. The absorption edge at 5 eV corresponds to the π plasmon coupling of CNTs and is only visible in this sample [28]. Moreover, a nonzero absorption is also observed throughout the visible region for this sample. The nonzero absorption in the red part of the visible spectrum implies that Zn vacancy-related defects, such as stacking faults, turn optically active under the influence of SPR of Ag NP.

The optical emission properties of the nanohybrids were assessed by means of room temperature PL measurements summarized in Figure 5. The employed 325 nm wavelength laser excitation corresponds to 3.81 eV photon energy, which closely matches the plasmonic absorption peak of the Ag NPs at 3.88 eV; i.e., this energy was just sufficient to engender the interband transition of plasmons of the Ag NP. In Figure 5a, the PL emission spectra of the Ag NPs show two main emission peaks centered at 2.7 and 2.12 eV. These emissions with large Stokes shifts of 1.1 and 1.68 eV, respectively, correspond to the electron and hole interband recombination processes in Ag NPs [31]. In the case of Ag clusters with a small average particle size of 2.8 nm, the manifestation of quantum confinement effects is expected [32], which generally consists of the breaking of the band structure into discrete states. The high-energy band at 2.7 eV can therefore be ascribed to the radiative recombination of electrons in the sp-bands with holes in the d-band of agglomerated Ag NP. The low-energy band at 2.1 eV is due to intra-band recombination in the sp-bands of the lowest unoccupied molecular level (LUMO) and highest occupied molecular level (HOMO) gap. However, on combining with CNT and at the same excitation wavelength, total quenching of the luminescence is observed in Figure 5d, suggesting an efficient charge transfer from Ag NP to the CNT. The PL spectra of the ZnO–CNT nanocomposites and hybrid ZnO–CNT–Ag nanostructures are placed alongside for comparison in Figure 5, highlighting the emission changes upon the incorporation of the Ag NPs. For the ZnO–CNT structures, the NBE and DLE emission bands are typical for ZnO nanomaterials [33]. On the other hand, the hybrid ZnO–CNT–Ag nanostructures demonstrate significant extension of emission into the UV region well above the optical band gap of ZnO. The shape of PL emission is typical for the degenerate semiconductor and accounts for both band gap narrowing (BGN) and BM effects. In particular, the low-energy edge of PL spectrum is determined by low-energy transitions from the electron states close to the renormalized gap (i.e., from the bottom of semifilled CB) to acceptor-bound and valence band holes, whereas the high-energy edge is given by the relatively washed out cut-off of the Fermi occupation function at 300 K.

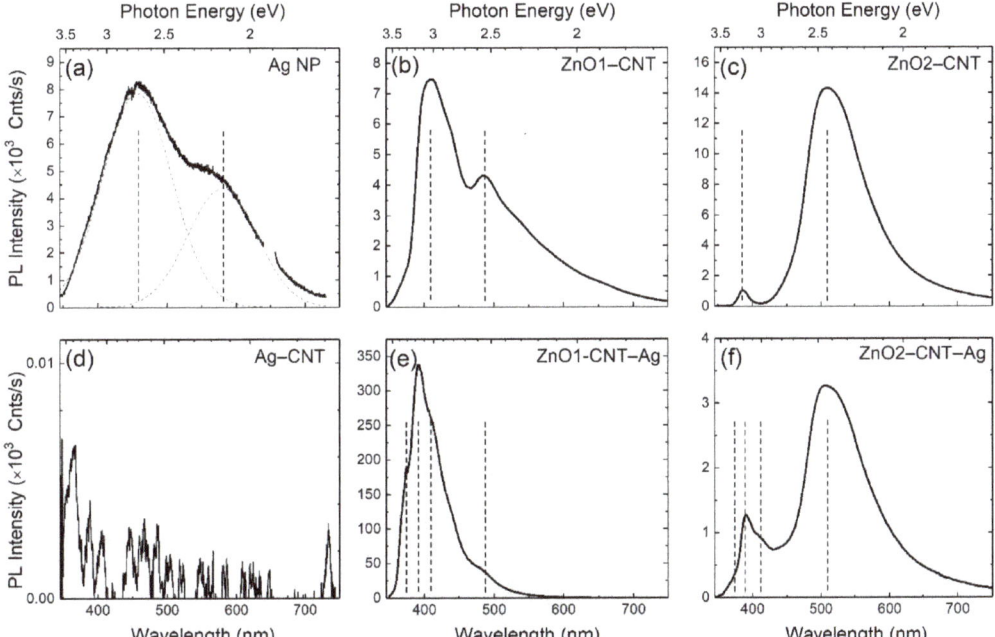

Figure 5. PL emission spectra of (**a**) pure Ag NPs, (**b**) ZnO1–CNT, (**c**) ZnO2–CNT, (**d**) Ag–CNT, (**e**) ZnO1-CNT–Ag and (**f**) ZnO2–CNT–Ag nanohybrids. Vertical dashed markers indicate major emission components discussed in the text.

The green emissions at around 2.5 and 2.3 eV represent the bulk or volume and depletion or surface regions of ZnO. The 2.5 eV emission pertains to a transition after a singly ionized oxygen vacancy captures a hole and becomes neutral. In the surface-related Vo emission, a hole is captured by the singly ionized oxygen vacancy, turning it into a doubly ionized oxygen vacancy. This acceptor state then accepts an electron from the conduction band or the depletion region. Nevertheless, for both samples of ZnO–CNT-Ag, there is a loss in the emission activity from the oxygen vacancy related defects when comparing NBE/DLE ratios of ZnO–CNT and ZnO–CNT–Ag (Figure 6). These defect levels of the volume- and surface-related oxygen vacancies lie close to the Fermi level of Ag. The emission in the red region due to transitions from Zn_i to O_i also shows the same tendency as the green emission. Some of the electrons from the NBE also undergo nonradiative de-excitation and are trapped in defect states such as O vacancies and Zn interstitials. The Ag NPs behave as sinks for these electrons, which are therefore spontaneously transferred to them. On photoexcitation, these electrons are then excited and transferred to higher empty levels of the conduction band, and the entire cyclic process of transfer from defect states of ZnO to the Fermi level of Ag and subsequently to the ZnO conduction band empty states continues. The photoexcitation of the electrons from the Fermi level of Ag can occur by the external excitation of the laser. Another possibility of photoexcitation via the photon emitted by recombination of e-h pairs of the NBE and DLE also exists. These excitations are energetic enough to excite the electrons from the Fermi level of Ag, which are then transferred to unoccupied states of the ZnO conduction band.

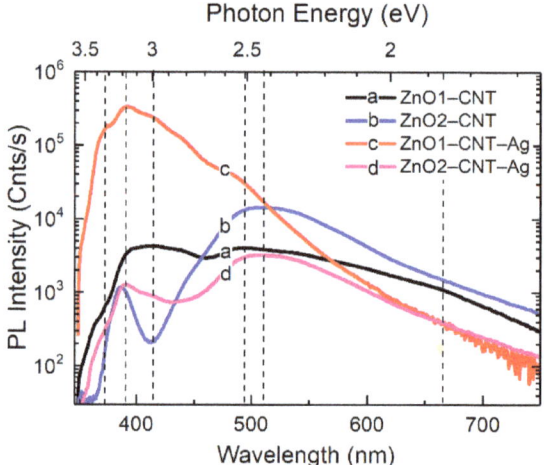

Figure 6. PL spectra at 300 K of (a) ZnO1–CNT, (b) ZnO2–CNT, (c) ZnO1–CNT–Ag and (d) ZnO2-CNT–Ag.

Figure 7 presents an outline of a possible model behind the optical phenomena in hybrid ZnO–CNT–Ag nanocomposites. In an ideal case, the downward band-bending model of Figure 7a corresponds to an Ohmic interfacial contact between ZnO and Ag. However, as previously described, absorption measurements (Tauc plots) indicate a blueshifted optical band gap with the Fermi level positioned within the conduction band at around $E_c + 0.3$ eV. Once an Ag NP comes into contact with a ZnO NP, the Fermi levels of both nanoparticles equalize, meaning an upward band-bending occurs at the surface of the ZnO NP. Furthermore, in a real case scenario, i.e., in air ambient, the upward band-bending in ZnO is also highly expected because of Fermi level pinning to always present surface defects, such as oxygen vacancies (V_o^{++}). This kind of upward bending at the ZnO surface implies Schottky barrier (SB) type contact between ZnO and Ag NP along with the presence of a depleted surface region (or space charge region), the width of which depends on several parameters including ZnO NP synthesis conditions as well as, their shape and size. In turn, the surface band-bending affects the intensity of both NBE and DLE, in particular the green luminescence component associated with the doubly ionized V_o^{++}.

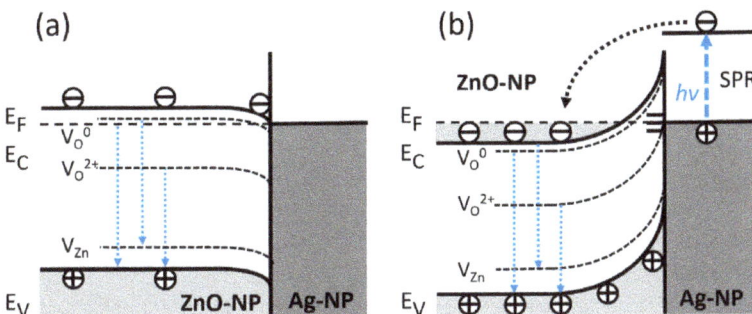

Figure 7. Sketch of optical phenomena in the nanohybrid structures at the interface of ZnO–Ag NP: (**a**) "Ideal" Ohmic contact model, where the interfacial band-bending is downwards because of the larger working function of ZnO than of Ag (~4.7–5.2 eV vs. 4.26 eV, respectively). (**b**) "Realistic" Schottky barrier model with upward band-bending at ZnO NP surfaces induced by interface states of adsorbents, such as hydroxyl groups and oxygen radicals.

In addition, for high interface state densities (Figure 7b), the SB height and band bending become independent of metal work function; i.e., the Fermi level at the surface is pinned by surface states within a narrow energy range in the band gap. In turn, the surface-pinned Fermi level leads to charge transfer that alters the carrier concentration within the surface space charge region, ionizing oxygen vacancy defect states involved in the DLE (green and red luminescence bands). Further, coupling with Ag NP leads to SPR-enhanced electron transfer to ZnO NP up to carrier concentrations causing a degenerate state with the Fermi level positioned within the conduction band, as confirmed by the BM shift in the absorption measurements.

4. Conclusions

In this work, we have successfully synthesized ZnO–CNT–Ag nanohybrids. Absorption studies demonstrated that CNTs serve as conducting pathways for the transfer of charges between ZnO and Ag nanoparticles. The plasmon–exciton interactions demonstrated an enhanced UV emission and a suppressed visible light emission in these structures. The widening of the optical band gap of ZnO was attributed to the SPR-induced Burstein–Moss shift [34]. For both ZnO hybrid samples, a nonzero absorption was also observed in the visible region. This suggests that they could be employed in applications such as visible light photodetection, photocatalysis and photovoltaics. Finally, the emission properties under visible light excitation will be assessed in a future work in order to more accurately evaluate the scope of such SPR-enhanced hybrid structures.

Author Contributions: P.R., A.G. and E.R. have contributed equally to the manuscript. All authors have read and agreed to the published version of the manuscript.

Funding: This research has been supported by the European Regional Development Fund project EQUiTANT TK134 (F180175TIBT). Partial financial support by the Research Council of Norway and the University of Oslo through the FUNDAMENT project (No. 251131, FRINATEK program) is gratefully acknowledged.

Acknowledgments: PRO-1 NANOSolutions is acknowledged for providing the silver metal nanoparticles. The authors thank M. Marques from CICECO, University of Aveiro, for CHN analysis. Laetitia Rapenne is acknowledged for access to TEM.

Conflicts of Interest: The authors declare no conflict of interest.

References

1. Janotti, A.; Van de Walle, C.G. Fundamentals of zinc oxide as a semiconductor. *Rep. Prog. Phys.* **2009**, *72*, 126501. [CrossRef]
2. Dai, T.-F.; Hsu, W.-C.; Hsu, H.-C. Improvement of photoluminescence and lasing properties in ZnO submicron spheres by elimination of surface-trapped state. *Opt. Express* **2014**, *22*, 27169–27174. [CrossRef]
3. Rauwel, P.; Salumaa, M.; Aasna, A.; Galeckas, A.; Rauwel, E. A Review of the Synthesis and Photoluminescence Properties of Hybrid ZnO and Carbon Nanomaterials. *J. Nanomater.* **2016**, *2016*, 5320625. [CrossRef]
4. Camarda, P.; Messina, F.; Vaccaro, L.; Agnello, S.; Buscarino, G.; Schneider, R.; Popescu, R.; Gerthsen, D.; Lorenzi, R.; Gelardi, F.M.; et al. Luminescence mechanisms of defective ZnO nanoparticles. *Phys. Chem. Chem. Phys.* **2016**, *18*, 16237–16244. [CrossRef] [PubMed]
5. Khokhra, R.; Bharti, B.; Lee, H.-N.; Kumar, R. Visible and UV photo-detection in ZnO nanostructured thin films via simple tuning of solution method. *Sci. Rep.* **2017**, *7*, 15032. [CrossRef]
6. Li, W.; Wang, G.; Chen, C.; Liao, J.; Li, Z. Enhanced Visible Light Photocatalytic Activity of ZnO Nanowires Doped with Mn(2+) and Co(2+) Ions. *Nanomaterials* **2017**, *7*, 20. [CrossRef]
7. Azarov, A.; Rauwel, P.; Hallén, A.; Monakhov, E.; Svensson, B.G. Extended defects in ZnO: Efficient sinks for point defects. *Appl. Phys. Lett.* **2017**, *110*, 022103. [CrossRef]
8. Azarov, A.; Vines, L.; Rauwel, P.; Monakhov, E.; Svensson, B.G. Silver migration and trapping in ion implanted ZnO single crystals. *J. Appl. Phys.* **2016**, *119*, 185705. [CrossRef]
9. Rauwel, E.; Galeckas, A.; Rauwel, P.; Sunding, M.F.; Fjellvåg, H. Precursor-Dependent Blue-Green Photoluminescence Emission of ZnO Nanoparticles. *J. Phys. Chem. C* **2011**, *115*, 25227–25233. [CrossRef]
10. Rauwel, E.; Galeckas, A.; Soares, M.R.; Rauwel, P. Influence of the Interface on the Photoluminescence Properties in ZnO Carbon-Based Nanohybrids. *J. Phys. Chem. C* **2017**, *121*, 14879–14887. [CrossRef]
11. Qi, K.; Cheng, B.; Yu, J.; Ho, W. Review on the improvement of the photocatalytic and antibacterial activities of ZnO. *J. Alloy. Compd.* **2017**, *727*, 792–820. [CrossRef]
12. Zhang, Q.; Xu, M.; You, B.; Zhang, Q.; Yuan, H.; Ostrikov, K. Oxygen Vacancy-Mediated ZnO Nanoparticle Photocatalyst for Degradation of Methylene Blue. *Appl. Sci.* **2018**, *8*, 353. [CrossRef]
13. Zhu, G.; Wang, H.; Yang, G.; Chen, L.; Guo, P.; Zhang, L. A facile synthesis of ZnO/CNT hierarchical microsphere composites with enhanced photocatalytic degradation of methylene blue. *RSC Adv.* **2015**, *5*, 72476–72481. [CrossRef]
14. Wang, Y.; Tang, W.; Zhang, L.; Zhao, J. Electron concentration dependence of optical band gap shift in Ga-doped ZnO thin films by magnetron sputtering. *Thin Solid Film.* **2014**, *565*, 62–68. [CrossRef]
15. Lai, H.H.-C.; Basheer, T.; Kuznetsov, V.L.; Egdell, R.G.; Jacobs, R.M.J.; Pepper, M.; Edwards, P.P. Dopant-induced bandgap shift in Al-doped ZnO thin films prepared by spray pyrolysis. *J. Appl. Phys.* **2012**, *112*, 083708. [CrossRef]
16. Obeid, M.M.; Jappor, H.R.; Al-Marzoki, K.; Al-Hydary, I.A.; Edrees, S.J.; Shukur, M.M. Unraveling the effect of Gd doping on the structural, optical, and magnetic properties of ZnO based diluted magnetic semiconductor nanorods. *RSC Adv.* **2019**, *9*, 33207–33221. [CrossRef]
17. Xie, G.C.; Fanga, L.; Peng, L.P.; Liu, G.B.; Ruan, H.B.; Wu, F.; Kong, C.Y. Effect of In-doping on the Optical Constants of ZnO Thin Films. *Phys. Procedia* **2012**, *32*, 651–657. [CrossRef]
18. Karthikeyan, B.; Sandeep, C.S.S.; Philip, R.; Baesso, M.L. Study of optical properties and effective three-photon absorption in Bi-doped ZnO nanoparticles. *J. Appl. Phys.* **2009**, *106*, 114304. [CrossRef]
19. Selloum, D.; Henni, A.; Karar, A.; Tabchouche, A.; Harfouche, N.; Bacha, O.; Tingry, S.; Rosei, F. Effects of Fe concentration on properties of ZnO nanostructures and their application to photocurrent generation. *Solid State Sci.* **2019**, *92*, 76–80. [CrossRef]
20. Bergum, K.; Hansen, P.-A.; Fjellvåg, H.; Nilsen, O. Structural, electrical and optical characterization of Ti-doped ZnO films grown by atomic layer deposition. *J. Alloys Compd.* **2014**, *616*, 618–624. [CrossRef]
21. Mia, M.N.H.; Pervez, M.F.; Hossain, M.K.; Reefaz Rahman, M.; Uddin, M.J.; Al Mashud, M.A.; Ghosh, H.K.; Hoq, M. Influence of Mg content on tailoring optical bandgap of Mg-doped ZnO thin film prepared by sol-gel method. *Results Phys.* **2017**, *7*, 2683–2691. [CrossRef]
22. Guidelli, E.J.; Baffa, O.; Clarke, D.R. Enhanced UV Emission From Silver/ZnO And Gold/ZnO Core-Shell Nanoparticles: Photoluminescence, Radioluminescence, And Optically Stimulated Luminescence. *Sci. Rep.* **2015**, *5*, 14004. [CrossRef] [PubMed]
23. Jiang, T.; Qin, X.; Sun, Y.; Yu, M. UV photocatalytic activity of Au@ZnO core–shell nanostructure with enhanced UV emission. *RSC Adv.* **2015**, *5*, 65595–65599. [CrossRef]
24. Mitsushio, M.; Miyashita, K.; Higo, M. Sensor properties and surface characterization of the metal-deposited SPR optical fiber sensors with Au, Ag, Cu, and Al. *Sens. Actuators A Phys.* **2006**, *125*, 296–303. [CrossRef]
25. Ruiz Peralta, M.D.L.; Pal, U.; Zeferino, R.S. Photoluminescence (PL) Quenching and Enhanced Photocatalytic Activity of Au Decorated ZnO Nanorods Fabricated through Microwave-Assisted Chemical Synthesis. *ACS Appl. Mater. Interfaces* **2012**, *4*, 4807–4816. [CrossRef]
26. Sharma, V.; Verma, D.; Okram, G.S. Influence of surfactant, particle size and dispersion medium on surface plasmon resonance of silver nanoparticles. *J. Phys. Condens. Matter* **2020**, *32*, 145302. [CrossRef]
27. Ghanbari, R.; Safaiee, R.; Sheikhi, M.H.; Golshan, M.M.; Horastani, Z.K. Graphene Decorated with Silver Nanoparticles as a Low-Temperature Methane Gas Sensor. *ACS Appl. Mater. Interfaces* **2019**, *11*, 21795–21806. [CrossRef]

8. Rauwel, P.; Galeckas, A.; Salumaa, M.; Ducroquet, F.; Rauwel, E. Photocurrent generation in carbon nanotube/cubic-phase HfO2 nanoparticle hybrid nanocomposites. *Beilstein J. Nanotechnol.* **2016**, *7*, 1075–1085. [CrossRef]
9. Riaz, A.; Ashraf, A.; Taimoor, H.; Javed, S.; Akram, M.A.; Islam, M.; Mujahid, M.; Ahmad, I.; Saeed, K. Photocatalytic and Photostability Behavior of Ag- and/or Al-Doped ZnO Films in Methylene Blue and Rhodamine B under UV-C Irradiation. *Coatings* **2019**, *9*, 202. [CrossRef]
10. Khurshid, F.; Jeyavelan, M.; Hudson, M.S.L.; Nagarajan, S. Ag-doped ZnO nanorods embedded reduced graphene oxide nanocomposite for photo-electrochemical applications. *R. Soc. Open Sci.* **2019**, *6*, 181764. [CrossRef]
11. Yang, T.-Q.; Peng, B.; Shan, B.-Q.; Zong, Y.-X.; Jiang, J.-G.; Wu, P.; Zhang, K. Origin of the Photoluminescence of Metal Nanoclusters: From Metal-Centered Emission to Ligand-Centered Emission. *Nanomaterials* **2020**, *10*, 261. [CrossRef]
12. Krajczewski, J.; Kołątaj, K.; Kudelski, A. Plasmonic nanoparticles in chemical analysis. *RSC Adv.* **2017**, *7*, 17559–17576. [CrossRef]
13. Vempati, S.; Mitra, J.; Dawson, P. One-step synthesis of ZnO nanosheets: A blue-white fluorophore. *Nanoscale Res. Lett.* **2012**, *7*, 470. [CrossRef]
14. Li, Z.; Xu, H. Nanoantenna effect of surface-enhanced Raman scattering: Managing light with plasmons at the nanometer scale. *Adv. Phys. X* **2016**, *1*, 492–521. [CrossRef]

Article

Numerical Study to Enhance the Sensitivity of a Surface Plasmon Resonance Sensor with BlueP/WS$_2$-Covered Al$_2$O$_3$-Nickel Nanofilms

Shivangani [1], Maged F. Alotaibi [2], Yas Al-Hadeethi [2], Pooja Lohia [1,*], Sachin Singh [3], D. K. Dwivedi [3,*], Ahmad Umar [4,5,*], Hamdah M. Alzayed [2], Hassan Algadi [5,6] and Sotirios Baskoutas [7]

1 Department of Electronics and Communication Engineering, Madan Mohan Malaviya University of Technology, Gorakhpur 273010, India; shivangani9673@gmail.com
2 Department of Physics, Faculty of Science, King Abdulaziz University, Jeddah 21589, Saudi Arabia; malhabrdi@kau.edu.sa (M.F.A.); yalhadeethi@kau.edu.sa (Y.A.-H.); hmalzayed@stu.kau.edu.sa (H.M.A.)
3 Photonics and Photovoltaic Research Lab, Department of Physics and Material Science, Madan Mohan Malaviya University of Technology, Gorakhpur 273010, India; sachin111iitp@gmail.com
4 Department of Chemistry, College of Science and Arts, Najran University, Najran 11001, Saudi Arabia
5 Promising Centre for Sensors and Electronic Devices (PCSED), Najran 11001, Saudi Arabia; hassan.algadi@gmail.com
6 Department of Electrical Engineering, College of Engineering, Najran University, Najran 11001, Saudi Arabia
7 Department of Materials Science, University of Patras, 265 04 Patras, Greece; bask@upatras.gr
* Correspondence: lohia.pooja6@gmail.com (P.L.); todkdwivedi@gmail.com (D.K.D.); ahmadumar786@gmail.com (A.U.)

Abstract: In the traditional surface plasmon resonance sensor, the sensitivity is calculated by the usage of angular interrogation. The proposed surface plasmon resonance (SPR) sensor uses a diamagnetic material (Al$_2$O$_3$), nickel (Ni), and two-dimensional (2D) BlueP/WS$_2$ (blue phosphorous-tungsten di-sulfide). The Al$_2$O$_3$ sheet is sandwiched between silver (Ag) and nickel (Ni) films in the Kretschmann configuration. A mathematical simulation is performed to improve the sensitivity of an SPR sensor in the visible region at a frequency of 633 nm. The simulation results show that an upgraded sensitivity of 332°/RIU is achieved for the metallic arrangement consisting of 17 nm of Al$_2$O$_3$ and 4 nm of Ni in thickness for analyte refractive indices ranging from 1.330 to 1.335. The thickness variation of the layers plays a curial role in enhancing the performance of the SPR sensor. The thickness variation of the proposed configuration containing 20 nm of Al$_2$O$_3$ and 1 nm of Ni with a monolayer of 2D material BlueP/WS$_2$ enhances the sensitivity to as high as 374°/RIU. Furthermore, it is found that the sensitivity can be altered and managed by means of altering the film portions of Ni and Al$_2$O$_3$

Keywords: surface plasmon resonance sensor; blue-phosphorus tungsten di-sulfide; Al$_2$O; nickel; sensitivity

1. Introduction

A method named surface plasmon resonance has arisen as an incredibly sensitive procedure for recognizing a very significant alteration in the refractive indexes of a detecting medium while communicating with the metal layer [1–3]. Enzyme detection, drug detection, medical diagnostics, and food safety are some of the biosensing applications of SPR-based biosensors [4–8]. Without any need for biomolecule labeling, a minute change in the refractive index (RI) can be detected in the detecting medium [9]. SPR is a highly sensitive technology that can detect very small fluctuations in the refractive index (RI) for biomolecule absorption of the order of 10^{-7} on the sensing interface. At the metals' dielectric contact, the collective oscillation of free electrons generates a transverse magnetically polarised electromagnetic wave known as a surface plasma wave (SPW). SPWs are made from metals with negative permittivity, such as gold, silver, copper, and aluminium, as well as dielectric materials, which can be liquid, gas, or solid. Researchers suggest that the Kretschmann arrangement brings about the productive coupling of light first from the crystal to the metal region, which is rooted in attenuated total reflection (ATR) [10]. The metallic regions of surface plasmons must be energized by the p-polarized light, while the s-polarised part is utilized as a source for the reference signal [11].

The SPR is acquired by the horizontal part of the evanescent wave (k_{ev}) being stage matched to the surface plasmon wave vector (k_{sp}). For example,:

$$k_{ev} = \sqrt[k_0]{\in p} k_p \sin \theta_{res} = k_{sp}$$

where the incident wave vector is addressed by $k_0 = \omega/c$, crystal permittivity is addressed by $\in p$, and the resonance point–angle is addressed by θ_{res}. Because of the total change of the p-polarized wave to the SP waves, the SPR condition causes a reduction in the depth of mirrored light (R).

In SPR sensors, normal metal layers of gold, silver, copper, and platinum are utilized. These are also plasmon-active metals that are used to generate surface plasmon waves in the sensing medium. The most promising metal is gold (Au), which has excellent optical properties, good chemical stability, and high oxidation and corrosion-resistant properties, although gold is the most expensive metal and lowers the biological molecules' absorption rate as compared to Ag metal [12–14]. The precision of an SPR sensor built from Ag film is better than that of an Au-film-based sensor. Because silver film is less costly than gold film, and as it was observed that silver has better sensitivity than Au metal, silver's SPR curve dip is narrower than gold's, meaning the sensitivity is improved. However, the chemical solidity problem of Ag should be alleviated, and the protecting layers must be explored for favorable optical properties [15–17]. As a result, this work introduces a novel aluminium oxide dielectric material to improve the SPR performance (Al_2O_3). It must be used to improve the SPR biosensor's performance parameters, such as the figure of merit (FOM), sensitivity, and so on. In fact, Al_2O_3 is widely used in mechanical areas due to its well-known properties, such as its superior corrosion resistance, high ductility, and high hardness, as well as in optical devices due to its high transparency and low refractive index. Certain 2D materials, such as graphene, WS_2, and the 2D heterostructure BlueP-WS_2, are the implicit aspirants used as a self-productive layer to give stability and boost the SPR sensor as it is slowly oxidized by air at room temperature and are considered corrosion resistant. Transition metal dichalcogenide (TMDC) substances have a curial position in SPR sensors. $PtSe_2$, Ti_3C_2Tx-Mxene, blue phosphorus, black phosphorus, and transition metal dichalcogenides (TMDC) are examples of substances that have stood out in the past twenty years due to their notable optical and electrical nature and have been used to make optoelectronic devices. Forty unique synthetic substances are presently incorporated in the TMDC family. MX_2 is another identifier for them, where M denotes metals such as tungsten, molybdenum, and niobium; and X denotes the chalcogen substances such as sulfur, selenium, and tellurium. MX_2's monolayer has three nuclear layers, with an alternate metal layer embedded between two chalcogen substance layers [18]. These nanomaterials collaborate with the metal layer and enhance the co-operation of the particles. If the oxidation of Al_2O_3 (2D material) can be minimized by means of coating it with another layer, we can use it as a high-sensitivity sensor. Because of its refractive index, Al_2O_3 ensure that the biosensor's performance is no longer affected. The homogeneity of the Al_2O_3 layer is a characteristic that leads to sensitivity enhancements [19]. Nickel (Ni), a ferromagnetic metal, is also gaining interest due to its notable magneto-optical and magnetic properties and being a good light absorber. Using inert magnetic metals minimizes the cost of the SPR sensor while simultaneously considerably improving its performance [20–22].

Moreover, because the TMDCs and blue phosphorene have a similar hexagonal crystal-like structure, BlueP/TMDCs can be formed [23]. Therefore BlueP/WS_2's heterostructure shows greater sensitivity. The sensing medium, also known as a sensing analyte, is the outermost layer in the proposed SPR biosensor design. There are 2 main SPR geometry configurations: the Kretschmann configuration and Otto's design. The Kretschmann configuration's benefits over Otto's arrangement have made it more widely applicable [24]. The SPR-based sensor's biosensing application covers the foundations of detecting the concentrations of biological things on a very small scale, such as bacteria, viruses, DNA, and proteins. Apart from biosensing and biomolecular analysis, the SPR sensor may also be used to detect nanostructured film depositions, as well as to quantify displacement and angular position [25–27].

In the present paper, calcium fluoride glass crystal (CaF_2) is taken for the proposed SPR biosensor since it gives maximal sensitivity and an enormous change in reflectivity when contrasted with different crystals (BK7, SF10, SF11, and so forth). A framework for an SPR sensor based on heterostructure containing Al_2O_3, BlueP/WS_2 with metal (Ag), and Ni (Nickle) is proposed in the present work to achieve improved sensitivity by altering the thickness of the Ni layers. The results demonstrate that adding the Al_2O_3 layer and BlueP/WS_2 to this structure boosts the sensitivity substantially. In this work, COMSOL Multiphysics 5.3a and MATLAB 2016a softwares are used to draw the plot of the reflectance curve, sensitivity curve, and electric field curve, which are calculated using these software programs.

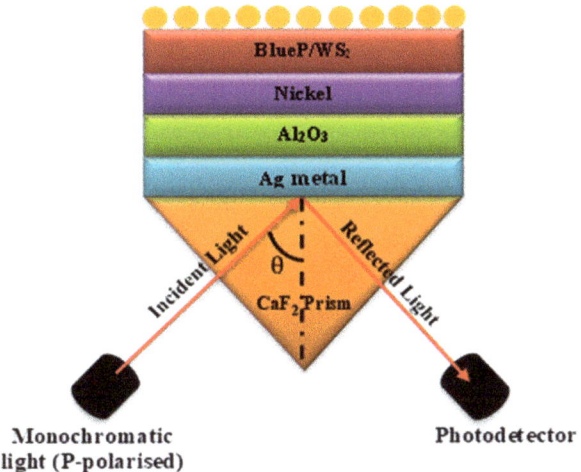

Figure 1. Sketch diagram of the proposed SPR biosensors.

2. Mathematical Modeling for the Proposed SPR Biosensor

2.1. Device Structure

In the present paper, the SPR sensor is composed of multiple layers primarily based on the Kretschmann configuration, in which the prism of CaF_2, Ag as a metal, Al_2O_3 (diamagnetic material), BlueP/WS_2, and SM on the top layer is used as shown in Figure 1. Table 1 refers to the width and different refractive indexes of the layers.

Table 1. Details of each layer of the proposed biosensor at 633 nm wavelength.

	Materials Used	Thickness (nm)	Refractive Index	References
1	CaF_2 prism	100	1.4329	[28]
2	Ag metal	50	$0.0803 + 1i \times 4.234$	[29]
3	Al_2O_3	20	1.7659	[30]
4	Nickel	1	$0.031957 + 1i \times 2.693$	[31]
5	BlueP/WS_2	0.75	$2.48 + 1i \times 0.170$	[32]
6	Sensing medium	300	1.330 to 1.335	This work

The structural diagram using Ag-Al$_2$O$_3$-Ni-BlueP/WS$_2$ is depicted in Figure 1 for this model. The 633 nm wavelength is utilized in SPR sensors for optimal results. The CaF$_2$ prism is employed in metal Ag with a refractive index of 1.4329 and a thickness of 50 nm, in Al$_2$O$_3$ with a thickness of 20 nm, in nickel with a thickness of 1 nm, and in BlueP/WS$_2$ with a thickness of 0.75 nm. Finally, the sensing medium has a refractive index of 1.330–1.335. The variation in the detecting medium is caused by adsorption, which occurs when biomolecules in the sensing medium contact with the BlueP/WS$_2$ layer causing the sensing medium's RI to alter. The first film is the CaF$_2$ prism, whose refractive index value may be computed using the Sellmeier relation.

$$n^2 = 1.33973 + \frac{0.69913 \times \lambda^2}{\lambda^2 - (0.09374)^2} + \frac{0.11994 \times \lambda^2}{\lambda^2 - (21.18)^2} + \frac{4.35181 \times \lambda^2}{\lambda^2 - (38.46)^2} \quad (1)$$

Here, 'λ' is the wavelength in nanometers. The RI of Ag is calculated using the Drude–Lorentz model:

$$n_{metal}(\lambda) = \left(1 - \frac{\lambda^2 \times \lambda_C}{\lambda_P^2(\lambda_C - \lambda \times i)}\right)^{\frac{1}{2}} \quad (2)$$

where λ_c and λ_P are the collision and plasma wavelengths, respectively. These are dispersion coefficients, while the used wavelength is 633 nm.

For Ag, λ_P = 145.41 nm and λ_C = 176.14 nm [33]. The transfer matrix method for the n-layer modeling and Fresnel equations is used throughout the numerical analysis. Using the reflectance curve, all the SPR sensor's performance characteristics in MATLAB software are evaluated. The graph is drawn using Origin software for all parameters and the FWHM values. Here, additionally compared distinct SPR sensors with the proposed model are mentioned in Table 2.

Table 2. Comparative study of the proposed SPR model.

Device Structure	Assembling of Films
design 1 (Conventional SPR)	CaF$_2$ crystal/Ag film/SM
design 2	CaF$_2$ crystal/Ag film/Al$_2$O$_3$/SM
design 3	CaF$_2$ crystal/Ag film/Al$_2$O$_3$/Ni/SM
design 4 (Proposed SPR)	CaF$_2$ crystal/Ag film/Al$_2$O$_3$/Ni/BlueP/WS$_2$/SM

2.2. Mathematical Expression for Reflectivity

The transfer matrix approach is utilized in the present work to obtain the reflection coefficient of the projected multilayer design (crystal, metal, Al$_2$O$_3$, Ni, and BlueP/WS$_2$ film). For the calculation of the reflectivity of the reflected light, the matrix approach for the N-film design is used. This method is quick and easy to use, and it does not involve any approximation. Along the z-axis, the layer thicknesses, dk, is considered. The kth layer's dielectric constant and RI are denoted by k and n, respectively. The tangential fields at $Z = Z_1 = 0$ are expressed in terms of the tangential field at $Z = Z_{N-1}$ using the boundary condition [34]:

$$\begin{bmatrix} U_1 \\ V_1 \end{bmatrix} = M \begin{bmatrix} U_{N-1} \\ V_{N-1} \end{bmatrix} \quad (3)$$

The places U_1 and V_1 address the tangential element of electric and magnetic fields, respectively, and the first and last layer of boundary are denoted by U_{N-1}, V_{N-1}, respectively. M_{ij} is the element for which the characteristics matrix is as follows [35]:

$$M_{ij} = \left(\prod_{k=2}^{N-1} M_k\right)_{ij} = \begin{bmatrix} M_{11} & M_{12} \\ M_{21} & M_{22} \end{bmatrix} \quad (4)$$

$$M_k = \begin{bmatrix} \cos\beta\kappa & (-i\sin\beta\kappa)/q_k \\ -iq_k\sin\beta\kappa & \cos\beta\kappa \end{bmatrix} \quad (5)$$

$$q_k = \left(\frac{u_k}{\varepsilon_k}\right)^{1/2} \cos\theta\kappa = \frac{(\varepsilon_k - \sin\theta_1 n_1^2)^{1/2}}{\varepsilon k} \quad (6)$$

and

$$q_k = \left(\frac{u_k}{\varepsilon_k}\right)^{1/2} \cos\theta\kappa = \frac{(\varepsilon_k - \sin\theta_1 n_1^2)^{1/2}}{\varepsilon k} \quad (7)$$

Following the mathematical steps, one can attain the reflection coefficient for p-polarized light, which is given below:

$$r_p = \frac{(M_{11} + M_{12}q_n)q_1 - (M_{21} + M_{22}q_n)}{(M_{11} + M_{12}q_n)q_1 + (M_{21} + M_{22}q_n)} \quad (8)$$

The multilayer configuration of the reflectivity R_p is given as:

$$R_p = |r_p|^2 \quad (9)$$

The conventional and three revised characteristics plots of the specific SPR sensor are shown in Figure 2. For the conventional SPR sensor shown in Figure 2a, the variation of the refractive index is Δn = 0.005, while the $\Delta\theta$ and sensitivity and full width at half maximum (FWHM) are 0.72(deg), 144°/RIU, and 1.04677, respectively. Further, Figure 2b–d presents a comparison with the SPR reflectance curve from Figure 2a. The reflectance curves of SPR designs 2 (Ag/Al$_2$O$_3$) and 3 (Ag/Al$_2$O$_3$/Ni) are displayed in Figure 2b,c, respectively. The angular shifts ($\Delta\theta$) for Figure 2b,c are 1.48 and 1.6, respectively, leading to sensitivities and FWHMs of 296° RIU^{-1} and 320° RIU^{-1}, and 2.05852 deg and 2.25936 deg, respectively. Figure 2d suggests a superior resonance angle, sensitivity, and FWHM in contrast to the other 3 designs. Figure 2d has the most extreme sensitivity of 397°/RIU among all other structures. Table 3 shows not only the maximum sensitivity but also the maximum resonance angle and a larger figure of merit compared to the other 3 designs.

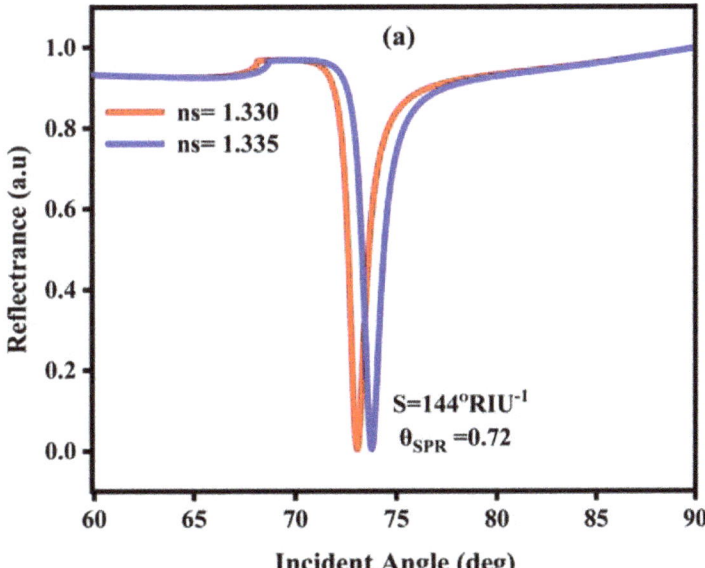

Figure 2. *Cont.*

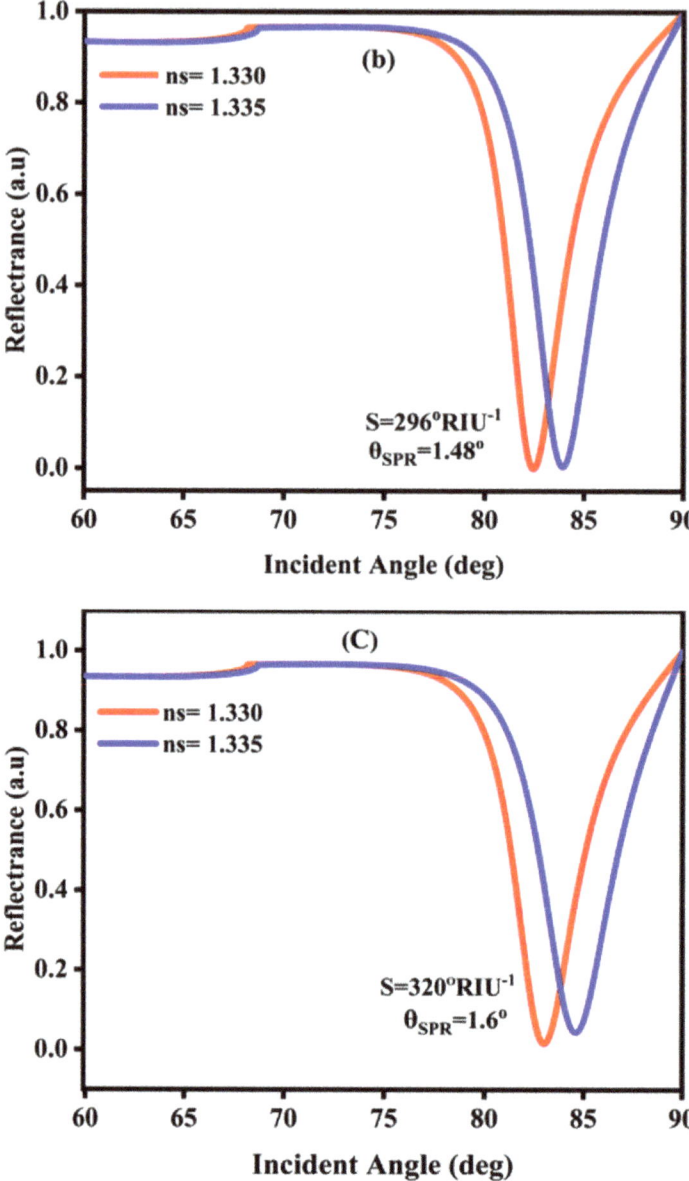

Figure 2. *Cont.*

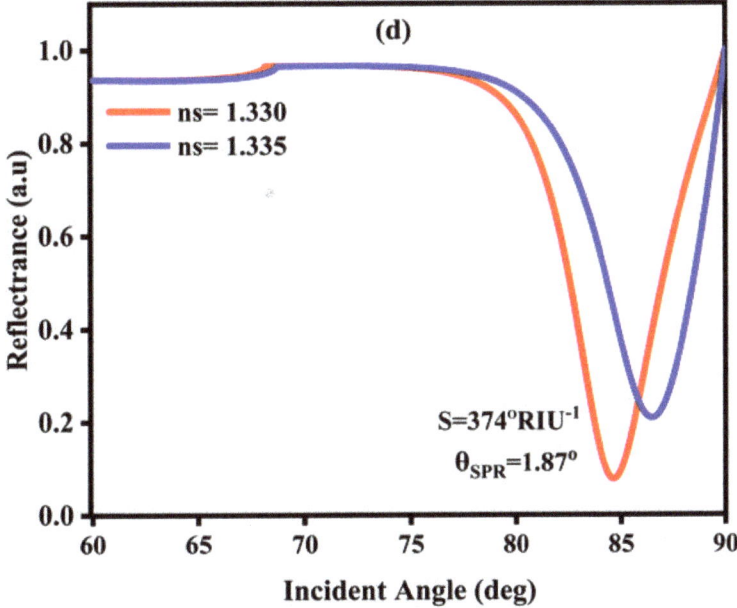

Figure 2. SPR reflectance curves: (**a**) design 1; (**b**) design 2; (**c**) design 3; (**d**) design 4.

Table 3. Comparative study of different proposed SPR sensors at a wavelength of 633 nm and $\Delta n = 0.005$.

Device Structure	$\Delta\theta$ (deg)	Sensitivity (° RIU^{-1})	FWHM (deg)	DA (deg^{-1})
Structure 1	0.72	144	1.04677	0.95531
Structure 2	1.48	296	2.05852	0.48578
Structure 3	1.6	320	2.25936	0.44260
Structure 4 (Proposed work)	1.87	374	2.75132	0.36346

Consequently, design 4 is considered the most reasonable decision among all designs.

3. Results and Discussion

The sensitivity properties of the biosensor with the changed Kretschmann design, which incorporates Al_2O_3 and Ni, are discussed here. To show how the sensitivity has reached the next level, from the reflectance curve the sensitivity is assessed according to the change in the resonance point. The excitement of the SPR causes a sharp drop in reflectance at a given point, which is clearly apparent. This event shows that the light is consumed by initiating the SPR in the biosensor arrangement, while the atom connection causes minor shifting in the refractive index of the sensor, which has a slight resonance dip around 1.87°. Thus, the design's sensitivity is achieved ($S_n = 374°$/RIU) by utilizing the connected computation articulation $S_n = S = \frac{\Delta\theta_{res}}{\Delta n}$ as displayed in Figure 2.

The effects of the thickness variation of Al_2O_3 and Ni on the performance of the projected SPR sensor are shown in Table 4. The execution constraints of the proposed SPR sensor for the variety of thicknesses of Al_2O_3 (14–20 nm) and Ni (1, 3, 5 nm) are displayed in Table 4. The greatest sensitivity of 374°/RIU is obtained with a thickness for Al_2O_3 of 20 nm and for Ni of 1 nm at a working wavelength of 633 nm.

Table 4. The optimized thickness values of Al$_2$O$_3$ and Ni with respect to the other parameters such as $\Delta\theta_{SPR}$, S, DA, FOM, and FWHM.

d(Al$_2$O$_3$) (nm)	d(Ni) (nm)	Δθ (deg)	S (°RIU^{-1})	DA (deg^{-1})	FWHM (deg)	FOM (RIU^{-1})
14	1	1.11	222	0.5190	1.9265	127.910
16	1	1.25	250	0.4753	2.1035	148.560
18	1	1.51	302	0.6471	1.5451	295.122
20	1	1.87	374	0.3635	2.7502	254.293
14	3	1.26	252	0.4399	2.2728	139.698
16	3	1.45	290	0.4000	2.2499	168.219
18	3	1.67	334	0.3498	2.8582	195.150
20	3	1.06	212	0.3048	3.2808	68.4950
14	5	1.44	288	0.3746	2.6690	155.381
16	5	1.61	322	0.3333	2.9995	172.834
18	5	1.34	268	0.2803	3.5672	100.670
20	5	0.09	18	0.2376	4.2070	0.38506

3.1. Use of CaF$_2$ Crystal

The refractive index directly affects the performance of the SPR sensor. Since the shift with the incident point in the reflectance bend is high and a sharp plunge is acquired, a CaF$_2$ crystal is utilized in the present situation. The sensitivity of this crystal material is incredible, with a lower RI than the CaF$_2$ crystal. Therefore, the CaF$_2$ glass crystal is at last utilized in the SPR sensor in the present theoretical investigation [36].

3.2. Performance Constraints of the SPR Sensor

The sensitivity, full width at half maximum, quality factor, detection accuracy, and limit of detection of the SPR sensor rely on certain variables. All of these constraints are dependent upon one another, and the reflectivity curve versus the incident angle determine the mathematical study of the SPR design.

3.3. Sensitivity (S)

The variation in resonance angle ($\Delta\theta_{res}$) with respect to the variation in the refractive index (Δn) decides the sensitivity and is defined as [37]:

$$S = \frac{\Delta\theta_{res}}{\Delta n} \left(\text{Unit} : {}^\circ\text{RIU}^{-1} \right) \tag{10}$$

The variation of the reflectance with the incident angle for the proposed SPR sensor is shown in Figure 3.

Figure 3 shows the shifts in resonance angle with the incident angle at different RI values (n_{SM} = 1.330–1.335). From Figure 3, the greatest alteration in resonance angle (1.87) is acquired for the present SPR design. The most extreme change in the resonance point shows the adjustment of the coupling state of the surface plasmon wave (SPW). The device's sensitivity should always be high. This means that the higher sensitivity sensor detects the minute variations in analyte (biomolecules) concentration, which shows that the sensor has superior sensing capabilities because it can easily detect minute RI variations in the structure.

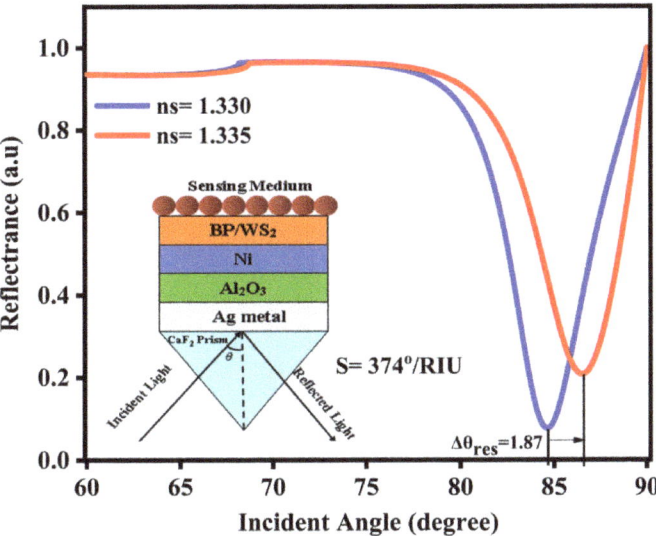

Figure 3. Variation of resonance angle vs. incident angle.

3.4. Quality Factor (QF)

The division of sensitivity with the full width at half maximum is defined as the quality factor. It is also called the figure of merit (FOM). On the other hand, the quality factor is the multiplication of the sensitivity by the detection accuracy. The figure of merit (FOM) is a quantity used to characterize the performance as the device increases [38].

$$QF = \frac{\Delta\theta_{res}}{\Delta n \times FWHM} \quad \left(\text{unit}: \text{RIU}^{-1}\right) \qquad (11)$$

The variation of the FOM with the different layer thicknesses of Al_2O_3 is shown in Figure 4. It can be observed that the FOM is at its maximum for the Ni thickness of 1 nm.

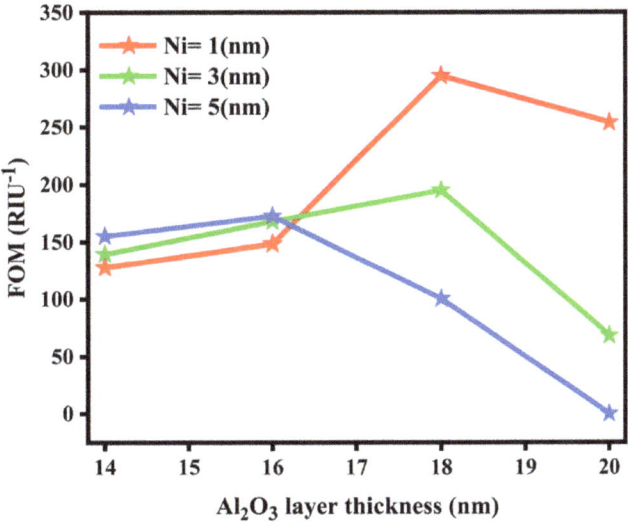

Figure 4. Variation of FOM vs. different layer thicknesses of Al_2O_3.

3.5. Full Width at Half Maximum (FWHM)

The incidence angle changes at the halfway point of the reflectance curve can be used to calculate the full width at half maximum (FWHM), and its values should be low to boost the FOM. The FWHM assumes a significant role in the sensor's execution because the majority of the parameters rely upon it. The width of the reflectance curve as the resonance angle shifts is measured by the FWHM. A low FWHM reduces the uncertainty in determining the resonance dip, and as a result improves the sensor's resolution. It is defined by [39]:

$$\text{FWHM} = \frac{1}{2}(\theta_{max} + \theta_{min}), \text{ (unit : degree)} \tag{12}$$

The FWHM with respect to the Ni thickness is shown in Figure 5.

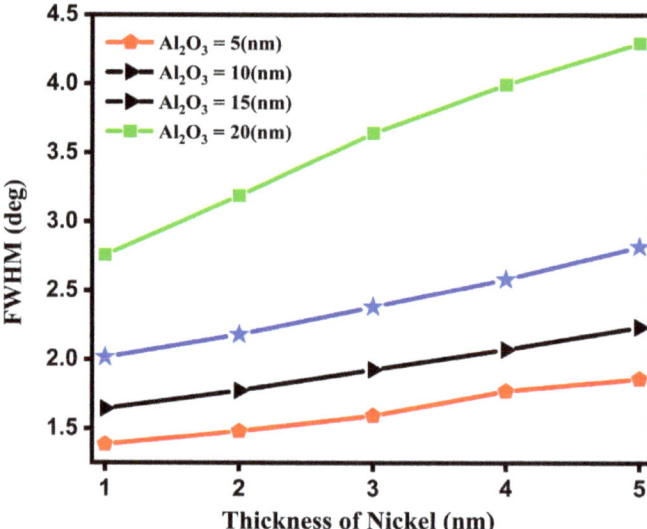

Figure 5. Variation of thickness of Ni vs. FWHM.

Figure 5 shows that the variations of Ni vs. FWHM give the best result for the FWHM at the 20 nm thickness of Al_2O_3 when the thickness of Ni varies from 1 nm to 5 nm. The FWHM is also used to calculate the quality factor and detection accuracy.

3.6. Detection Accuracy (DA)

This is conversely connected with the full width at half maximum. It is also termed the signal-to-noise ratio, and the ratio of the signal to the noise should be high as possible to better the device quality. Basically, it shows how the noise level is impacting the device structure. It is defined as [30]:

$$\text{DA} = \left[\frac{1}{\text{FWHM}}\right], \left(\text{degree}^{-1}\right) \tag{13}$$

The detection accuracy vs. layer thickness plot for Al_2O_3 is shown in Figure 6. It can be observed that the detection accuracy decreases with the layer thickness.

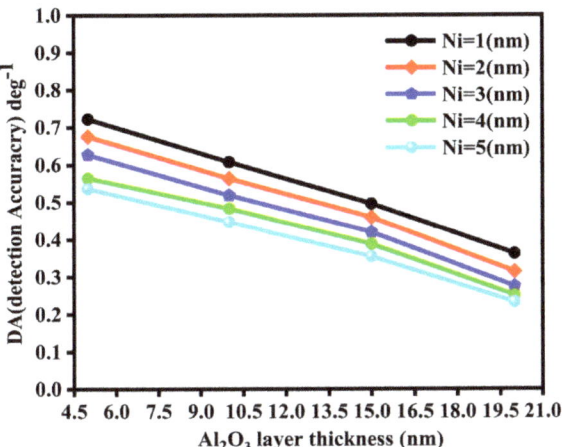

Figure 6. Plot of DA vs. different layer thickness for Al$_2$O$_3$.

3.7. Limit of Detection (LOD)

This is the difference in biomolecule fixation or analyte concentration in the detecting area, and it is also defined as the proportion of progress in the RI (Δn) with the change in resonance angle (Δθres) [30]:

$$\text{LOD} = \frac{\Delta n}{\Delta \theta_{res}} \times 0.001^0 \qquad (14)$$

where 0.001^0 is a very small shift in the detecting medium.

3.8. The Impact of the Refractive Index on the Reflectance Curves of Different SPR Sensor Structures

The curves of SPR reflectivity for the sensor in the detecting medium with the RI range from 1.330 to 1.335. The proposed SPR curve for the changing RI of an analyte (n_s = 0.005) with adjustment of the resonance angle (1.87°), sensitivity (374°/RIU), quality factor (56.211 RIU^{-1}), and DA (0.281 deg^{-1}) for the recommended design might be found in the SPR reflectance bends. The sensitivity and resonance angle shift are significantly bigger than with conventional sensors, as displayed in Table 3. The variation of the resonance angle with the RI of the sensing medium for the conventional and proposed sensor designs are plotted in Figure 7.

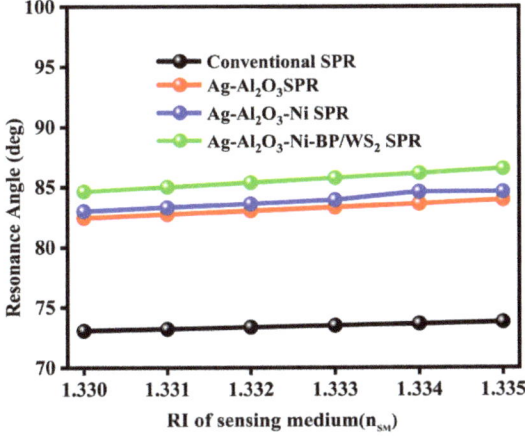

Figure 7. Variation of the resonance angle with Ri for different sensor designs.

3.9. Optimization of the Thicknesses of Al_2O_3 and Metal (Ni)

The refractive index of Al_2O_3 is small, so it should be selected because it minimizes the losses in performance of the biosensor, which is another characteristic contributing to the sensor's improvement [32]. Since the Al_2O_3 layer has small damping properties with a higher penetration rate of the surface plasmon in the sensing medium, this property helps to stop the corrosion phenomena and enhances the performance of the SPR sensor [40]. On the other hand, the cost of the SPR sensor is reduced by using the ferromagnetic material nickel (Ni) because it has incredible magneto-optical characteristics, magnetic qualities and small optical losses. At the optimum thickness of Ni (1 nm), the molecular absorption of light increases with minimum reflectance. The Ni layer is also used as a protective layer, which helps to raise the sensitivity of the SPR sensor [41]. Figure 8 shows a graph demonstrating the fluctuations in Al_2O_3 and Ni thickness and the varying sensitivities at different thicknesses. Figure 8a shows that the maximum sensitivity is achieved with a layer of Al_2O_3 of 20 nm. For metal, the best sensitivity is 374°/RIU (Ag). To determine the best sensitivity for the suggested SPR sensor, the thickness of Ni is optimized, as Figure 8b illustrates, whereby the sensitivity changes with the changes in Ni thickness. The optimal thickness ranges from 1 to 5 nm, with a maximum sensitivity of 374°/RIU and minimal reflection (R_{min}). Consequently, the proposed SPR sensor shows the best results when the thickness of the Ni layer is 1 nm.

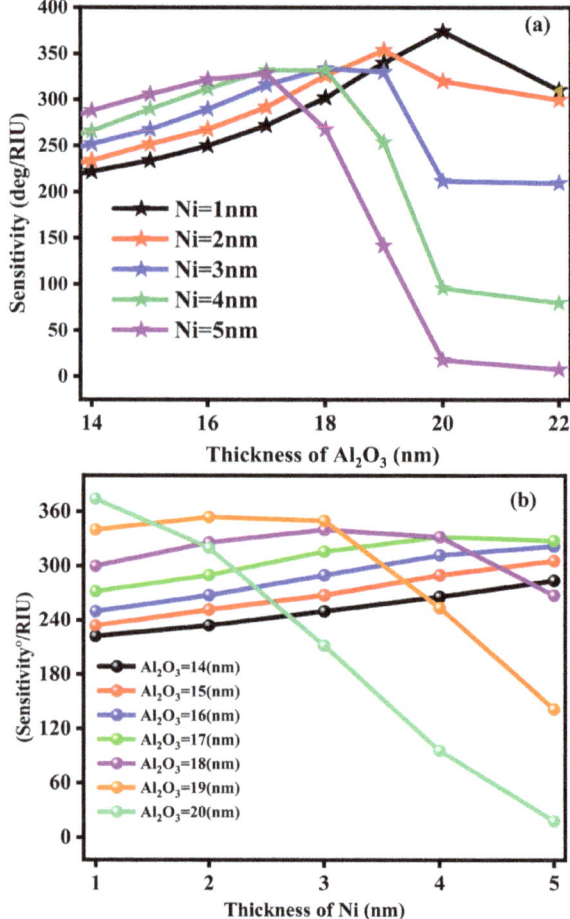

Figure 8. (**a**) Variation of sensitivity vs. Al_2O_3 (nm). (**b**) Variation of sensitivity vs. Ni (nm).

3.10. Parameter Analysis of the Projected SPR Biosensor

The design (CaF$_2$/Ag/Al$_2$O$_3$/Ni/BlueP/WS$_2$/sensing medium) arrangement has the most elevated sensitivity (374°RIU^{-1}) for sensor applications. It is additionally advantageous for energizing surface plasmons by changing over from the crystal-directed mode to the surface plasmon polariton (SPP) mode proficiently. The performance parameters, sensitivity, and QF show increments with high RI values. However, the DA (detection accuracy) diminishes, which causes the increase in FWHM increments with a huge change in the reflectivity bend. The FWHM and DA have an opposite relationship.

3.11. Clarification of Transvers Magnetic (TM) Field and Penetration Depth

The transverse magnetic (TM) field plot is discussed in this part. Utilizing the COMSOL Multiphysics programming software, the recommended SPR biosensor's transverse magnetic (TM) field variation is shown in Figure 9. The TM field likewise helps with the estimation of a vital measurement, the sensor's penetration depth. The distance of the electric field force diminishes to 1/e, which is called the penetration depth, and its value is 108.25 nm. Subsequently, when compared with past SPR sensor calculations, the proposed sensor has the best penetration depth, and along these lines is more delicate. To show the electromagnetic field, the effective mode index (EMI = 1.0926 − 0.31363 × i) is utilized [42].

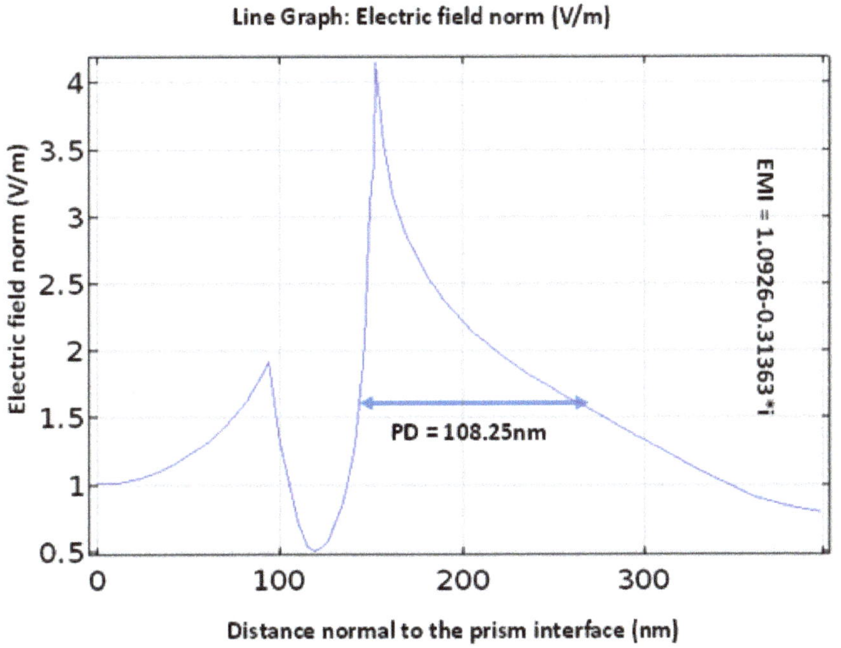

Figure 9. Penetration depth variation and transverse magnetic field.

4. Conclusions

In the work, the sensitivity was improved by utilizing a layer of Al$_2$O$_3$ over the metals silver (Ag) and nickel (Ni) on the top. A layer of 2D material, BlueP/WS$_2$, was utilized to upgrade the sensitivity and safeguard the device from corrosion. The greatest sensitivity was found for Ag metal (374°/RIU) and the primary arrangement for the most extreme sensitivity was CaF$_2$/Ag/Al$_2$O$_3$/Ni/BlueP/WS$_2$/SM. The performance qualities of the heterostructure-based SPR sensors, such as FWHM, detection accuracy (DA), and LOD, showed great correlations with traditional sensors for the appropriate scope of the RI from 1.330 to 1.335. From the above study, the projected SPR sensor configuration has incredible sensitivity and could be utilized in the biosensing field.

Author Contributions: Data curation: S., Y.A.-H., P.L., S.S., D.K.D. and A.U.; Formal analysis: M.F.A., P.L., H.M.A., H.A. and S.B.; Methodology: S., M.F.A., S.S. and H.M.A.; Project administration: Y.A.-H., D.K.D. and A.U.; Supervision: S.B.; Writing—original draft, P.L. and H.A. All authors have read and agreed to the published version of the manuscript.

Funding: This research work was funded by Institutional Fund Projects under grant no. (IFPIP: 1663-130-1442).

Data Availability Statement: Not applicable.

Acknowledgments: Authors gratefully acknowledge technical and financial support from the Ministry of Education and King Abdulaziz University, DSR, Jeddah, Kingdom of Saudi Arabia.

Conflicts of Interest: The authors declare no conflict of interest.

References

1. Schuller, J.A.; Barnard, E.; Cai, W.; Jun, Y.C.; White, J.S.; Brongersma, M.L. Plasmonics for extreme light concentration and manipulation. *Nat. Mater.* **2010**, *9*, 193–204. [CrossRef] [PubMed]
2. Gramotnev, D.K.; Bozhevolnyi, S.I. Plasmonics beyond the diffraction limit. *Nat. Photonics* **2010**, *4*, 83–91. [CrossRef]
3. Zhixun, L.; Yunying, S.; Yunfei, Y.; Yuanyuan, F.; Lvqing, B. Electro-Optic Hybrid Logic Gate Derived from a Silicon-Based Hybrid Surface Plasmon Polariton Waveguide. *J. Nanoelectron. Optoelectron.* **2022**, *17*, 298–304. [CrossRef]
4. Mao, Y.; Zhu, Y.; Jia, C.; Zhao, T.; Zhu, J. A Self-Powered Flexible Biosensor for Human Exercise Intensity Monitoring. *J. Nanoelectron. Optoelectron.* **2021**, *16*, 699–706. [CrossRef]
5. Syubaev, S.A.; Zhizhchenko, A.Y.; Pavlov, D.; Gurbatov, S.O.; Pustovalov, E.V.; Porfirev, A.P.; Khonina, S.N.; Kulinich, S.; Rayappan, J.B.B.; Kudryashov, S.I.; et al. Plasmonic Nanolenses Produced by Cylindrical Vector Beam Printing for Sensing Applications. *Sci. Rep.* **2019**, *9*, 19750. [CrossRef]
6. Yi, Z.; Liang, C.; Chen, X.; Zhou, Z.; Tang, Y.; Ye, X.; Yi, Y.; Wang, J.; Wu, P. Dual-Band Plasmonic Perfect Absorber Based on Graphene Metamaterials for Refractive Index Sensing Application. *Micromachines* **2019**, *10*, 443. [CrossRef]
7. Tabassum, R.; Kant, R. Recent trends in surface plasmon resonance based fiber–optic gas sensors utilizing metal oxides and carbon nanomaterials as functional entities. *Sens. Actuators B Chem.* **2020**, *310*, 127813. [CrossRef]
8. Saad, Y.; Selmi, M.; Gazzah, M.H.; Bajahzar, A.; Belmabrouk, H. Performance enhancement of a cop-per-based optical fiber SPR sensor by the addition of an oxide layer. *Optik* **2019**, *190*, 19. [CrossRef]
9. Homola, J.; Yee, S.S.; Gauglitz, G. Surface plasmon resonance sensors. *Sens. Actuators B Chem.* **1999**, *54*, 3–15. [CrossRef]
10. Kim, S.-K. Computational Sub-10 nm Plasmonic Nanogap Patterns by Block Copolymer Self-Assembly. *J. Nanoelectron. Optoelectron.* **2021**, *16*, 1063–1066. [CrossRef]
11. Kowalczyk, S.W.; Tuijtel, M.; Donkers, S.P.; Dekker, C. Unraveling Single-Stranded DNA in a Solid-State Nanopore. *Nano Lett.* **2010**, *10*, 1414–1420. [CrossRef]
12. West, P.; Ishii, S.; Naik, G.; Emani, N.K.; Shalaev, V.; Boltasseva, A. Searching for better plasmonic materials. *Laser Photon. Rev.* **2010**, *4*, 795–808. [CrossRef]
13. Naik, G.V.; Shalaev, V.M.; Boltasseva, A. Alternative Plasmonic Materials: Beyond Gold and Silver. *Adv. Mater.* **2013**, *25*, 3264–3294. [CrossRef] [PubMed]
14. Maurya, J.B.; Prajapati, Y.K.; Singh, V.K.; Saini, J.P.; Tripathi, R. Performance of graphene–MoS2 based surface plasmon resonance sensor using Silicon layer. *Opt. Quantum Electron.* **2015**, *47*, 3599–3611. [CrossRef]
15. Prajapati, Y.K.; Pal, S.; Saini, J.P. Effect of a Metamaterial and Silicon Layers on Performance of Surface Plasmon Resonance Biosensor in Infrared Range. *Silicon* **2018**, *10*, 1451–1460. [CrossRef]
16. Xiao, L.; Youji, L.; Feitai, C.; Peng, X.; Ming, L. Facile synthesis of mesoporous titanium dioxide doped by Ag-coated graphene with enhanced visible-light photocatalytic performance for methylene blue degradation. *RSC Adv.* **2017**, *7*, 25314–25324. [CrossRef]
17. Tang, N.; Li, Y.; Chen, F.; Han, Z. In situ fabrication of a direct Z-scheme photocatalyst by immobilizing CdS quantum dots in the channels of graphene-hybridized and supported mesoporous titanium nanocrystals for high photocatalytic performance under visible light. *RSC Adv.* **2018**, *8*, 42233–42245. [CrossRef]
18. Zhang, X.; Teng, S.Y.; Loy, A.C.M.; How, B.S.; Leong, W.D.; Tao, X. Transition Metal Dichalcogenides for the Application of Pollution Reduction: A Review. *Nanomaterials* **2020**, *10*, 1012. [CrossRef]
19. Sharma, N.K. Performances of different metals in optical fibre-based surface plasmon resonance sensor. *Pramana* **2012**, *78*, 417–427. [CrossRef]
20. Ordal, M.A.; Bell, R.J.; Alexander, R.W.; Long, L.L.; Querry, M.R. Optical properties of fourteen metals in the infrared and far infrared: Al, Co, Cu, Au, Fe, Pb, Mo, Ni, Pd, Pt, Ag, Ti, V, and W. *Appl. Opt.* **1985**, *24*, 4493–4499. [CrossRef]
21. Ehrenreich, H.; Philipp, H.R.; Olechna, D.J. Optical Properties and Fermi Surface of Nickel. *Phys. Rev. (Ser. I)* **1963**, *131*, 2469–2477. [CrossRef]
22. Shukla, S.; Sharma, N.K.; Sajal, V. Theoretical Study of Surface Plasmon Resonance-based Fiber Optic Sensor Utilizing Cobalt and Nickel Films. *Braz. J. Phys.* **2016**, *46*, 288–293. [CrossRef]

3. Peng, Q.; Wang, Z.; Sa, B.; Wu, B.; Sun, Z. Electronic structures and enhanced optical properties of blue phosphorene/transition metal dichalcogenides van der Waals heterostructures. *Sci. Rep.* **2016**, *6*, 31994. [CrossRef] [PubMed]
4. Chen, H.; Chen, Z.; Yang, H.; Wen, L.; Yi, Z.; Zhou, Z.; Dai, B.; Zhang, J.; Wu, X.; Wu, P. Multi-mode surface plasmon resonance absorber based on dart-type single-layer graphene. *RSC Adv.* **2022**, *12*, 7821–7829. [CrossRef]
5. Long, F.; Zhang, Z.; Wang, J.; Yan, L.; Zhou, B. Cobalt-nickel bimetallic nanoparticles decorated graphene sensitized imprinted electrochemical sensor for determination of octylphenol. *Electrochimica Acta* **2015**, *168*, 337–345. [CrossRef]
6. Wu, X.; Zheng, Y.; Luo, Y.; Zhang, J.; Yi, Z.; Wu, X.; Cheng, S.; Yang, W.; Yu, Y.; Wu, P. A four-band and polarization-independent BDS-based tunable absorber with high refractive index sensitivity. *Phys. Chem. Chem. Phys.* **2021**, *23*, 26864–26873. [CrossRef]
7. Alagdar, M.; Yousif, B.; Areed, N.F.; Elzalabani, M. Highly sensitive fiber optic surface plasmon resonance sensor employing 2D nanomaterials. *Appl. Phys. A* **2020**, *126*, 522. [CrossRef]
8. Naresh, V.; Lee, N. A Review on Biosensors and Recent Development of Nanostructured Materials-Enabled Biosensors. *Sensors* **2021**, *21*, 1109. [CrossRef]
9. Hasib, M.H.H.; Nur, J.N.; Rizal, C.; Shushama, K.N. Improved Transition Metal Dichalcogenides-Based Surface Plasmon Resonance Biosensors. *Condens. Matter* **2019**, *4*, 49. [CrossRef]
10. Gupta, B.; Sharma, A.K. Sensitivity evaluation of a multi-layered surface plasmon resonance-based fiber optic sensor: A theoretical study. *Sens. Actuators B Chem.* **2005**, *107*, 40–46. [CrossRef]
11. Ouyang, Q.; Zeng, S.; Jiang, L.; Qu, J.; Dinh, X.-Q.; Qian, J.; He, S.; Coquet, P.; Yong, K.-T. Two-Dimensional Transition Metal Dichalcogenide Enhanced Phase-Sensitive Plasmonic Biosensors: Theoretical Insight. *J. Phys. Chem. C* **2017**, *121*, 6282–6289. [CrossRef]
12. Shah, K.; Sharma, N.K. SPR based Fiber Optic Sensor Utilizing Thin Film of Nickel. In *AIP Conference Proceedings*; AIP Publishing LLC: College Park, MD, USA, 2018; Volume 2009, p. 020040.
13. Han, L.; Hu, Z.; Pan, J.; Huang, T.; Luo, D. High-sensitivity goos-hänchen shifts sensor based on bluep-tmdcs-graphene heterostructure. *Sensors* **2020**, *20*, 3605. [CrossRef] [PubMed]
14. Tabassum, R.; Mishra, S.K.; Gupta, B.D. Surface plasmon resonance-based fiber optic hydrogen sulphide gas sensor utilizing Cu–ZnO thin films. *Phys. Chem. Chem. Phys.* **2013**, *15*, 11868–11874. [CrossRef] [PubMed]
15. Yamamoto, M. Surface plasmon resonance (SPR) theory: Tutorial. *Rev. Polarogr.* **2002**, *48*, 209–237. [CrossRef]
16. Verma, R.; Gupta, B.D.; Jha, R. Sensitivity enhancement of a surface plasmon resonance based biomolecules sensor using graphene and silicon layers. *Sens. Actuators B Chem.* **2011**, *160*, 623–631. [CrossRef]
17. Vasimalla, Y.; Pradhan, H.S. Modeling of a novel K5 prism-based surface Plasmon resonance sensor for urea detection employing Aluminum arsenide. *J. Opt.* **2022**, 1–12. [CrossRef]
18. Meshginqalam, B.; Barvestani, J. Performance Enhancement of SPR Biosensor Based on Phosphorene and Transition Metal Dichalcogenides for Sensing DNA Hybridization. *IEEE Sens. J.* **2018**, *18*, 7537–7543. [CrossRef]
19. Sharma, A.K.; Kaur, B. Analyzing the effect of graphene's chemical potential on the performance of a plasmonic sensor in infrared. *Solid State Commun.* **2018**, *275*, 58–62. [CrossRef]
20. AlaguVibisha, G.; Nayak, J.K.; Maheswari, P.; Priyadharsini, N.; Nisha, A.; Jaroszewicz, Z.; Rajesh, K.; Jha, R. Sensitivity enhancement of surface plasmon resonance sensor using hybrid configuration of 2D materials over bimetallic layer of Cu–Ni. *Opt. Commun.* **2020**, *463*, 125337. [CrossRef]
21. Alagdar, M.; Yousif, B.; Areed, N.F.; ElZalabani, M. Improved the quality factor and sensitivity of a surface plasmon resonance sensor with transition metal dichalcogenide 2D nanomaterials. *J. Nanoparticle Res.* **2020**, *22*, 189. [CrossRef]
22. Agarwal, S.; Prajapati, Y.K.; Maurya, J.B. Effect of metallic adhesion layer thickness on surface roughness for sensing application. *IEEE Photonics Technol. Lett.* **2016**, *28*, 2415–2418. [CrossRef]

Significance of Hydroxyl Groups on the Optical Properties of ZnO Nanoparticles Combined with CNT and PEDOT:PSS

Keshav Nagpal [1], Erwan Rauwel [1], Elias Estephan [2], Maria Rosario Soares [3] and Protima Rauwel [1,*]

1. Institute of Forestry and Engineering, Estonian University of Life Sciences, 51014 Tartu, Estonia
2. LBN, University of Montpellier, 34193 Montpellier, France
3. CICECO, University of Aveiro, 3810-193 Aveiro, Portugal
* Correspondence: protima.rauwel@emu.ee

Abstract: We report on the synthesis of ZnO nanoparticles and their hybrids consisting of carbon nanotubes (CNT) and polystyrene sulfonate (PEDOT:PSS). A non-aqueous sol–gel route along with hydrated and anhydrous acetate precursors were selected for their syntheses. Transmission electron microscopy (TEM) studies revealed their spherical shape with an average size of 5 nm. TEM also confirmed the successful synthesis of ZnO-CNT and ZnO-PEDOT:PSS hybrid nanocomposites. In fact, the choice of precursors has a direct influence on the chemical and optical properties of the ZnO-based nanomaterials. The ZnO nanoparticles prepared with anhydrous acetate precursor contained a high amount of oxygen vacancies, which tend to degrade the polymer macromolecule, as confirmed from X-ray photoelectron spectroscopy and Raman spectroscopy. Furthermore, a relative increase in hydroxyl functional groups in the ZnO-CNT samples was observed. These functional groups were instrumental in the successful decoration of CNT and in producing the defect-related photoluminescence emission in ZnO-CNT.

Keywords: ZnO; ZnO-CNT; ZnO-PEDOT:PSS; nanoparticles; hybrids; hydroxyl groups; non-aqueous sol–gel; surface defects; photoluminescence

1. Introduction

Hybrid nanocomposites combining organic and inorganic counterparts have a multitude of applications, e.g., light-emitting diodes (LED), solar cells, and photodetectors [1–3]. Organic materials consist of polymers possessing remarkable properties, such as easy processing, flexibility, and good conductivity [4]. However, their high cost and lack of stability are obstacles for practical devices. On the other hand, inorganic materials present higher structural, chemical, and functional stability, as well as a high charge mobility, making them suitable for optoelectronic applications [5]. Therefore, the combination of organic with inorganic materials provides robust multifunctional nanocomposites with applications in flexible electronic and photonic devices [6–8].

Conducting polymers such as poly (3,4-ethylenedioxythiophene) poly(styrenesulfonate) (PEDOT:PSS) are already being incorporated into organic thin film transistors, organic LED, organic solar cells, capacitors, batteries, and thermoelectric devices, as well as technologies such as touch screens and electronic papers [9,10]. In addition, PEDOT:PSS is mechanically stable and highly flexible. Various combinations of PEDOT:PSS with inorganic materials, such as SnO_2, TiO_2, CdS, CdSe, ZnO and metal nanostructures, have been investigated to that end [11,12]. Among these inorganic nanomaterials, ZnO is promising due its wide band gap of 3.37 eV, large exciton binding energy of 60 meV, high chemical stability, and remarkable electrical and optical properties [13]. Moreover, the high surface-to-volume ratio of ZnO nanoparticles implies a spontaneous presence of surface defects, including oxygen vacancies (V_O), oxygen interstitials (O_i), and zinc interstitials (Zn_i). Therefore, in addition to the UV emission, known as the near-band emission (NBE), ZnO nanoparticles emit within the entire visible spectrum, also known as defect-level emission (DLE) [14]. The latter depends on both the surface and the volume defects introduced during the synthesis of the nanoparticles [15,16]. For example, ZnO nanoparticles prepared by aqueous sol–gel routes tend to emit higher NBE and a negligible DLE [17]. On the other hand, for ZnO nanoparticles prepared by non-aqueous sol–gel routes, the emission depends on the presence of hydrates in the precursor [18,19]. In fact, hydrates in the precursor contribute to the enhancement of NBE due to improved oxidation of ZnO during synthesis [18]. On the other hand, adsorption of hydroxyl groups on the surface of ZnO nanoparticles has been shown to increase the visible PL emission [20]. The chemisorption of oxygen radicals from air on the surface of ZnO nanoparticles also augments green emission from them [21]. In general, small nanoparticles possess a high surface-to-volume ratio, and therefore harbor higher amounts of surface defects. For larger ZnO nanoparticles, defects can be both surface and volume related [22].

Recently, ZnO-CNT nanohybrids have attracted considerable interest due to their high stability and superior photonic, electrochemical and electromagnetic properties, which originate from interfacial effects. In a previous study, we successfully passivated ZnO surface states by combining them with CNT via sonication [17]. In this work, we carry out ZnO nanoparticle synthesis via a non-aqueous sol–gel route with hydrated (zinc acetate dihydrate ($Zn(CH_3CO_2)_2.2H_2O$) and anhydrous (zinc acetate anhydrous ($Zn(CH_3CO_2)_2$) precursors. In a study by Šarić et al., it was shown that with similar precursors, ZnO precipitation could be promoted through an esterification reaction that generates water upon the addition of acetic acid [23]. In our reaction, only absolute ethanol is used as a solvent. It plays a crucial role in controlling the size and shape of ZnO nanoparticles, and consequently, in the formation of various surface defects. Due to the addition of sodium hydroxide in this work, the basic character of the solution prevents any esterification reaction and in turn, no water molecules are formed. This route therefore enables the formation of very small ZnO nanoparticles functionalized with hydroxyl groups, promoting the decoration of CNT with ZnO. We then decorated CNT with ZnO nanoparticles in order to create a hybrid nanocomposite. Subsequently, we fabricated a second type of hybrid nanocomposite consisting of ZnO-PEDOT:PSS and compared the evolution of the surface defects to ZnO-CNT. These optical properties are discussed in terms of synthesis conditions, crystal structure, chemical properties, and the morphology of ZnO nanoparticles and their hybrids. The objective is to use these hybrid materials in LED. Therefore, finding a way to control these surface defects or trap states for LED applications is a priority, as they are detrimental to device properties.

2. Materials and Methods

2.1. Synthesis

2.1.1. ZnO

Two different zinc precursors, $Zn(CH_3CO_2)_2.2H_2O$ (99.5%, Fisher Scientific, Loughborough, UK) and $Zn(CH_3CO_2)_2$ (99.9%, Alfa Aesar, Kandel, Germany), were used for the synthesis of ZnO nanoparticles via non-aqueous sol–gel routes. Sodium hydroxide (NaOH) (99.9%, Aldrich) was used as a reducing agent. All the chemicals used were of analytic reagent grade. To prepare 0.05 M solutions of zinc precursors, 219.5 mg of $Zn(CH_3CO_2)_2.2H_2O$ or 183.48 mg of $Zn(CH_3CO_2)_2$ were dissolved in 20 mL absolute ethanol in a beaker placed in a water bath. The solutions were maintained at 65 °C under continuous magnetic stirring until the precursors were completely dissolved in absolute ethanol. Furthermore, a solution of 0.10 M NaOH in 20 mL absolute ethanol was prepared. The NaOH solution was added dropwise to the zinc precursor solutions. Thereafter, the mixtures were maintained at 65 °C for 2 h after which they were cooled to ambient temperature. White ZnO precipitates settled at the bottom of the reaction vessel. The resulting solutions containing ZnO nanoparticles were then centrifuged at 4500 rpm for 6 min, followed by drying for 24 h in air at 60 °C. This resulted in an agglomeration of ZnO nanoparticles in the form of a pellet, which is typical after drying nanoparticles synthesized via sol–gel routes. These pellets were thereafter crushed using a pestle and mortar to obtain a very fine powder of ZnO nanoparticles.

2.1.2. ZnO-CNT Hybrids

For the preparation of ZnO-CNT hybrids, firstly, a solution of CNT was prepared by mixing 4 mg of CNT in 50 mL absolute ethanol and sonicating until a homogenous mixture was obtained. As before, 0.05 M zinc precursor solutions were prepared in 20 mL absolute ethanol. To prepare 0.10 M NaOH solutions, ~80 mg of NaOH was added to 19 mL absolute ethanol, in which 1 mL CNT mixture (~0.08 mg) was added. The final mixtures were sonicated and added dropwise to zinc precursor solutions. Thereafter, the reaction was completed as described earlier, and ZnO-CNT hybrid pellets were obtained. These pellets were gently crushed to obtain fine black powders of ZnO-CNT nanohybrids.

2.1.3. ZnO-PEDOT:PSS Hybrids

For the preparation of ZnO-PEDOT:PSS hybrids, 20 mg of the as-synthesized ZnO nanoparticles were taken, to which 400 mg of PEDOT:PSS (as purchased) was added. The mixtures were sonicated for 1 h and dried at 70 °C for 24 h. The dried mixtures were further gently crushed to obtain blue powders with agglomerated particles of ZnO-PEDOT:PSS nanohybrids.

2.2. Characterization

X-ray diffraction patterns were collected in Bragg–Brentano geometry using a Bruker D8 Discover diffractometer (Bruker AXS, Germany) with CuKα1 radiation (λ = 0.15406 nm) selected by a Ge (111) monochromator and LynxEye detector. Transmission electron microscopy (TEM) was carried out on a Tecnai G2 F20 (Netherlands) is a 200 kV field emission gun (FEG) for high-resolution and analytical TEM/STEM. It provided a point-to-point resolution of 2.4 Å. XPS measurements were performed at room temperature with a SPECS PHOIBOS 150 hemispherical analyzer (SPECS GmbH, Berlin, Germany)) with a base pressure of 5×10^{-10} mbar using monochromatic Al K alpha radiation (1486.74 eV) as excitation source operated at 300 W. The energy resolution as measured by the FWHM of the Ag $3d_{5/2}$ peak for a sputtered silver foil was 0.62 eV. The spectra were calibrated with respect to the C1s at 284.8 eV. The optical absorbance of ZnO, ZnO-CNT and ZnO-PEDOT:PSS nanohybrids was determined using a NANOCOLOR UV-VIS II spectrometer (MACHEREY-NAGEL, Germany) in the 200–900 nm region. The band gap of ZnO, ZnO-CNT and ZnO-PEDOT:PSS nanohybrids was subsequently calculated using Tauc plots. PL spectroscopy was carried out at room temperature with an excitation wavelength of 365 nm of an LSM-365A LED (Ocean Insight, USA) with a specified output power of 10 mW. The emission was collected by FLAME UV-Vis spectrometer (Ocean optics, USA) with a spectral resolution 1.34 nm. Optical images of ZnO were taken under a UV lamp ZLUV220 (China) with an excitation source of 365 nm. Raman spectra were collected using a WITec Confocal Raman Microscope System alpha 300R (WITec Inc., Ulm, Germany). Excitation in confocal Raman microscopy is generated by a frequency-doubled Nd:YAG laser (New-port, Irvine, CA, USA) at a wavelength of 532 nm, with 50 mW maximum laser output power in a single longitudinal mode. The system was equipped with a Nikon (Otawara, Japan) objective with a X20 magnification and a numerical aperture NA = 0.46. The acquisition time of a single spectrum was set to 0.5 s.

3. Results

Table 1 provides a list of ZnO, ZnO-CNT and ZnO-PEDOT:PSS samples synthesized in this work. Samples ZnO-D and ZnO-A correspond to ZnO nanoparticles synthesized from Zn(CH$_3$CO$_2$)$_2$·2H$_2$O and Zn(CH$_3$CO$_2$)$_2$ precursors, respectively. Samples ZnO-D-CNT, ZnO-A-CNT, ZnO-D-PEDOT:PSS and ZnO-A-PEDOT:PSS correspond to the hybrids of samples ZnO-D and ZnO-A with CNT and PEDOT:PSS, respectively. In this study, the terms ZnO samples refer to samples ZnO-D and ZnO-A; ZnO-CNT hybrids refer to samples ZnO-D-CNT and ZnO-A-CNT and ZnO-PEDOT:PSS hybrids refer to samples ZnO-D-PEDOT:PSS and ZnO-A-PEDOT:PSS.

Table 1. List of ZnO-CNT hybrids and ZnO-PEDOT:PSS hybrids synthesized in this work.

Sample Name	CNT (wt%)	PEDOT:PSS (wt%)
ZnO-D-CNT	~1	-
ZnO-A-CNT	~1	-
ZnO-D-PEDOT:PSS	-	~95
ZnO-A-PEDOT:PSS	-	~95

3.1. Structure and Morphology

The XRD patterns of ZnO samples ZnO-D and ZnO-A are shown in Figure 1. The peaks (100), (002), (101), (102), and (110) correspond to the hexagonal Wurtzite structure (a = 3.25 Å and c = 5.20 Å) of ZnO (JCPDS, Card Number 36-1451). No secondary phases are visible in the XRD patterns, indicating that single-phase ZnO nanoparticles were formed. In addition, XRD patterns illustrate that both samples ZnO-D and ZnO-A exhibit very small particle sizes due to broader XRD peaks. The size of nanoparticles was estimated using the Scherrer equation [24].

$$D = \frac{0.9\lambda}{\beta \cos\theta} \tag{1}$$

where D is particle size, λ (=0.15406 nm) is the wavelength of incident X-ray beam, β is FWHM in radians, and θ is Bragg's diffraction angle. Size calculation was carried out by considering the highest-intensity (101) peak. The calculated ZnO nanoparticle sizes of samples ZnO-D and ZnO-A were ~9 nm and ~5 nm, respectively. However, the actual size and shape of ZnO nanoparticles were confirmed from TEM studies as discussed below.

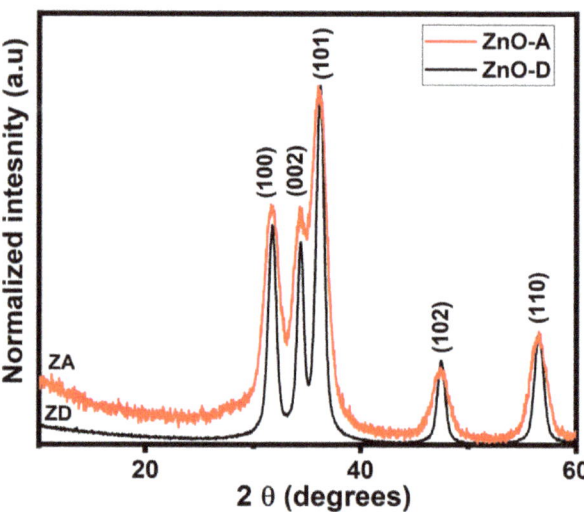

Figure 1. Normalized XRD patterns of samples ZnO-D and ZnO-A.

The morphological features of the as-synthesized ZnO samples, ZnO-CNT hybrids and ZnO-PEDOT:PSS hybrids were studied by TEM, as shown in Figure 2. TEM images in Figure 2a,b consist of overviews of the as-grown samples ZnO-D and ZnO-A, respectively. The micrographs reveal spherical nanoparticles of uniform size that tend to agglomerate. With the help of size distribution histograms of the as-synthesized ZnO samples, we estimate an average nanoparticle size of ~5.2 nm and ~4.8 nm for ZnO-D and ZnO-A, respectively. Figure 2c is a high-resolution TEM (HRTEM) image of sample ZnO-A, where two ZnO nanoparticles are oriented along the [0001] zone axis of the basal plane of the Wurtzite structure. Figure 2d,e are low-magnification TEM images of the samples ZnO-D-CNT and ZnO-A-CNT, respectively. ZnO nanoparticles dominate the TEM images due to the low wt% (~1 wt%) of CNT in the samples. HRTEM images of ZnO-CNT are presented in Figure 2g,h. The walls of the CNT are clearly visible along with nanoparticles decorating them. We observe that the presence of CNT does not alter the crystallinity or the size distribution of the nanoparticles, and average sizes of ~5.7 nm and ~4.7 nm were retained. The micrographs therefore clearly indicate successful decoration of the nanoparticles on the walls of the CNT. In our study, the nanotubes were functionalized by sonication in pure ethanol; hence, the most likely functional groups present are carboxyl (COOH) that can be broken down into carbonyl (C-O) and hydroxyl (OH) [25]. These functional groups promote a covalent bonding between the CNT and ZnO nanoparticles, necessary for the decoration of CNT. Figure 2f,i are the TEM and scanning transmission electron microscopy (STEM) images of ZnO-D-PEDOT:PSS and ZnO-A-PEDOT:PSS samples, respectively. PEDOT:PSS appears as flakes without any noticeable agglomeration of ZnO nanoparticles in the polymer layer. However, some areas of PEDOT:PSS are more densely packed with ZnO nanoparticles. The insets of Figure 2f,i are high-magnification TEM images of the samples emphasizing on their homogeneous distribution in the PEDOT:PSS matrix.

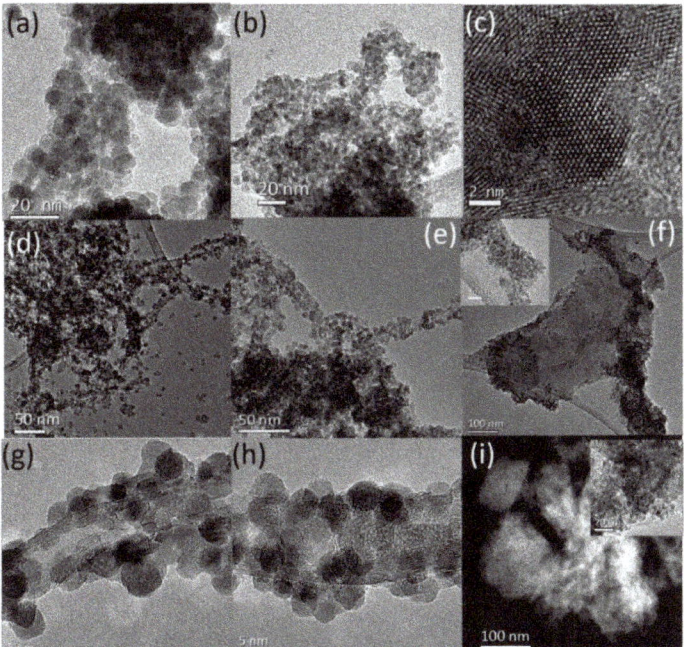

Figure 2. Overview TEM images of samples (**a**) ZnO-D, (**b**) ZnO-A, (**c**) HRTEM image of ZnO-A nanoparticles. Overview TEM images of (**d**) ZnO-D-CNT, (**e**) ZnO-A-CNT, (**f**) ZnO-D-PEDOT:PSS. HRTEM images of (**g**) ZnO-D-CNT, (**h**) ZnO-A-CNT and (**i**) STEM image of ZnO-A-PEDOT:PSS.

The high-resolution XPS spectra of the C 1s and O 1s regions of the as-synthesized and hybrid ZnO nanoparticles are shown in Figures 3 and 4, respectively. For the C 1s spectra of the as-synthesized samples in Figure 3a,d, the photoelectron peak at 284.8 eV corresponds to adventitious carbon [26]. Several carbon bonds are present in the samples, such as C-OH, O=C-O originating from the NaOH and acetate precursors used in the syntheses [27]. Both ZnO-A and ZnO-D contain oxygen and hydroxyl groups that are chemisorbed. The C 1s region of ZnO-CNT in Figure 3b,e manifests an additional peak corresponding to sp^2

hybridization of C atoms in the CNT at a binding energy of 283 eV. In addition, the C-OH peak is relatively more intense for the ZnO-A-CNT sample compared to the as-synthesized ZnO sample in Figure 3d, which could indicate an increase of hydroxyl groups or oxygen vacancies [18]. In fact, sonication of CNT in ethanol engenders a breakdown of the sidewalls which then produces C-dangling bonds [28]. After sonication and during the initial stage of synthesis, CNT were mixed in ethanol and heated to a temperature of 65 °C for 2 h during which a solution of NaOH was added dropwise. Considering the hydroxyl rich conditions, the attachment of OH groups to C-dangling bonds is likely. For the ZnO-D-based samples in Figure 3a–c, the relative intensities of the various peaks in the C1s region are similar unlike the ZnO-A-based samples. In addition, in Figure 3f, a decrease in the O=C-O and C-OH peak intensities relative to the C-C peak for sample ZnO-A-PEDOT:PSS is observed indicating an oxygen-deficient or -reduced PEDOT:PSS polymer.

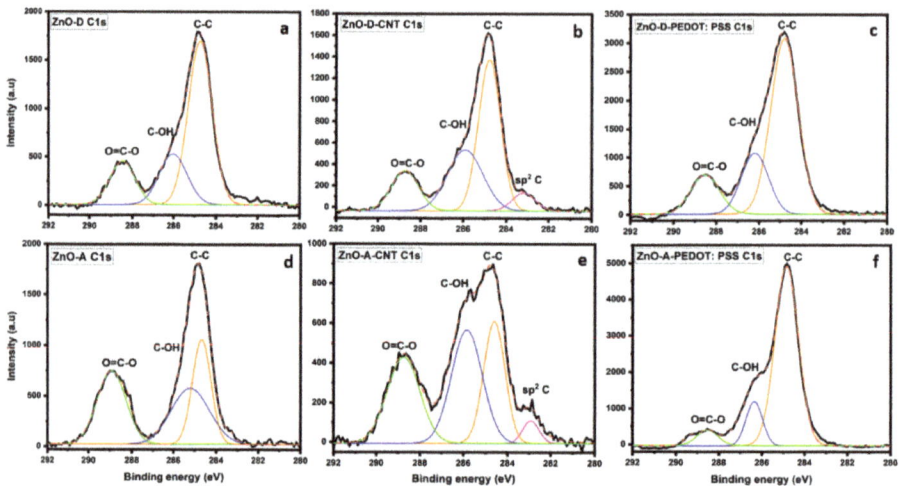

Figure 3. High-resolution XPS spectra of the C1s region of (**a**) ZnO-D, (**b**) ZnO-D-CNT, (**c**) ZnO-D-PEDOT:PSS, (**d**) ZnO-A, (**e**) ZnO-A-CNT and (**f**) ZnO-A-PEDOT:PSS.

The high-resolution spectra of the O 1s region of the ZnO samples, ZnO-CNT hybrids and ZnO-PEDOT:PSS hybrids in Figure 4 consist of several peaks, including lattice oxygen peak of ZnO or the Zn-O bond. Additionally, for the samples ZnO-D (Figure 3a), ZnO-D-CNT (Figure 3b), ZnO-A (Figure 3d) and ZnO-A-CNT (Figure 3e), photoelectron peaks that correspond to hydroxyl groups are also visible. In particular, the photoelectron peak at around 531.5 eV is attributed to Zn-OH bonds as well as oxygen vacancies [22]. TEM analysis estimated an average ZnO nanoparticle size of 5 nm, implying a very high surface-to-volume ratio. In such small nanoparticles, surface oxygen vacancies are prevalent. Since the C 1s region contains oxygen or hydroxyl components and the O 1s region contains carbon and hydroxyl components, it therefore suggests that hydroxyl groups are responsible for the decoration of CNT with ZnO. This directly implies that hydroxyl groups enable the anchoring of ZnO on CNT surface through covalent bonding with carbon, as there is no indication of Zn-C bonds. TEM images clearly indicate that ZnO nanoparticles grow directly on the CNT sidewalls through Zn-O/OH-C bonds. Furthermore, the O 1s region of the ZnO-PEDOT:PSS hybrids of Figure 4c,f display additional peaks, along with differences in relative intensities of peaks compared to ZnO and ZnO-CNT hybrids. In these samples, the characteristics of the PEDOT:PSS polymer is more dominant. In fact two peaks—C-O-C of PEDOT at 532.7 eV and O=S of PSS at 531.7 eV—are visible as well as a third peak of Zn-O [29]. In general, the ZnO-D lattice, i.e., as-synthesized ZnO-D or ZnO-D in the nanohybrids, shows a more stable oxygen component, when considering the C 1s spectra of Figure 5a–c, where the relative intensities of O=C-O, C-OH and C-C peaks are rather constant. However, for the ZnO-A-PEDOT:PSS, the PEDOT peak is less intense than the PSS peak. In fact, PEDOT:PSS macromolecule consists of PEDOT that is positively charged, highly conductive, and hydrophobic. On the other hand, PSS is negatively charged, insulating and hydrophilic. If we consider that the nanoparticles were dispersed in an aqueous solution of PEDOT:PSS, then the adsorption of hydroxyl groups on the surface of the ZnO nanoparticles is inevitable. From the relative intensities of various peaks of the O1s region in Figure 4a,d, ZnO-A tends to adsorb a higher quantity of hydroxyl groups than ZnO-D. Consequently, the surface of ZnO-A is more electronegative with a propensity to the positively charged PEDOT. The O 1s region of ZnO-A-PEDOT:PSS consists of a less intense C-O-C peak and a highly intense O=S peak compared to ZnO-D-PEDOT:PSS. The reduction in the relative intensity of the C-O-C peak suggests that either PEDOT was removed or degraded on adding ZnO-A. In addition, the shift in the Zn-O and C-O-C peaks to higher binding energies confirms the formation of a covalent bond between the C of PEDOT and OH groups present on the ZnO surface. The higher binding energy of the Zn-O peak along with an increase in its intensity indicates that the configuration for the lattice oxygen of ZnO-A becomes more stable.

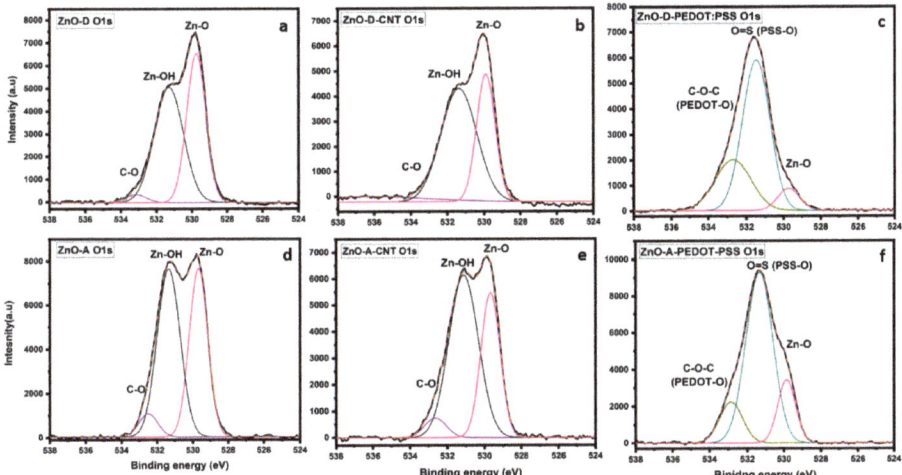

Figure 4. High-resolution XPS spectra of the O 1s region, (**a**) ZnO-D, (**b**) ZnO-D-CNT, (**c**) ZnO-D-PEDOT:PSS, (**d**) ZnO-A, (**e**) ZnO-A-CNT and (**f**) ZnO-A-PEDOT:PSS.

The vibrational properties of the ZnO nanoparticles and their hybrids were investigated using Raman spectroscopy. The results obtained from Raman spectroscopy complement those obtained via XPS. In fact, chemical and structural changes can be evaluated simultaneously on ZnO and CNT or PEDOT:PSS using Raman spectroscopy. Figure 5a compares the different vibrational modes obtained from these samples in the range of 100–800 cm^{-1}. The first-order phonon modes obtained at ~440 cm^{-1}, ~585 cm^{-1} and ~667 cm^{-1}, correspond to E_{2H}, E_1 (LO) and E_2 (TO) modes, respectively [21]. Other modes obtained at ~320 cm^{-1} and 506 cm^{-1} are multiphonon scattering modes that correspond to the $E_{2H}-E_{2L}$ and E_1(TO) + E_{2L} modes, respectively [30]. The E_{2H}, $E_{2H}-E_{2L}$, E_1 (LO) modes involve the oxygen component of ZnO. More specifically, the E_{2H} at 440 cm^{-1} corresponds to lattice oxygen, whereas the E_1 (LO) corresponds to oxygen-related defects [31]. For all the samples, the E_{2H} mode intensities are high, implying that the ZnO lattice structure is unaffected on hybridizing with CNT or PEDOT:PSS. However, the relative intensity of the E_1 (LO) band increases in the nanocomposites, indicating an increased number of surface defects [32]. The attachment of ZnO on the sidewalls of the CNT through hydroxyl functional groups indicates that the interfacial region and, therefore, the surface of ZnO are highly defective. Additionally, the E_{2H} peak for ZnO-PEDOT:PSS samples has shifted to a higher wavenumber of 445 cm^{-1} owing to chemical interactions between PEDOT and ZnO. This peak is more intense for ZnO-A-PEDOT:PSS than ZnO-D-PEDOT:PSS, which once again supports that ZnO-A has a more stable lattice configuration in PEDOT:PSS.

Raman signatures lower than 300 cm^{-1} are assigned to the vibrations of Zn_i, and those above 300 cm^{-1} are assigned to the vibrations of oxygen atoms [33]. The peak a 275 cm^{-1} has been attributed to Zn_i or Zn_i clustering [34,35]. The intensity of this mode increases relative to the other modes in the CNT-based nanocomposites and is the highest for PEDOT:PSS-based nanocomposites. This suggests that the amount of Zn_i is higher than Vo in the hybrid samples. Another mode at 526 cm^{-1} is observed for the PEDOT:PSS nanocomposites, corresponding to the combination of Vo and Zn_i [34]. A lower-intensity peak at the same localization is also visible in the ZnO-CNT-based samples. In general the relative intensity of this combined mode increases in the hybrid samples owing to an increase in Zni. In Figure 5b, Raman bands from 1200–1800 cm^{-1} of pristine CNT are compared to those of ZnO-CNT. The D-band at 1341 cm^{-1} for pristine CNT redshifts for ZnO-CNT to ~1351 cm^{-1}. A similar redshift in the G-band from 1579 cm^{-1} to ~1592 cm^{-1} is also observed. These redshifts further confirm the presence of oxygen or hydroxy groups on the CNT surface [36]. The (*) marked peaks in ZnO-CNT samples are assigned to C-O bond vibrations from the acetate precursor used during synthesis [21]. Infrared spectroscopy studies of hydrogen adsorption on ZnO suggest that OH and H are adsorbed simultaneously [37]. In fact, the dissociation of hydrogen followed by its adsorption manifests as a change in the corresponding vibrational frequency, including stretching vibrations of Zn-H and O-H, which are very different from the free hydroxyl group vibrational frequency [38]. However, hydrogen adsorption is more likely on prismatic surfaces, implying that facetted ZnO nanoparticles would be more susceptible to hydrogen adsorption [39]. However, for successful hydrogen adsorption, firstly, a more acidic environment is required when working in aqueous media, or a high pressure when working in gaseous media. In addition, the nanoparticles presented in this study are spherical and not facetted. In our case, the NaOH-rich conditions provide a basic environment that is advantageous to the adsorption of hydroxyl groups, further promoted by the presence of V_o.

Figure 5c shows the Raman spectra of ZnO-PEDOT:PSS samples in the range (900–1700 cm^{-1}), where the contributions from PSS and PEDOT vibrational modes are the most significant. Two typical PSS vibrational modes at 988 cm^{-1} and 1097 cm^{-1} are observed [40]. The vibrational modes of PEDOT observed at 1263 cm^{-1}, 1369 cm^{-1} 1436 cm^{-1} and 1517 cm^{-1} correspond to C_α-C_α, C_β-C_β, symmetrical C_α=C_β and asymmetrical C_α=C_β stretching vibrational modes, respectively. In the ZnO-D-PEDOT:PSS samples, the symmetrical vibrational mode at 1436 cm^{-1} is redshifted compared to the pristine PEDOT:PSS (~1440 cm^{-1}) [41]. However, this mode is slightly more redshifted in the ZnO-A-PEDOT:PSS, suggesting a slightly higher benzoid (coil) to quinoid (linear structural transition [41,42]. The PEDOT chains of linear conformation tend to increase the conductivity of the polymer due to a stronger covalent bonding with ZnO. Additionally asymmetrical C_α=C_β bonds of PEDOT have similar intensities for both samples, whereas C_α-C_α and C_β-C_β bonds for sample ZnO-D-PEDOT:PSS are more intense than sample ZnO-A-PEDOT:PSS. On the other hand, the asymmetrical C_α=C_β bond of PEDOT at 1517 cm^{-1} is more intense for ZnO-A-PEDOT:PSS samples. This implies that the bonds in the PEDOT chain have undergone structural modification provoking a breakdown in symmetry. This again suggests that PEDOT was degraded or removed from the macromolecule upon combining with ZnO-A.

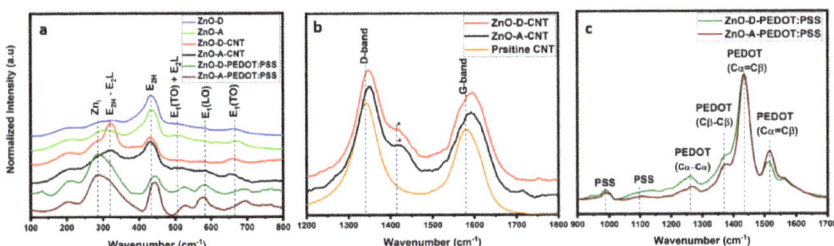

Figure 5. Normalized Raman spectra of (**a**) ZnO-D, ZnO-A, ZnO-D-CNT, ZnO-A-CNT, ZnO-D-PEDOT:PSS and ZnO-A-PEDOT:PSS from 100-800 cm^{-1}, (**b**) pristine CNT, ZnO-D-CNT and ZnO-A-CNT from 1200-1800 cm^{-1}, and (**c**) ZnO-D-PEDOT:PSS and ZnO-A-PEDOT:PSS from 900–1600 cm^{-1}.

3.2. Optical Properties

The band gaps of the ZnO samples, ZnO-CNT hybrids and ZnO-PEDOT:PSS hybrids were calculated via UV-Vis absorption spectroscopy followed by Tauc plots, presented in Figure 6. The band gaps of these samples range from 3.11 to 3.3 eV, which correspond to the theoretical band gap of ZnO, implying that the absorbance in the nanocomposites is dominated by ZnO. Depending on the synthesis routes, variations in the band gaps of ZnO have been observed [43]. The absorption spectra of ZnO samples revealed a sharp shoulder at ~3.3 eV, stretching down to 2.0 eV, whereas, a broader shoulder at ~3.3 eV stretching down to ~1.5 eV was observed for the CNT hybrids [44].

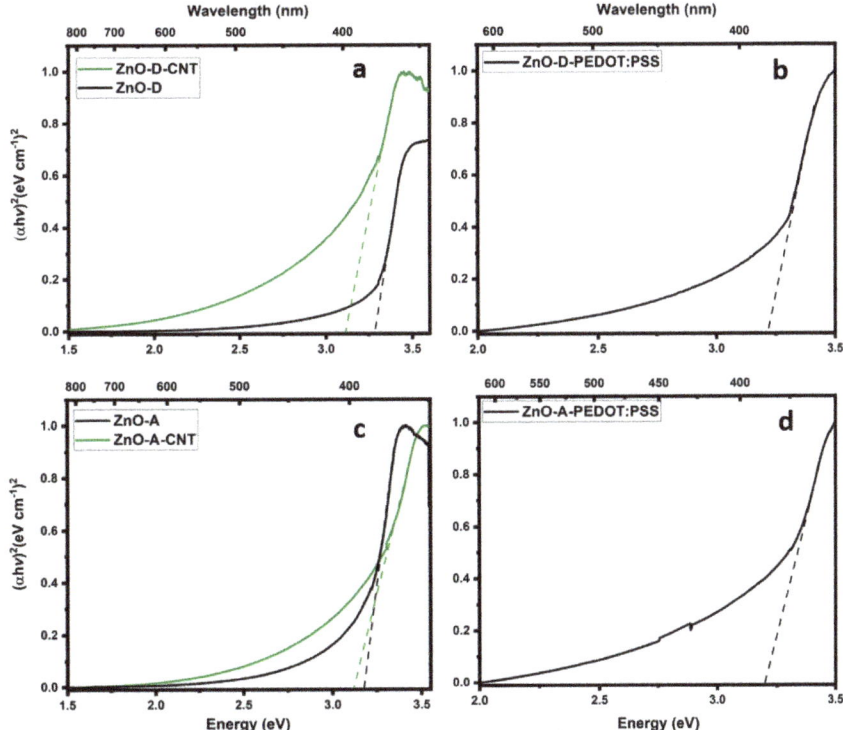

Figure 6. Tauc plot of (**a**) ZnO-D and ZnO-D-CNT, (**b**) ZnO-D-PEDOT:PSS, (**c**) ZnO-A and ZnO-A-CNT, and (**d**) ZnO-A-PEDOT:PSS.

The emission properties of the as-synthesized ZnO samples, ZnO-CNT hybrids, and ZnO-PEDOT:PSS hybrids were investigated at room temperature. A 365 nm (3.4 eV) excitation source was used to induce band-to-band transitions in these samples with band gaps between 3.11 eV and 3.3 eV. The PL spectra in Figure 7a–f present typical PL emission characteristics of ZnO nanoparticles, consisting of the NBE and DLE [45]. The similarities in emission peak localizations indicate that the emissions mainly originate from ZnO nanoparticles, for both the freestanding and hybrid nanocomposites. However, there are significant changes in the overall quantum efficiencies and intensities of certain emission peaks of the hybrid samples. This suggests that interfacial bonding between ZnO nanoparticles and CNT or PEDOT:PSS via OH groups plays an important role in excitonic separation and recombination. The probable origin of the DLE is the combination of several point defects, such as oxygen interstitials (O_i), oxygen vacancies (V_O), zinc vacancies (V_{Zn}), zinc interstitials (Zn_i), and their complexes [22,46] that are related to the presence of hydrates in the ZnO precursor. In addition, the NBE to DLE ratio is useful in evaluating the crystalline quality of ZnO. Moreover, an average nanoparticle size of 5 nm indicates a high surface-to-volume ratio, which in turn denotes a high amount of surface defects. These surface traps also consist of chemisorbed species, allowing additional radiative or non-radiative recombination mechanisms, which would alter the quantum efficiency of the DLE.

The doubly ionized oxygen vacancy, i.e., V_O^{++}, or surface oxygen vacancy at 2.2 eV is dominant in all the samples due to the high surface-to-volume ratio. The 2.2 eV transition is associated with the capture of a hole by V_O^+ from surface charges to form V_O^{++} [47]. The single ionized oxygen vacancy V_O^+ or volume oxygen vacancy emits at ~2.5 eV. Both types of vacancies produce green luminescence in ZnO. Additionally, the intensity of the green emission can be strongly influenced by free carriers on the surface, especially for nanoparticles with very small sizes [47,48]. Since PL measurements were performed in air, it is likely that hydroxyl groups or oxygen molecules are adsorbed on the surface of the nanoparticles. The chemisorbed oxygen species provoke an upward band bending in the as-synthesized ZnO nanoparticles (Figure 7g), which allows V_O^+ to convert into V_O^+ through the tunneling of surface-trapped holes to deep levels. Therefore, the observed dominant green emission in ZnO-D and ZnO-A samples is mainly surface related. In a previous study, the chemisorption of hydroxyl groups/oxygen species was suppressed by covering the nanoparticles with CNT [17]. In that study, non-functionalized CNT were used, and a successful passivation of surface states was obtained. In the present case, the CNT were functionalized with OH functional groups, as discussed previously. Therefore, in the present case, the upward band bending is enhanced (Figure 7h). The increased upward band bending leads to further increase in the depletion region size, whereupon the probability of electron capture at the defect sites increases. This mechanism also reduces the probability of band-to-band transitions and the NBE is diminished.

For ZnO-PEDOT:PSS samples, a complete coverage of ZnO with PEDOT:PSS is visible in the TEM images. In general, there is a reduction in the overall emission compared to the as-synthesized and ZnO-CNT samples, due to the low amount of ZnO nanoparticles. However, the NBE-to-DLE ratio is higher in these samples. The increase in NBE can be attributed to the reduced surface hydroxyl groups, relative to the as-synthesized and ZnO-CNT samples, leading to lower upward band bending (Figure 7i). More particularly, the NBE-to-DLE ratio is higher for the ZnO-D-PEDOT:PSS than ZnO-A-PEDOT:PSS. An increase in DLE for the latter can be attributed to the higher amount of hydroxyl groups present on the ZnO-A sample, as assessed on the basis of the XPS studies, leading to a slightly higher upward band bending than for ZnO-D. Additionally, the NBE of both types of hybrid sample, i.e., CNT and PEDOT:PSS, has redshifted, suggesting an increased amount of Zn_i, further corroborating the Raman spectroscopy results. Finally, the red emission at ~1.75 eV is the least significant component in the PL spectra, associated mainly with V_{Zn}-related defects [49].

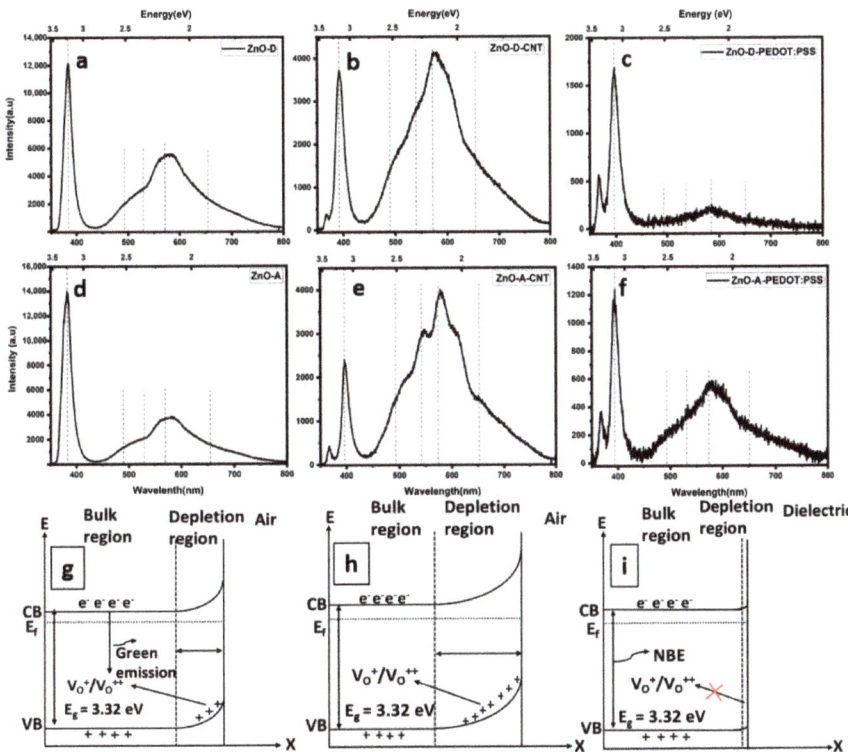

Figure 7. PL emission spectra of (**a**) ZnO-D, (**b**) ZnO-D-CNT, (**c**) ZnO-D-PEDOT:PSS, (**d**) ZnO-A, (**e**) ZnO-A-CNT and (**f**) ZnO-A-PEDOT:PSS. Schematics of upward band bending (**g**) chemisorbed Oxygen species, (**h**) CNT decorated with ZnO through hydroxyl groups, (**i**) ZnO-PEDOT:PSS nanohybrids.

4. Conclusions

In this study, we successfully synthesized ZnO nanoparticles and their hybrids containing CNT and PEDOT:PSS. The effect of hydroxyl groups on the optical properties of ZnO nanoparticles and their hybrids was investigated. The ZnO nanoparticles display optical properties that are both bandgap and defect related. In addition, the choice of precursor immensely influences the overall properties of the nanoparticles and their hybrids. During the synthesis of the hybrid nanocomposites, hydroxyl groups adhere to the surface of the ZnO nanoparticles, and in turn, intensify the defect related emission. These hydroxyl groups are necessary for the successful decoration of the CNT and the incorporation of ZnO in the PEDOT:PSS matrix. Additionally, the linear transformation of PEDOT from the coil structure implies a more conductive polymer, which would enhance the I-V characteristics of the nanocomposite. However, ZnO-A tends to degrade the PEDOT:PSS macromolecule by removing or degrading the conducting PEDOT, which could prove detrimental to the electrical properties of the nanocomposite. Future research consists of incorporating these nanoparticles and their hybrids in LED applications. This present work aids in understanding the modification of the physical and chemical properties of the ZnO nanoparticles when hybridized to PEDOT:PSS and CNT. On the basis of this work, we conclude that ZnO-D and its hybrid nanocomposites, synthesized with hydrate precursors show higher stability and are likely to offer better electrical conductivity when used in LED.

Author Contributions: Conceptualization: K.N., P.R.; methodology: K.N., P.R., E.R.; validation, P.R., E.R. and E.E.; formal analysis, K.N., E.E., M.R.S.; investigation, K.N., P.R., E.R.; resources, E.E., P.R., E.R.; data curation, K.N.; writing—original draft preparation, K.N.; writing—review and editing,

K.N., P.R. and E.R.; supervision, P.R. and E.R.; project administration, P.R.; funding acquisition, P.R. and E.R. All authors have read and agreed to the published version of the manuscript.

Funding: This research has been supported by the European Regional Development Fund project grant number TK134 "EQUiTANT" and T210013TIBT "PARROT mobility program". We thank EU H2020 research and innovation program under grant agreement no. 1029 supporting the Transnational Access Activity within the framework NFFA-Europe.

Data Availability Statement: Not applicable.

Conflicts of Interest: The authors declare no conflict of interest.

References

1. Huang, J.; Yin, Z.; Zheng, Q. Applications of ZnO in organic and hybrid solar cells. *Energy Environ. Sci.* **2011**, *4*, 3861–3877. [CrossRef]
2. Periyayya, U.; Kang, J.H.; Ryu, J.H.; Hong, C.-H. Synthesis and improved luminescence properties of OLED/ZnO hybrid materials. *Vacuum* **2011**, *86*, 254–260. [CrossRef]
3. Nagpal, K.; Rauwel, E.; Ducroquet, F.; Rauwel, P. Assessment of the optical and electrical properties of light-emitting diodes containing carbon-based nanostructures and plasmonic nanoparticles: A review. *Beilstein J. Nanotechnol.* **2021**, *12*, 1078–1092. [CrossRef] [PubMed]
4. Shirota, Y.; Kageyama, H. 1—Organic materials for optoelectronic applications: Overview. In *Handbook of Organic Materials for Electronic and Photonic Devices*, 2nd ed.; Ostroverkhova, O., Ed.; Woodhead Publishing: Sawston, UK, 2019; pp. 3–42.
5. Yu, K.J.; Yan, Z.; Han, M.; Rogers, J.A. Inorganic semiconducting materials for flexible and stretchable electronics. *NPJ Flex. Electron.* **2017**, *1*, 4. [CrossRef]
6. Könenkamp, R.; Word, R.C.; Godinez, M. Ultraviolet Electroluminescence from ZnO/Polymer Heterojunction Light-Emitting Diodes. *Nano Letters* **2005**, *5*, 2005–2008. [CrossRef]
7. Biju, V.; Itoh, T.; Anas, A.; Sujith, A.; Ishikawa, M. Semiconductor quantum dots and metal nanoparticles: Syntheses, optical properties, and biological applications. *Anal. Bioanal. Chem.* **2008**, *391*, 2469–2495. [CrossRef]
8. Yang, P.; Yan, H.; Mao, S.; Russo, R.; Johnson, J.; Saykally, R.; Morris, N.; Pham, J.; He, R.; Choi, H.-J. Controlled Growth of ZnO Nanowires and Their Optical Properties. *Adv. Funct. Mater.* **2002**, *12*, 323–331. [CrossRef]
9. Huseynova, G.; Hyun Kim, Y.; Lee, J.-H.; Lee, J. Rising advancements in the application of PEDOT:PSS as a prosperous transparent and flexible electrode material for solution-processed organic electronics. *J. Inf. Disp.* **2020**, *21*, 71–91. [CrossRef]
10. Sun, K.; Zhang, S.; Li, P.; Xia, Y.; Zhang, X.; Du, D.; Isikgor, F.H.; Ouyang, J. Review on application of PEDOTs and PEDOT:PSS in energy conversion and storage devices. *J. Mater. Sci. Mater. Electron.* **2015**, *26*, 4438–4462. [CrossRef]
11. Park, Y.; Müller-Meskamp, L.; Vandewal, K.; Leo, K. PEDOT:PSS with embedded TiO2 nanoparticles as light trapping electrode for organic photovoltaics. *Appl. Phys. Lett.* **2016**, *108*, 253302. [CrossRef]
12. Sharma, B.K.; Khare, N.; Ahmad, S. A ZnO/PEDOT:PSS based inorganic/organic hetrojunction. *Solid State Commun.* **2009**, *149*, 771–774. [CrossRef]
13. Lin, B.; Fu, Z.; Jia, Y. Green luminescent center in undoped zinc oxide films deposited on silicon substrates. *Appl. Phys. Lett.* **2001**, *79*, 943–945. [CrossRef]
14. Raji, R.; Gopchandran, K.G. ZnO nanostructures with tunable visible luminescence: Effects of kinetics of chemical reduction and annealing. *J. Sci. Adv. Mater. Devices* **2017**, *2*, 51–58. [CrossRef]
15. Polsongkram, D.; Chamninok, P.; Pukird, S.; Chow, L.; Lupan, O.; Chai, G.; Khallaf, H.; Park, S.; Schulte, A. Effect of synthesis conditions on the growth of ZnO nanorods via hydrothermal method. *Phys. B Condens. Matter* **2008**, *403*, 3713–3717. [CrossRef]
16. Wang, J.; Chen, R.; Xiang, L.; Komarneni, S. Synthesis, properties and applications of ZnO nanomaterials with oxygen vacancies: A review. *Ceram. Int.* **2018**, *44*, 7357–7377. [CrossRef]
17. Nagpal, K.; Rapenne, L.; Wragg, D.S.; Rauwel, E.; Rauwel, P. The role of CNT in surface defect passivation and UV emission intensification of ZnO nanoparticles. *Nanomater. Nanotechnol.* **2022**, *12*, 18479804221079419. [CrossRef]
18. Rauwel, E.; Galeckas, A.; Rauwel, P.; Sunding, M.F.; Fjellvåg, H. Precursor-Dependent Blue-Green Photoluminescence Emission of ZnO Nanoparticles. *J. Phys. Chem. C* **2011**, *115*, 25227–25233. [CrossRef]
19. Niederberger, M. Nonaqueous Sol–Gel Routes to Metal Oxide Nanoparticles. *Acc. Chem. Res.* **2007**, *40*, 793–800. [CrossRef]
20. Kim, S.; Somaratne, R.M.D.S.; Whitten, J.E. Effect of Adsorption on the Photoluminescence of Zinc Oxide Nanoparticles. *J. Phys. Chem. C* **2018**, *122*, 18982–18994. [CrossRef]
21. Sharma, A.; Singh, B.P.; Dhar, S.; Gondorf, A.; Spasova, M. Effect of surface groups on the luminescence property of ZnO nanoparticles synthesized by sol–gel route. *Surf. Sci.* **2012**, *606*, L13–L17. [CrossRef]
22. Sahai, A.; Goswami, N. Probing the dominance of interstitial oxygen defects in ZnO nanoparticles through structural and optical characterizations. *Ceram. Int.* **2014**, *40*, 14569–14578. [CrossRef]
23. Šarić, A.; Gotić, M.; Štefanić, G.; Dražić, G. Synthesis of ZnO particles using water molecules generated in esterification reaction. *J. Mol. Struct.* **2017**, *1140*, 12–18. [CrossRef]

Leitão Muniz, F.; Miranda, M.; Morilla-Santos, C.; Sasaki, J. The Scherrer equation and the dynamical theory of X-ray diffraction. *Acta Crystallogr. Sect. A Found. Adv.* **2016**, *72*, 385–390. [CrossRef] [PubMed]

Hosseini Largani, S.; Akbarzadeh Pasha, M. The effect of concentration ratio and type of functional group on synthesis of CNT–ZnO hybrid nanomaterial by an in situ sol–gel process. *Int. Nano Lett.* **2017**, *7*, 25–33. [CrossRef]

Chen, X.; Wang, X.; Fang, D. A review on C1s XPS-spectra for some kinds of carbon materials. *Fuller. Nanotub. Carbon Nanostructures* **2020**, *28*, 1048–1058. [CrossRef]

Aïssa, B.; Fauteux, C.; El Khakani, M.A.; Daniel, T. Structural and photoluminescence properties of laser processed ZnO/carbon nanotube nanohybrids. *J. Mater. Res.* **2009**, *24*, 3313–3320. [CrossRef]

Rossell, M.D.; Kuebel, C.; Ilari, G.; Rechberger, F.; Heiligtag, F.J.; Niederberger, M.; Koziej, D.; Erni, R. Impact of sonication pretreatment on carbon nanotubes: A transmission electron microscopy study. *Carbon* **2013**, *61*, 404–411. [CrossRef]

Mengistie, D.A.; Ibrahem, M.A.; Wang, P.-C.; Chu, C.-W. Highly Conductive PEDOT:PSS Treated with Formic Acid for ITO-Free Polymer Solar Cells. *ACS Appl. Mater. Interfaces* **2014**, *6*, 2292–2299. [CrossRef]

Zeferino, R.S.; Flores, M.B.; Pal, U. Photoluminescence and Raman Scattering in Ag-doped ZnO Nanoparticles. *J. Appl. Phys.* **2011**, *109*, 014308. [CrossRef]

Gao, Q.; Dai, Y.; Li, C.; Yang, L.; Li, X.; Cui, C. Correlation between oxygen vacancies and dopant concentration in Mn-doped ZnO nanoparticles synthesized by co-precipitation technique. *J. Alloys Compd.* **2016**, *684*, 669–676. [CrossRef]

Naeem, M.; Qaseem, S.; Gul, I.H.; Maqsood, A. Study of active surface defects in Ti doped ZnO nanoparticles. *J. Appl. Phys.* **2010**, *107*, 124303. [CrossRef]

Khachadorian, S.; Gillen, R.; Choi, S.; Ton-That, C.; Kliem, A.; Maultzsch, J.; Phillips, M.R.; Hoffmann, A. Effects of annealing on optical and structural properties of zinc oxide nanocrystals. *Phys. Status Solidi B* **2015**, *252*, 2620–2625. [CrossRef]

Wang, J.B.; Huang, G.J.; Zhong, X.L.; Sun, L.Z.; Zhou, Y.C.; Liu, E.H. Raman scattering and high temperature ferromagnetism of Mn-doped ZnO nanoparticles. *Appl. Phys. Lett.* **2006**, *88*, 252502. [CrossRef]

Gluba, M.A.; Nickel, N.H.; Karpensky, N. Interstitial zinc clusters in zinc oxide. *Phys. Rev. B* **2013**, *88*, 245201. [CrossRef]

Montanheiro, T.L.d.A.; de Menezes, B.R.C.; Ribas, R.G.; Montagna, L.S.; Campos, T.M.B.; Schatkoski, V.M.; Righetti, V.A.N.; Passador, F.R.; Thim, G.P. Covalently γ-aminobutyric acid-functionalized carbon nanotubes: Improved compatibility with PHBV matrix. *SN Appl. Sci.* **2019**, *1*, 1177. [CrossRef]

Tsyganenko, A.A.; Lamotte, J.; Saussey, J.; Lavalley, J.C. Bending vibrations of OH groups resulting from H_2 dissociation on ZnO. *J. Chem. Soc. Faraday Trans. 1 Phys. Chem. Condens. Phases* **1989**, *85*, 2397–2403. [CrossRef]

Scarano, D.; Bertarione, S.; Cesano, F.; Vitillo, J.G.; Zecchina, A. Plate-like zinc oxide microcrystals: Synthesis and characterization of a material active toward hydrogen adsorption. *Catal. Today* **2006**, *116*, 433–438. [CrossRef]

Dent, A.L.; Kokes, R.J. Hydrogenation of ethylene by zinc oxide. I. Role of slow hydrogen chemisorption. *J. Phys. Chem.* **1969**, *73*, 3772–3780. [CrossRef]

Chang, S.H.; Chiang, C.H.; Kao, F.S.; Tien, C.L.; Wu, C.G. Unraveling the Enhanced Electrical Conductivity of PEDOT:PSS Thin Films for ITO-Free Organic Photovoltaics. *IEEE Photonics J.* **2014**, *6*, 1–7. [CrossRef]

Lee, H.; Kim, Y.; Cho, H.; Lee, J.-g.; Kim, J.H. Improvement of PEDOT:PSS linearity via controlled addition process. *RSC Adv.* **2019**, *9*, 17318–17324. [CrossRef]

Wang, Y.; Sun, K.; Fu, J.; Chen, R.; Li, M.; Zang, Z.; Liu, X.; Li, B.; Gong, H.; Ouyang, J. Enhancement of Conductivity and Thermoelectric Property of PEDOT:PSS via Acid Doping and Single Post-Treatment for Flexible Power Generator. *Adv. Sustain. Syst.* **2018**, *2*, 1800085. [CrossRef]

Davis, K.; Yarbrough, R.; Froeschle, M.; White, J.; Rathnayake, H. Band gap engineered zinc oxide nanostructures via a sol–gel synthesis of solvent driven shape-controlled crystal growth. *RSC Adv.* **2019**, *9*, 14638–14648. [CrossRef] [PubMed]

Rauwel, P.; Galeckas, A.; Salumaa, M.; Ducroquet, F.; Rauwel, E. Photocurrent generation in carbon nanotube/cubic-phase HfO2 nanoparticle hybrid nanocomposites. *Beilstein J. Nanotechnol.* **2016**, *7*, 1075–1085. [CrossRef] [PubMed]

Kumar Jangir, L.; Kumari, Y.; Kumar, A.; Kumar, M.; Awasthi, K. Investigation of luminescence and structural properties of ZnO nanoparticles, synthesized with different precursors. *Mater. Chem. Front.* **2017**, *1*, 1413–1421. [CrossRef]

Oudhia, A.; Choudhary, A.; Sharma, S.; Aggrawal, S.; Dhoble, S.J. Study of defect generated visible photoluminescence in zinc oxide nano-particles prepared using PVA templates. *J. Lumin.* **2014**, *154*, 211–217. [CrossRef]

Vanheusden, K.; Warren, W.L.; Seager, C.H.; Tallant, D.R.; Voigt, J.A.; Gnade, B.E. Mechanisms behind green photoluminescence in ZnO phosphor powders. *J. Appl. Phys.* **1996**, *79*, 7983–7990. [CrossRef]

Rauwel, P.; Galeckas, A.; Rauwel, E. Enhancing the UV Emission in ZnO–CNT Hybrid Nanostructures via the Surface Plasmon Resonance of Ag Nanoparticles. *Nanomaterials* **2021**, *11*, 452. [CrossRef]

Przezdziecka, E.; Guziewicz, E.; Jarosz, D.; Snigurenko, D.; Sulich, A.; Sybilski, P.; Jakiela, R.; Paszkowicz, W. Influence of oxygen-rich and zinc-rich conditions on donor and acceptor states and conductivity mechanism of ZnO films grown by ALD—Experimental studies. *J. Appl. Phys.* **2020**, *127*, 075104. [CrossRef]

Optimizing the PMMA Electron-Blocking Layer of Quantum Dot Light-Emitting Diodes

Mariya Zvaigzne [1,*], Alexei Alexandrov [2], Anastasia Tkach [1], Dmitriy Lypenko [2], Igor Nabiev [1,3,4] and Pavel Samokhvalov [1,*]

1. Laboratory of Nano-Bioengineering, Institute of Engineering Physics for Biomedicine, National Research Nuclear University MEPhI (Moscow Engineering Physics Institute), 31 Kashirskoe Highway, 115409 Moscow, Russia; taa025@campus.mephi.ru (A.T.); igor.nabiev@univ-reims.fr (I.N.)
2. Laboratory of Electronic and Photonic Processes in Polymeric Nanomaterials, A.N. Frumkin Institute of Physical Chemistry and Electrochemistry of the Russian Academy of Sciences, 31, bld.4, Leninsky Prospect, 119071 Moscow, Russia; klays007@gmail.com (A.A.); lypenko@phyche.ac.ru (D.L.)
3. Laboratory of Immunopathology, I.M. Sechenov First Moscow State Medical University, 8-2 Trubetskaya Str., 119991 Moscow, Russia
4. Laboratoire de Recherche en Nanosciences, Université de Reims Champagne-Ardenne, 51 rue Cognacq Jay, 51100 Reims, France
* Correspondence: mazvajgzne@mephi.ru (M.Z.); pssamokhvalov@mephi.ru (P.S.)

Abstract: Quantum dots (QDs) are promising candidates for producing bright, color-pure, cost-efficient, and long-lasting QD-based light-emitting diodes (QDLEDs). However, one of the significant problems in achieving high efficiency of QDLEDs is the imbalance between the rates of charge-carrier injection into the emissive QD layer and their transport through the device components. Here we investigated the effect of the parameters of the deposition of a poly (methyl methacrylate) (PMMA) electron-blocking layer (EBL), such as PMMA solution concentration, on the characteristics of EBL-enhanced QDLEDs. A series of devices was fabricated with the PMMA layer formed from acetone solutions with concentrations ranging from 0.05 to 1.2 mg/mL. The addition of the PMMA layer allowed for an increase of the maximum luminance of QDLED by a factor of four compared to the control device without EBL, that is, to 18,671 cd/m^2, with the current efficiency increased by an order of magnitude and the turn-on voltage decreased by ~1 V. At the same time, we have demonstrated that each particular QDLED characteristic has a maximum at a specific PMMA layer thickness; therefore, variation of the EBL deposition conditions could serve as an additional parameter space when other QDLED optimization approaches are being developed or implied in future solid-state lighting and display devices.

Keywords: quantum dots; QDLED; electron-blocking layer; PMMA

1. Introduction

Fluorescent semiconductor nanocrystals (NCs) or quantum dots (QDs) have plenty of advantageous properties, such as the possibility of tuning the luminescence wavelength by varying the physical size of the NCs and the capacity for forming stable colloidal solutions, which makes it possible to obtain coatings by inexpensive solution-process methods, and make QDs promising materials in optoelectronic, bioimaging, lighting, and other applications [1–5]. Today, light-emitting devices (LEDs) based on organic compounds (OLEDs) prevail in commercial lighting and display appliances. Quantum dots outperform traditional organic dyes in terms of the width of the absorption spectrum, molar extinction, and photostability. Thus, quantum dots are expected to be promising candidates to overcome the material stability issues typical of OLEDs, such as drastic efficiency roll-off at high current densities and mediocre operational lifetimes. Moreover, due to their inorganic nature, QDs are much more thermally stable materials, which makes it possible to increase the brightness of QD-based LEDs by increasing the current density in the device.

In addition, QDs have quite narrow fluorescence and electroluminescence spectra and hence, are promising components of displays or illuminators with a wide color gamut. To fully exploit the superior properties of QDs, a number of QD-based LED (QDLED) structures with different device and material configurations have been developed [6]. A typical QDLED-emitting layer represents a thin QD film sandwiched between two charge transport layers, and its interaction with them may cause luminescence quenching through various nonradiative pathways, such as the Auger recombination [7,8], QD charging [7] and charge and/or energy transfer from QDs to the charge-transport materials [7,9], and so forth.

One of the main shortcomings of QD-based LEDs is imbalance between the rates of charge carrier injection and transport [7], which leads to the formation of excess charges (electrons or holes) in the emitting QD layer and quenching of QD radiation due to the aforementioned nonradiative processes. This phenomenon leads to a significant decrease in radiation efficiency, especially at high current densities, and, hence, to overall low performance of the QDLEDs. In most modern QDLED configurations [10], this imbalance mainly results from a larger potential barrier for the injection of holes into the QD layer than for the injection of electrons, as well as a higher mobility of electrons in the electron transport layer (ETL), usually based on ZnO, compared to the hole mobility in organic hole transport layers (HTL) [11]. One of the approaches to solving this problem is the introduction of an electron-blocking layer (EBL). The EBL materials and methods of its integration into the QDLED structure vary widely. For example, it has been shown that the addition of a 4,4,4 tris(N-carbazolyl)-triphenylamine (TcTa) EBL between the hole-transporting layer and the light-emitting QD layer [12] or a combined hole-transporting and electron-blocking layer of deoxyribonucleic acid (DNA) complexed with cetyltrimetylammonium (CTMA) [13] enhances efficiency due to the reduction of electron overflow and improvement of hole injection. Another approach is to insert an EBL between the ETL and the QD layer to restrict the flow of electrons. In the framework of this approach, Al_2O_3 [14] and poly(methyl methacrylate) (PMMA) [15] have been demonstrated to be effective materials for the EBL. However, a thin film of poly (methyl methacrylate) (PMMA) is more often used as an EBL [15,16], because the positions of its energy levels provide a high potential barrier for electron injection into the emitting layer for most types of QDs. In addition, PMMA is soluble, such as in acetone, which makes it possible to apply PMMA onto the underlying QD layer without partially dissolving or deforming the latter.

The first introduction of a PMMA electron blocking layer into the QDLED structure was reported by Dai et al. [15], who fabricated QDLEDs with a 6 nm PMMA insulating layer between a CdSe/CdS core/shell QD layer and a ZnO ETL to optimize the charge balance in the device. They compared the hole mobility in an HTL consisting of poly TPD ($1 \cdot 10^{-4}$ $cm^2 \cdot V^{-1} \cdot s^{-1}$) and PVK ($2.5 \cdot 10^{-6}$ $cm^2 \cdot V^{-1} \cdot s^{-1}$) with the electron mobility in an ETL based on a ZnO nanocrystal film ($\sim 1.8 \cdot 10^{-3}$ $cm^2 \cdot V^{-1} \cdot s^{-1}$), and concluded that the insertion of the PMMA layer may lead to excess electron injection into the QD emissive layer. To confirm this assumption, they measured and compared the current densities of the electron-only devices (ITO/Al/QDs/ZnO/Al) and hole-only devices (ITO/PEDOT:PSS/poly-TPD/PVK/QDs/Pd). In this case, the addition of the PMMA layer between the ETL and the QD layer to optimize charge balance did not cause any considerable changes in either the turn-on voltage or the brightness in comparison with the control QDLEDs without the PMMA layer. However, for equally bright QDLEDs, the current density in the control device was much greater, which indicated that the efficiency of this device was substantially lowered by the excess electron current. Thus, the efficiency of the EBL-based approach in terms of enhancing the performance of QDLEDs was confirmed. In addition, the stability of the devices without the PMMA layers was relatively poor, and it was improved 20-fold by adding the PMMA EBL.

Rahmati et al. [16] presented a new QDLED architecture with multiple PMMA EBLs sandwiched between a pair or more of QD layers. The authors developed QDLED structures with one, two, and three PMMA layers and showed that a device containing two

PMMA and three QD layers had the best current efficiency of 17.8 cd·A^{-1} and a luminance of 194,038 cd·m^{-2}. The substantial improvement of QDLED performance was mainly attributed to the addition of the PMMA EBL, which reduced the backward electron leakage from the active QD region and enhanced electron confinement, leading to an increased electron concentration in the QD active layers and a higher radiative recombination rate. It is worth noting that the aforementioned QDLED configuration where the EBL is sandwiched between a pair of QD EMLs could be employed in the design of white QDLEDs by combining isolated blue/green/red QD layers separated by two PMMA spacers into a complex emissive layer [16].

Although it has been demonstrated that adding a PMMA layer to the QDLED structure may significantly improve the performance of QDLEDs, this approach needs further optimization and detailed study to rationally exploit the PMMA electron-blocking capacity. Here, we studied the correlations between the parameters of the PMMA EBL deposition, such as the PMMA solution concentration, and the most important performance characteristics of QDLEDs.

2. Materials and Methods

2.1. Synthesis of CdSe/ZnS/CdS/ZnS (CdSe/MS) QDs

The synthesis of CdSe cores with a diameter of 2.3 nm was carried out by the hot injection technique using cadmium hexadecylphosphonate and trioctylphosphine as precursors at a temperature of 240 °C; the procedure is described in more detail in [17]. After the synthesis, the separation and purification of the CdSe cores were carried out by means of re-precipitation of nanocrystals and subsequent gel permeation chromatography, after which their surface was treated with oleylamine in the presence of sodium borohydride to replace the hexadecylphosphonic acid residues with oleylamine, which facilitated further growth of inorganic shells. Then, after additional purification steps, the CdSe cores were placed into a reaction mixture of 1-octadecene and oleylamine (1:1, v/v) for growing the shells. After accurate quantification of CdSe cores in the reaction solution using the approach described in [18], we calculated the quantities of precursors required for obtaining QDs with the desired shell structure. The shells were grown at 170 °C in an argon atmosphere at an average growth rate of 1 monolayer per 30 min. After the synthesis and isolation of QDs from the crude solution, the organic ligands were replaced with hexadecylammonium palmitate (HDA-PA), which reduced the sensitivity of the optical properties of QDs to atmospheric exposure during long-term storage. The luminescence and absorption spectra, as well as the transmission electron microscopy (TEM) image of the obtained QDs are shown in Figures S1 and S2, provided in the Supporting Information (SI) file.

2.2. Fabrication of QDLED Devices

Glass substrates with an indium tin oxide (ITO) layer were preliminarily cleaned by treatment in an ultrasonic bath and then in oxygen plasma. Then, a hole-injecting layer of PEDOT:PSS (poly(3,4-ethylenedioxythiophene): poly(styrene sulfonate)) was deposited on the substrates by spin-coating at 2000 rpm, followed by annealing at 110 °C for 10 min. The film thickness was 40 nm. The substrates coated with a PEDOT: PSS layer were transferred into a glove box containing argon (O_2 < 1 ppm, H_2O < 1 ppm). Next, hole transport layers of poly-TPD (poly(N,N'-bis-4-butylphenyl-N,N'-bisphenyl) benzidine, solution in chlorobenzene, 8 mg/mL) and PVK (poly (vinylcarbazole), solution in o-xylene, 1.5 mg/mL) were deposited alternately by spin-coating at 2000 rpm. Layers of poly-TPD (30 nm) and PVK (5 nm) were annealed at 100 °C for 10 min before applying the next layer. Then, a QD layer was applied from a solution in n-octane (20 mg/mL) by spin-coating at 2500 rpm and annealed at 100 °C for 10 min. The QD film thickness was 40 nm. The following PMMA (Sigma Aldrich, Saint Louis, MO, USA, average M_w ~ 120,000 Da) layer was applied by spin-coating from a solution in acetone at 3500 rpm and then annealed at 100 °C for 10 min. The concentration of the acetone solution of PMMA varied from 1.2 mg/mL to 0.05 mg/mL to obtain different blocking-layer thicknesses. A 50 nm electron-transport layer (ETL) was applied from solutions of ZnO nanoparticles in isopropyl alcohol (25 mg/mL) by spin-coating at 1000 rpm followed by annealing at 60 °C for 10 min. Finally, an aluminum cathode with a thickness of 80 nm was deposited onto the ETL through a shadow mask by thermal evaporation in vacuum ($2 \cdot 10^{-6}$ mbar).

2.3. Instrumental Methods

The luminescence spectra were measured using an Agilent Cary Eclipse spectrofluorimeter. The absorption spectra were measured using an Agilent Cary 60 UV-Vis spectrophotometer. Transmission electron microscopy (TEM) images were obtained on a JEOL JEM-2100F (JEOL Ltd., Tokyo, Japan) instrument operated at 200 kV acceleration voltage. TEM specimens were prepared by drop-casting a solution of QDs in hexane onto carbon/Formvar-coated 200 mesh copper TEM grids. The voltage–current and voltage–brightness characteristics were measured with a Keithley 2601 SourceMeter 2601 (Keithley Instruments, Inc., Solon, OH, USA), a Keithley 485 picoampermeter (Keithley Instruments, Inc., Solon, OH, USA), and a TKA-04/3 luxmeter–brightness meter (Scientific Instruments "TKA", St. Petersburg, Russia). The preparation of QDLED samples and measurements of their characteristics were performed at room temperature in an argon atmosphere. The film thicknesses were determined by ellipsometry using an MII-4 interferometer ("LOMO", St. Petersburg, Russia) and by means of a MultiMode V (Bruker Corporation, Billerica, MA, USA) atomic force microscope.

3. Results and Discussion

Deposition of an ultimately thin PMMA charge-blocking layer on top or within the emissive QD layer by means of solution processes requires the minimum possible distortion of the QD layer to achieve efficient performance of the QDLED. Thus, selection of the appropriate solvent for PMMA is of utmost importance. PMMA can be dissolved in a number of organic nonpolar (toluene, chloroform, etc.) and weakly polar (acetone) solvents. In most QDLED configurations, QDs capped with long-chain aliphatic ligands are deposited from nonpolar solvents, such as octane and toluene. Therefore, acetone becomes the solvent of choice for deposition of PMMA because, being polar, it cannot dissolve the underlying QDs and, at the same time, being only mildly polar, it cannot cause severe wrapping or cracking of the underlying thin film of QDs. On the other hand, acetone is quite volatile and has a low viscosity; therefore, it is hard to control the parameters of its deposition by spin-coating other than the concentration of PMMA in solution.

The structure of the fabricated QDLED devices is illustrated in Figure 1 along with the schematic of the flat-band energy level diagram of the layers in the device.

We used QDs with the core/multishell structure to eliminate the negative effects on QD fluorescence caused by Auger recombination and surface-trapping [19–21], and because our previous studies revealed that this type of QD possessed the optimal characteristics of photostability due to the suppression of charge transfer [4]. In order to increase the efficiency of hole injection into the QD-emitting layer, we added a poly-TPD/PVK bilayer-structured hole-injection layer between QDs and PEDOT:PSS. This configuration creates a gradual step transition between the hole energy levels of QDs and PEDOT:PSS hole-transport layer. A thin layer of ZnO nanoparticles was deposited as an electron transport layer (ETL) because ZnO proved to be the most favorable ETL material due to its high transparency, low work function, and high electron mobility [6,10,15]. To investigate the effect of the PMMA layer preparation routine on the QDLED performance, a series of devices were fabricated. To do this, we varied the concentration of the PMMA solution in acetone from 0.05 to 1.2 mg/mL. Unfortunately, we were unable to measure the exact thickness of the PMMA EBL using the available AFM instrumentation, because the measurement error was higher than the measured value. Otherwise, we may roughly estimate the thickness range of PMMA EBL in our devices as 0.13–3 nm, corresponding to the lower and upper limits of the solution concentrations, respectively. The details of this estimation are given in the SI, the results of the calculation are given in the Table S1. Yet, in the following sections, we prefer to stick to the known experimental concentration values rather than our rough thickness estimations.

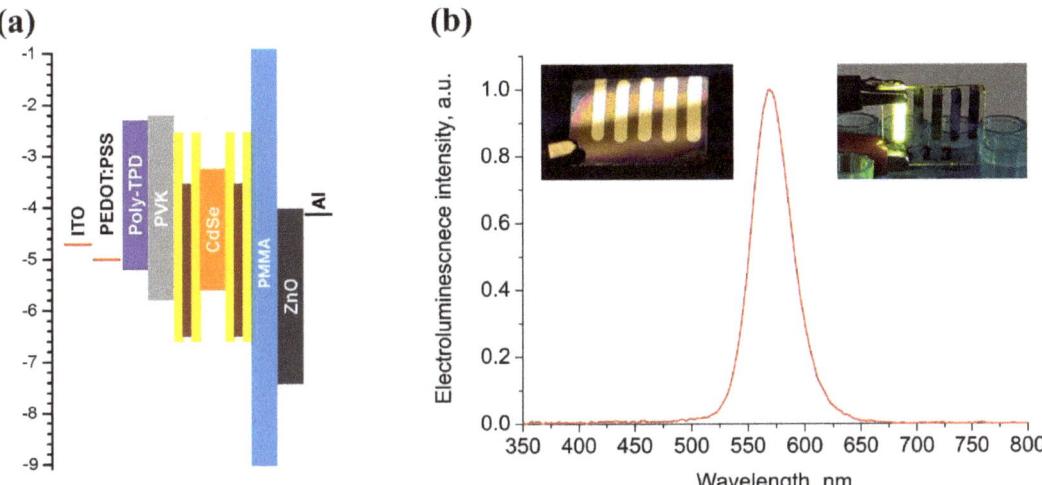

Figure 1. Flat-band energy level diagram of the fabricated QDLEDs (**a**) and electroluminescence spectrum of QDLED with the PMMA electron-blocking layer deposited from a solution with a PMMA concentration of 0.2 mg/mL (**b**). The insets in panel (**b**) show photographs of the device under ambient light (left) and operated at 6 V (right).

Figure 2a shows the current density–voltage and luminance–voltage characteristics of the devices under investigation. As can be seen, the addition of a PMMA blocking layer in most cases led to an accelerated rise and an overall increase in the current density at all voltages relative to the structure without a blocking layer. Only when we applied the EBL of PMMA from a solution with the highest concentration (1.2 mg/mL) did we observe a drop in this characteristic. These results might be counterintuitive at first glance, because blocking of electron flux through the device by a potential barrier created by the PMMA layer led to an overall increase in the current flow through the whole device. However, previous studies showed that the imbalance between the electron and hole currents led to the accumulation of excess electrons inside the device and interfacial charging [7,11], which, in turn, acts as a counter-driving force for electron currents and leads to less efficient electron transport and injection. EBL, in this case, diminishes the charge flow imbalance and allows the device to operate in the optimal regime, since the PMMA interlayer provides quite a high energy barrier of around 3 eV against electron flow from the ETL (Figure 1). Thus, PMMA EBL can block excess electron flow from ETL to the QD light-emitting layer by reducing the electron current density, which leads to the improved charge carrier balance inside the emissive QD layer, as it was shown for a number of other EBL materials [22–25].

As can be seen from our data, this optimal charge carrier balance was achieved when EBL was deposited from a solution with a PMMA concentration in the range of 0.1–0.4 mg/mL. In this range, higher PMMA concentrations yield devices with a lower current density but similar luminance. On the other hand, when the PMMA concentration in deposition solution was lowered to 0.05 mg/mL, we observed current leakage even at low voltages (0–2.5 V), which suggested a short circuit due to disturbance of the emissive layer during EBL deposition.

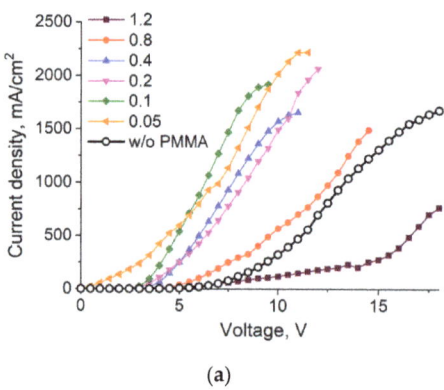
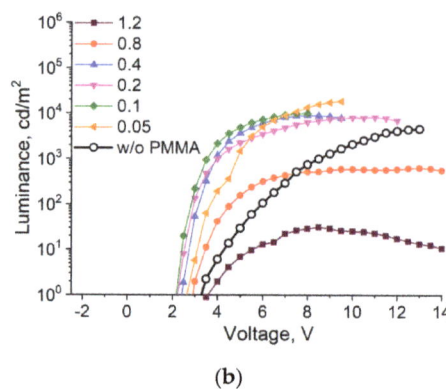

(a)   (b)

Figure 2. Current density (**a**) and luminance (**b**) versus voltage characteristics of QDLED samples employing a PMMA EBL deposited from PMMA solutions in acetone with different concentrations.

A similar trend was observed for the luminance–voltage characteristics (Figure 2b). In this case, the luminance saturation plateau was reached faster, and the brightness values were higher in QDLED structures fabricated with a blocking PMMA layer. As an exception, QDLED samples employing a PMMA EBL deposited from solutions with concentrations of 0.8 and 1.2 mg/mL exhibited only minor, if any, improvement of this characteristic. In the case of a 0.8 mg/mL solution, sharper growth was observed, but the brightness value did not exceed that for the device without an EBL. These effects may arise from hindered injection of electrons into the emitting layer due to an increase in the thickness of the potential barrier and, as a consequence, a decrease in the probability of carrier tunneling.

The performance parameters of all fabricated QDLED devices are summarized in Table 1. The lowest turn-on voltage of 2.1 V was observed for the two lowest PMMA solution concentrations, 0.1 and 0.05 mg/mL. At the same time, the QDLED structure without a PMMA layer had one of the highest turn-on voltage values, 3.3 V. In general, a distinct minimum was observed in the plot of the turn-on voltage versus the PMMA solution concentration (Figure 3).

Table 1. Summary of the performance parameters of the fabricated QDLED devices with and without a PMMA layer deposited from PMMA solutions with different concentrations.

PMMA Solution Concentration, mg/mL	1.2	0.8	0.4	0.2	0.1	0.05	w/o PMMA
Turn-on voltage, V	3.6	2.9	2.5	2.3	2.1	2.1	3.3
Maximum current efficiency, cd/A	0.04	0.18	0.95	0.63	0.73	0.99	0.49
Luminance, cd/m^2	33	635	9093	8146	9969	18,671	4472

In terms of the maximum current efficiency, the QDLED with an EBL deposited from a 0.4 mg/mL PMMA solution turned out to be the optimal one (Table 1). An increase in PMMA concentration led to a sharp drop of the current efficiency, while its decrease also resulted in a 1.5-fold lower current efficiency. For concentrations of 0.2 and 0.1 mg/mL, there were no significant differences in either current efficiency or turn-on voltage. However, the brightness steadily increased with decreasing PMMA concentration in the EBL deposition procedure. Thus, the maximum brightness in our experiment was 18 671 cd/m^2, obtained in the case of QDLEDs with an EBL fabricated using a 0.05 mg/mL PMMA solution. This luminance value was four times higher than that for devices without a blocking layer. Figures 4 and 5 show the dependences of the current efficiency and luminance at 9 V on the PMMA solution concentration.

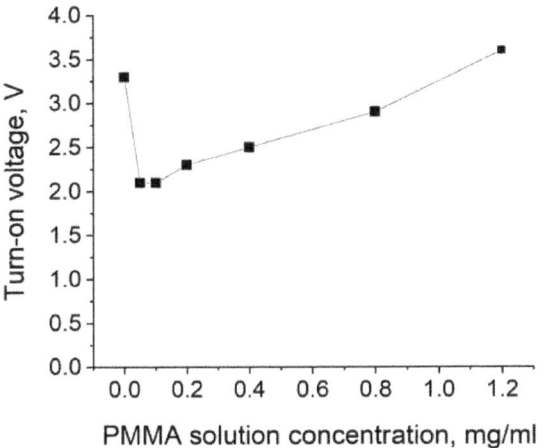

Figure 3. Effect of PMMA solution concentration on the turn-on voltage value of the QDLED device.

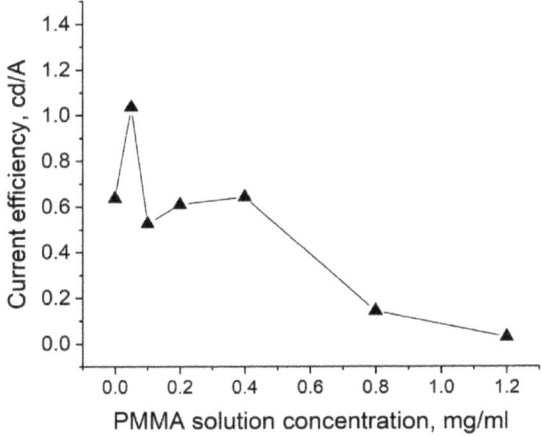

Figure 4. Effect of the PMMA solution concentration on the current efficiency of the QDLED device at 9 V.

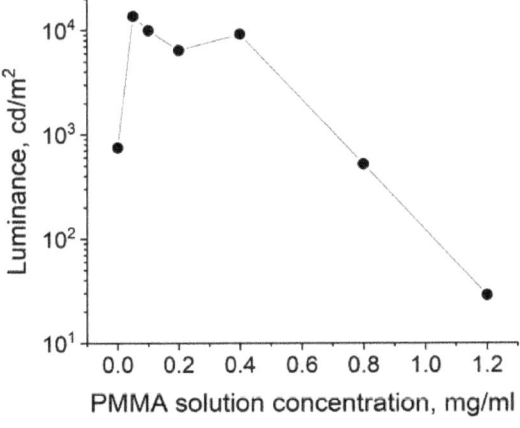

Figure 5. Effect of the PMMA solution concentration on the luminance of the QDLED device at 9 V.

As can be seen, in the case of current efficiency, the apparent maximum is observed for the device where the EBL had the minimum thickness, when it was deposited from a solution with a PMMA concentration of 0.05 mg/mL, and for the device without EBL. Additionally, a local maximum was observed for the device whose EBL was formed from a 0.4 mg/mL PMMA solution. This result suggests that either excess electron injection or over-blocking of the electron current deteriorates charge balance in the QDLEDs and thereby degrades current efficiency values.

Regarding the luminance (Figure 5), addition of even the thinnest PMMA layer to the QDLED led to a drastic increase in this characteristic, apparently due to reducing the probability of the formation of excess charges in the QD emissive layer and preventing the luminescence quenching via nonradiative processes. However, further increase in the concentration of PMMA in the EBL deposition solution resulted in deterioration of this characteristic. This may have been due to the hindered injection of electrons into the emitting layer as a result of an increased thickness of the potential barrier and, as a consequence, a decreased probability of carrier tunneling.

Our findings show that the most important characteristics of QDLEDs can be substantially improved by careful adjustment of the PMMA EBL deposition parameters, such as PMMA solution concentration. Notably, among the QDLEDs studied here, there was no obvious best device in terms of the turn-on voltage, current efficiency, and luminance. Therefore, the addition of the PMMA as an EBL alone should not be considered as a single treatment to improve all the QDLED characteristics, but it may be quite effective if applied along with other optimization approaches. In this case, the PMMA layer deposition parameters should be adjusted according to the requirements of each specific QDLED structure. Our results may be helpful as guidance for the preparation of a PMMA EBL in order to adjust specific QDLED parameters.

4. Conclusions

An electron-blocking layer of poly(methyl methacrylate) was added to the standard QDLED structure in order to improve the brightness characteristics and current efficiency. It has been shown that the concentration of the PMMA solution during layer deposition plays a significant role in achieving high QDLED efficiency. Specifically, at a concentration as high as 1.2 mg/mL, the characteristics of the current efficiency and brightness of the QDLEDs dropped significantly relative to a similar device without an EBL. This may be due to the hindered injection of electrons into the emitting layer due to an increase in the thickness of the potential barrier and, as a consequence, a decrease in the probability of carrier-tunneling.

At the same time, a low concentration of the initial PMMA solution leads to a sharp improvement of the characteristics of the QDLEDs, both in terms of brightness and current efficiency and in terms of lowering the turn-on voltage. In terms of current efficiency, the QDLED sample with an EBL deposited from a 0.4 mg/mL PMMA solution turned out to be the optimal one. Apparently, this was why the resultant EBL provided a better balance of the inflow of charge carriers into the QD layer. In the case of the minimum concentration of the PMMA solution, the brightness of the LEDs produced was 18,671 cd/m^2, which is four times higher than these values for devices without a blocking layer due to reducing the number of charged QDs and probability of nonradiative processes.

Supplementary Materials: The following are available online at https://www.mdpi.com/article/10.3390/nano11082014/s1. Figure S1: Luminescence and absorption spectra of the synthesized CdSe/ZnS/CdS/ZnS QDs; Figure S2: TEM image of the synthesized CdSe/ZnS/CdS/ZnS QDs; Formulas regarding the estimation of the thickness of PMMA electron blocking layers; Table S1: Estimated PMMA EBL layer thickness deposited from PMMA solutions in acetone with different concentrations.

Author Contributions: Conceptualization, I.N. and P.S.; methodology, M.Z., A.A., A.T. and D.L.; validation, D.L.; writing—original draft preparation, M.Z.; writing—review and editing, P.S. and

I.N.; project administration, P.S. The manuscript was written through contributions of all authors. All authors have given approval to the final version of the manuscript.

Funding: This work was supported by the Russian Science Foundation (grant no. 18-19-00588-П) in its part related to the synthesis of nanomaterials and QDLEDs preparation and by the Ministry of Science and Higher Education of Russian Federation (grant no. FSWU-2020-0035) in its part related to the characterization and validation of created samples.

Data Availability Statement: The data presented in this study are available on request from the corresponding authors.

Acknowledgments: I.N. acknowledges supports from the Université de Reims Champagne-Ardenne and the Ministry of Higher Education, Research and Innovation of French Republic. We thank Vladimir Ushakov for the help with technical preparation of the manuscript.

Conflicts of Interest: The authors declare no conflict of interest.

References

1. Ananthakumar, S.; Ramkumar, J.; Moorthy Babu, S. Synthesis of thiol modified CdSe nanoparticles/P3HT blends for hybrid solar cell structures. *Mater. Sci. Semicond. Process.* **2014**, *22*, 44–49. [CrossRef]
2. Aubert, T.; Soenen, S.J.; Wassmuth, D.; Cirillo, M.; Van Deun, R.; Braeckmans, K.; Hens, Z. Bright and stable CdSe/CdS@SiO$_2$ nanoparticles suitable for long-term cell labeling. *ACS Appl. Mater. Interfaces* **2014**, *6*, 11714–11723. [CrossRef]
3. Lim, J.; Jeong, B.G.; Park, M.; Kim, J.K.; Pietryga, J.M.; Park, Y.S.; Klimov, V.I.; Lee, C.; Lee, D.C.; Bae, W.K. Influence of shell thickness on the performance of light-emitting devices based on CdSe/Zn$_{1-X}$Cd$_X$S core/shell heterostructured quantum dots. *Adv. Mater.* **2014**, *26*, 8034–8040. [CrossRef] [PubMed]
4. Dayneko, S.; Lypenko, D.; Linkov, P.; Sannikova, N.; Samokhvalov, P.; Nikitenko, V.; Chistyakov, A. Application of CdSe/ZnS/CdS/ZnS Core-multishell Quantum Dots to Modern OLED Technology. *Mater. Today Proc.* **2016**, *3*, 211–215. [CrossRef]
5. Linkov, P.; Krivenkov, V.; Nabiev, I.; Samokhvalov, P. High Quantum Yield CdSe/ZnS/CdS/ZnS Multishell Quantum Dots for Biosensing and Optoelectronic Applications. *Mater. Today Proc.* **2016**, *3*, 104–108. [CrossRef]
6. Sun, Y.; Jiang, Y.; Sun, X.W.; Zhang, S.; Chen, S. Beyond OLED: Efficient Quantum Dot Light-Emitting Diodes for Display and Lighting Application. *Chem. Rec.* **2019**, *19*, 1729–1752. [CrossRef] [PubMed]
7. Anikeeva, P.O.; Madigan, C.F.; Halpert, J.E.; Bawendi, M.G.; Bulović, V. Electronic and excitonic processes in light-emitting devices based on organic materials and colloidal quantum dots. *Phys. Rev. B Condens. Matter Mater. Phys.* **2008**, *78*, 085434. [CrossRef]
8. Bae, W.K.; Park, Y.-S.; Lim, J.; Lee, D.; Padilha, L.A.; McDaniel, H.; Robel, I.; Lee, C.; Pietryga, J.M.; Klimov, V.I. Controlling the influence of Auger recombination on the performance of quantum-dot light-emitting diodes. *Nat. Commun.* **2013**, *4*, 2661. [CrossRef]
9. Sun, Y.; Wang, W.; Zhang, H.; Su, Q.; Wei, J.; Liu, P.; Chen, S.; Zhang, S. High-Performance Quantum Dot Light-Emitting Diodes Based on Al-Doped ZnO Nanoparticles Electron Transport Layer. *ACS Appl. Mater. Interfaces* **2018**, *10*, 18902–18909. [CrossRef]
10. Alexandrov, A.; Zvaigzne, M.; Lypenko, D.; Nabiev, I.; Samokhvalov, P. Al-, Ga-, Mg-, or Li-doped zinc oxide nanoparticles as electron transport layers for quantum dot light-emitting diodes. *Sci. Rep.* **2020**, *10*, 7496. [CrossRef]
11. Jia, H.; Wang, F.; Tan, Z. Material and device engineering for high-performance blue quantum dot light-emitting diodes. *Nanoscale* **2020**, *12*, 13186–13224. [CrossRef]
12. Peng, H.; Wang, W.; Chen, S. Efficient Quantum-Dot Light-Emitting Diodes With 4,4,4-Tris(N-Carbazolyl)-Triphenylamine (TcTa) Electron-Blocking Layer. *IEEE Electron. Device Lett.* **2015**, *36*, 369–371. [CrossRef]
13. Sun, Q.; Subramanyam, G.; Dai, L.; Check, M.; Campbell, A.; Naik, R.; Grote, J.; Wang, Y. Highly Efficient Quantum-Dot Light-Emitting Diodes with DNA-CTMA as a Combined Hole-Transporting and Electron-Blocking Layer. *ACS Nano* **2009**, *3*, 737–743. [CrossRef]
14. Jin, H.; Moon, H.; Lee, W.; Hwangbo, H.; Yong, S.H.; Chung, H.K.; Chae, H. Charge balance control of quantum dot light emitting diodes with atomic layer deposited aluminum oxide interlayers. *RSC Adv.* **2019**, *9*, 11634–11640. [CrossRef]
15. Dai, X.; Zhang, Z.; Jin, Y.; Niu, Y.; Cao, H.; Liang, X.; Chen, L.; Wang, J.; Peng, X. Solution-processed, high-performance light-emitting diodes based on quantum dots. *Nature* **2014**, *515*, 96–99. [CrossRef]
16. Rahmati, M.; Dayneko, S.; Pahlevani, M.; Shi, Y. Highly Efficient Quantum Dot Light-Emitting Diodes by Inserting Multiple Poly(methyl methacrylate) as Electron-Blocking Layers. *Adv. Funct. Mater.* **2019**, *29*, 1906742. [CrossRef]
17. Krivenkov, V.; Samokhvalov, P.; Zvaigzne, M.; Martynov, I.; Chistyakov, A.; Nabiev, I. Ligand-Mediated Photobrightening and Photodarkening of CdSe/ZnS Quantum Dot Ensembles. *J. Phys. Chem. C* **2018**, *122*, 15761–15771. [CrossRef]
18. Jasieniak, J.; Smith, L.; Embden, J.; Mulvaney, P.; Califano, M. Re-examination of the Size-Dependent Absorption Properties of CdSe Quantum Dots. *J. Phys. Chem. C* **2009**, *113*, 19468–19474. [CrossRef]
19. Li, L.; Reiss, P.; Protie, M. Core/Shell Semiconductor Nanocrystals. *Small* **2009**, *5*, 154–168. [CrossRef]

20. Pietryga, J.M.; Park, Y.-S.; Lim, J.; Fidler, A.F.; Bae, W.K.; Brovelli, S.; Klimov, V.I. Spectroscopic and Device Aspects of Nanocrystal Quantum Dots. *Chem. Rev.* **2016**, *116*, 10513–10622. [CrossRef]
21. Shang, Y.; Ning, Z. Colloidal quantum-dots surface and device structure engineering for high-performance light-emitting diodes. *Natl. Sci. Rev.* **2017**, *4*, 170–183. [CrossRef]
22. Zhang, H.; Sui, N.; Chi, X.; Wang, Y.; Liu, Q.; Zhang, H.; Ji, W. Ultrastable Quantum-Dot Light-Emitting Diodes by Suppression of Leakage Current and Exciton Quenching Processes. *ACS Appl. Mater. Interfaces* **2016**, *8*, 31385–31391. [CrossRef]
23. Jin, X.; Chang, C.; Zhao, W.; Huang, S.; Gu, X.; Zhang, Q.; Li, F.; Zhang, Y.; Li, Q. Balancing the Electron and Hole Transfer for Efficient Quantum Dot Light-Emitting Diodes by Employing a Versatile Organic Electron-Blocking Layer. *ACS Appl. Mater. Interfaces* **2018**, *10*, 15803–15811. [CrossRef] [PubMed]
24. Fu, Y.; Jiang, W.; Kim, D.; Lee, W.; Chae, H. Highly Efficient and Fully Solution-Processed Inverted Light-Emitting Diodes with Charge Control Interlayers. *ACS Appl. Mater. Interfaces* **2018**, *10*, 17295–17300. [CrossRef]
25. Ding, K.; Chen, H.; Fan, L.; Wang, B.; Huang, Z.; Zhuang, S.; Hu, B.; Wang, L. Polyethylenimine Insulativity-Dominant Charge-Injection Balance for Highly Efficient Inverted Quantum Dot Light-Emitting Diodes. *ACS Appl. Mater. Interfaces* **2017**, *9*, 20231–20238. [CrossRef]

Communication

Synthesis of Group II-VI Semiconductor Nanocrystals via Phosphine Free Method and Their Application in Solution Processed Photovoltaic Devices

Mingyue Hou [1], Zhaohua Zhou [1], Ao Xu [1], Kening Xiao [1], Jiakun Li [1], Donghuan Qin [1,2,*], Wei Xu [1,2,*] and Lintao Hou [3,*]

[1] School of Materials Science and Engineering, South China University of Technology, Guangzhou 510640, China; 201830140083@mail.scut.edu.cn (M.H.); 201830320430@mail.scut.edu.cn (Z.Z.); 202020118759@mail.scut.edu.cn (A.X.); 201830320324@mail.scut.edu.cn (K.X.); 201830640170@mail.scut.edu.cn (J.L.)
[2] State Key Laboratory of Luminescent Materials & Devices, Institute of Polymer Optoelectronic Materials & Devices, South China University of Technology, Guangzhou 510640, China
[3] Guangdong Provincial Key Laboratory of Optical Fiber Sensing and Communications, Guangzhou Key Laboratory of Vacuum Coating Technologies and New Energy Materials, Siyuan Laboratory, Department of Physics, Jinan University, Guangzhou 510632, China
* Correspondence: qindh@scut.edu.cn (D.Q.); xuwei@scut.edu.cn (W.X.); thlt@jnu.edu.cn (L.H.)

Abstract: Solution-processed CdTe semiconductor nanocrystals (NCs) have exhibited astonishing potential in fabricating low-cost, low materials consumption and highly efficient photovoltaic devices. However, most of the conventional CdTe NCs reported are synthesized through high temperature microemulsion method with high toxic trioctylphosphine tellurite (TOP-Te) or tributylphosphine tellurite (TBP-Te) as tellurium precursor. These hazardous substances used in the fabrication process of CdTe NCs are drawing them back from further application. Herein, we report a phosphine-free method for synthesizing group II-VI semiconductor NCs with alkyl amine and alkyl acid as ligands. Based on various characterizations like UV-vis absorption (UV), transmission electron microscope (TEM), and X-ray diffraction (XRD), among others, the properties of the as-synthesized CdS, CdSe, and CdTe NCs are determined. High-quality semiconductor NCs with easily controlled size and morphology could be fabricated through this phosphine-free method. To further investigate its potential to industrial application, NCs solar cells with device configuration of ITO/ZnO/CdSe/CdTe/Au and ITO/ZnO/CdS/CdTe/Au are fabricated based on NCs synthesized by this method. By optimizing the device fabrication conditions, the champion device exhibited power conversion efficiency (PCE) of 2.28%. This research paves the way for industrial production of low-cost and environmentally friendly NCs photovoltaic devices.

Keywords: group II-VI semiconductor nanocrystals; phosphine free method; solution process; solar cells

1. Introduction

Group II-VI semiconductor nanocrystals (NCs), such as CdS, CdSe, CdTe, etc., have recently attracted much attention due to their tunable direct bandgap and potential application in optoelectronic devices like photodetectors, photovoltaic devices, and light emitting diodes (LEDs) [1–8]. As the desire for clean and renewable energy increases, photovoltaic devices have become one major research hotspot. In this case, solution processed CdTe NCs solar cells is a promising candidate for next generation commercial photovoltaic device for their low cost, low consumption of materials, simple fabricating process, and being suitable for roll-to-roll printing techniques in industrial mass production [9–11]. The controllable synthesis process of high-quality II-VI NCs with uniform morphology, composition, and desired crystal structure is of great significance for fabricating efficient photovoltaic devices [12,13]. To date, group II-VI NCs in most of the reported researches are fabricated by the well-known hot injection method. In such a method, cadmium precursor is first dispersed in a high boiling point solvent such as 1-Octadecene (ODE), Trioctylphosphine oxide (TOPO), and oleic acid (OA), among others. After that, Se or Te precursors (chalcogen elements dissolved into alkylphosphines, such as trioctylphosphine (TOP) or tributylphosphine (TBP)) are quickly injected into the cadmium precursor solution at high temperature (>200 °C) [14–17]. CdS, CdSe, or CdTe NCs with homogeneous size, crystalline, and morphology can be obtained by varying the reaction temperature, ligands, and precursor concentration. However, the alkylphosphines is highly toxic and expensive, making this method a highly environmentally harmful process. Comparing to the S or Se, preparation of the Te precursors is a great challenge for the insolubility and strong metallicity of the Te element [18–20]. As a result, the first step towards a green synthesis method for CdS, CdSe, and CdTe NCs is to prepare phosphine-free S, Se, and Te precursors. Previous researches have reported several ways to obtain a phosphine-free Te precursor. One is based on the water phase green synthesis method, which had been well developed in recent years [21–23]. In this method, NaHTe was selected as a Te precursor, which was prepared by adding Te powder and NaHB to deionized water and refluxed under N_2 flow. Water-soluble CdTe or CdSe NCs can be fabricated based on Se and Te precursors obtained by this process. However, photovoltaic devices fabricated from such water-soluble CdTe/CdSe NCs showed lower devices performance, when compared to similar NC solar cells based on organic phase synthesized CdTe/CdSe NCs [24–34]. In addition, as the reactivity of Te precursor is too high, it is difficult to control the size and morphology of the as-synthesized NCs. Recently, Webber et al. [35] developed a novel binary diamine-dithiol solvent mixture. This mixture is capable of dissolving the VI group element, including S, Se, Te, and other six metals. High-quality telluride semiconductor thin films can be formed by using these precursors. Following this, Yao et al. [19] reported a phosphine-free Se and Te precursor prepared by dissolving SeO_2 and TeO_2 in dodecanethiol solvent by sonication under low temperature. It was found that SeO_2 and TeO_2 are reduced by dodecanethiol to elemental Se and Te, respectively. After combining disulfides, various metal chalcogenide NCs, including CdSe, CdTe, and other tellurite semiconductors can be synthesized. Later, Wu's group [36] reported a phosphine-free Te precursor prepared by dissolving TeO_2 into dodecanethiol and oleylamine mixture. CdTe, PbTe, $FeTe_2$, and Cu_2Te NCs with homogeneous size and morphology were obtained by controlling reaction conditions, such as temperature, and precursor concentration. Although the II-VI group semiconductor NCs was fabricated successfully via a phosphine-free environmentally friendly way, there are still no reports on photovoltaic devices' application based on these semiconductor NCs.

In this research, we demonstrate a low-cost, efficient phosphine-free method for synthesis of CdS, CdSe, and CdTe NCs in organic solvents. Dodecanethiol and oleyl amine were selected as complex ligands for the Te precursor, while Se or S powder was used directly for the synthesis of CdSe or CdS NCs. The morphology and structure of the NCs were characterized by TEM and XRD, respectively, while the UV-vis absorption was used to investigate the optical properties of NCs. Based on the semiconductor NCs synthesized by phosphine-free process, solar cells with configuration of ITO/ZnO/CdS/CdTe/Au and ITO/ZnO/CdSe/CdTe/Au were fabricated through a layer-by-layer solution process. The effects of active layer thickness and annealing temperature on the device's performance are investigated and discussed in detail. Champion devices with PCE of 1.08% and 2.28% were fabricated when adopting CdSe and CdS as n-type materials, respectively. Using this low-cost and low-toxicity synthesis method, this research has shown the potential of manufacturing environmentally friendly, low-cost, and large-area solution-processed CdTe NCs solar cells.

2. Experiment Procedure

2.1. Materials

Anhydrous $CdCl_2$ (99.99%), TeO_2 powder (99%), Se powder (99%), sublimed sulfur (AR), 2,2′-Dithiobisbenzothiazole (98%), Tetraethylthiuram disulfide (97%), 1-Dodecanethiol (DDT, 98%), 3-Mercaptopropionic acid (99%), Zinc acetate dehydrate (99.99%), Ethanolamine (99%), 2-methoxyethanol (99.8%), Oleic acid (OA, AR), Oleylamine (OLA, 80%), and 1-Octadecene (ODE, 90%) were purchased from Alfa Aesar. All other chemicals and solvents were used as received.

2.2. Synthesis Method of CdTe NCs

In a typical synthesis procedure, 1 mmol TeO_2 powder is dissolved in DDT (3 mL) and stirred for 3 min at room temperature, obtaining a yellow solution. OLA (3 mL) is injected into the above TeO_2 mixture under N_2 atmosphere; the solution turns black rapidly and the OLA-Te precursor is obtained. 2 mmol $CdCl_2$ is dissolved in OLA and loaded into a three-neck bottle. The mixture is heated to 220 °C under N_2 atmosphere and this temperature is maintained for 15 min until a clear solution is obtained. After that, the mixture is rapidly heated to 240 °C and the OLA-Te precursor is quickly injected into the mixture. To investigate the growth dynamics of CdTe NCs, aliquots of solution are taken from the reaction flask at regular time intervals and diluted with toluene for UV-vis absorption (UV) and photoluminescence (PL) measurements. The final product was dispersed and washed three times with methanol and toluene.

2.3. Synthesis of CdSe NCs and CdS NCs

The fabrication of CdSe and CdS NCs are based on previous reported methods [37,38] with optimal process. For CdSe NCs fabrication, cadmium myristate was selected as a single cadmium salt. In a typical synthetic procedure for the CdSe NCs, 1 mmol of cadmium myristate and 10.0 mL ODE were loaded into a flask. Following this, the reaction system was heated to 170 °C and kept at this temperature for 10 min to dissolve the cadmium myristate. The reaction was then cooled down to room temperature and 2 mmol Se power was introduced into the reaction mixture. The mixture was then heated up to 230 °C under N_2 flow. 0.35 mL OA was injected into the flask after 2 min. Finally, the reaction was kept at 230 °C for 30 min and cooled down to room temperature. The final product was dispersed and washed three times with acetone, methanol, and toluene. Fabrication of CdS NCs: In this case, cadmium myristate was selected as cadmium salt, while cadmium acetate was used in the literature [38]. Moreover, no myristic acid was used in this case. In a typical process, 2 mmol cadmium myristate, 0.125 mmol tetraethylthiuram disulfide, 0.375 mmol 2,2′-dithiobisbenzothiazole, 1 mmol sulfur, and 50 mL ODE were loaded into a three-necked flask. The mixture was heated to 140 °C under N_2 flow and maintained at this temperature for 1 h to obtain a clear solution. The mixture was then heated to 240 °C at a rate of 15°/min and kept at this temperature for 30 min and then cooled down to room temperature. The purify method for CdS NCs is the same as that for CdSe NCs. There are no phosphine mixtures used in the fabrication of CdS NCs and CdSe NCs listed above.

2.4. Device Fabrication and Characterization

The CdTe NCs are dispersed in toluene with a concentration of 45 mg/mL. The CdSe NCs are dispersed into a chlorobenzene/pyridine (with volume ratio of 9:1) solvent, with a concentration of 43 mg/mL, while the CdS NCs are dispersed into a pyridine/n-propanol (with volume ratio of 1:1) solvent with a concentration of 36 mg/mL. In a typical process, solar cells with configuration of ITO/ZnO/CdSe/CdTe/Au or ITO/ZnO/CdS/CdTe/Au are fabricated by a layer-by-layer solution process. The Zn^{2+} precursor was prepared by dissolved 3.2925 g zinc acetate dehydrate into 30 mL 2-methoxyethanol with 0.905 mL ethanolamine and refluxed at 80 °C for 1 h under N_2 flow. The mixture was collected into a vial after being cooled down to room temperature. Firstly, several drops of the Zn^{2+} precursor solution are deposited on ITO substrates and spin-casted at 3000 rpm for

30 s. Then the substrate is transferred to a hotplate and annealed at 400 °C for 10 min to eliminate any organic solvent and impurity. Then, several drops of CdSe or CdS NCs solution are deposited on the ITO/ZnO substrate and spin-casted at 3000 rpm for 20 s. After that, the substrate is transferred to a hot plate and annealed at 150 °C for 10 min and 350 °C (CdSe NCs) or 380 °C (CdS NCs) for 40 s. Several drops of the CdTe NCs solution (dissolved into toluene at a concentration of 48 mg/mL) are deposited on the ITO/ZnO/CdSe (or CdS) substrate and spin-casted at 1100 rpm for 20 s. The substrate is then transferred to a hotplate and annealed at 150 °C for 3 min. Following this, the substrate is dipped into saturated $CdCl_2$/methanol solution for 10 s and rinsed with n-propyl and quickly putted on the hotplate at 360 °C for 40 s. Finally, five layers of CdTe NCs are deposited on the ITO/ZnO/CdS (or CdSe) via a similar layer-by-layer sintering process. The detail fabricating process can be found in our previously reported works [13]. Finally, Au (~80 nm) back contact electrodes are deposited via thermal vacuum evaporation through a shadow mask with an active area of 0.16 cm^2. The morphology of NC is characterized by transmission electron microscopy (TEM, FEI Tecnai G2 F20, FEI, OR, USA) and atomic force microscopy (AFM, DI/MutiMode, Veeco Instruments Inc., Plainview, MN, USA), while the structure is characterized by X-ray diffraction (XRD, X'pert Pro M, Philips, Amsterdam, The Netherlands). A step profiler (Dektak XT10, Bruker, Karlsruhe, Germany) is used to measure the thickness of the NC thin film. The optical properties and electronic properties of the NC devices are investigated by using a solar simulator (XES-40S1, San-Ei Electric Co., Ltd., Osaka, Japan), UV-Vis (UV-5100B, Shanghai Metash Instrument Co., Ltd., Shanghai, China) and photo-luminescence (PL, FL3-2IHR, HORIBA Instruments Inc., Irvine, CA, USA), and external quantum efficiency (EQE, Solar Cell Scan100, Zolix Instruments Co., Ltd., Beijing, China).

3. Result and Discussion

In this research, CdS, CdSe, and CdTe NCs are synthesized using phosphine-free receipt. In the case of CdSe and CdS NCs, ODE is selected as a non-coordinating solvent and the synthesis processes are non-injection methods. As shown in Figure 1a, Se powder is used as a Se precursor, while cadmium carboxyl (cadmium myristate) is used as a Cd^{2+} precursor. At a temperature of 230 °C, H^+ would be supplied in situ by ODE and react with Se to form H_2Se, as reported before [37]. CdSe NCs will be formed because of the reaction between H_2Se and cadmium myristate. As myristic acid is generated and acts as ligands capping on the surface of CdSe NCs during the reaction, NCs with homogeneous size are obtained. In the case of synthesizing CdS NCs, as the activity of sulfur is low when comparing to the Cd ion, the addition of the 2,2'-Dithiobisbenzothiazole initiator ensures the activity of S precursor and guarantees the formation of high-quality CdS NCs (Figure 1b). For CdTe NCs (Figure 1c), however, a similar non-injection method will not work. Because of the low activity of Te element, Te powder cannot be used directly as a Te precursor for CdTe NCs' fabrication. An efficient phosphine-free Te precursor is essential to fabricate high-quality CdTe NCs. To overcome this difficulty, a Te precursor is prepared by reducing TeO_2 powder with DDT in the presence of OLA to generate a soluble alkylammonium telluride at room temperature. CdTe NCs can then be prepared by injecting this phosphine-free Te precursor into the $CdCl_2$/OLA mixture at a temperature of 240 °C and refluxed at 220 °C for 30 min. It should be noticed that the NCs fabricated by this phosphine-free receipt is both economical and environmentally-friendly. The chemicals prices for fabricating 1 g CdS, CdSe, and CdTe NCs under different methods are presented in Tables S1 and S2. Using our phosphine-free receipt, the cost for fabricating 1g CdS, CdSe and CdTe NCs is $58.96, $16.90, and $18.33, respectively, while the cost of NCs synthesized by conventional methods are $94.11, $71.63, and $61.29 respectively. It is obvious that our phosphine-free receipt can significantly decrease the cost of NC photovoltaic products and is suitable for industrial mass production.

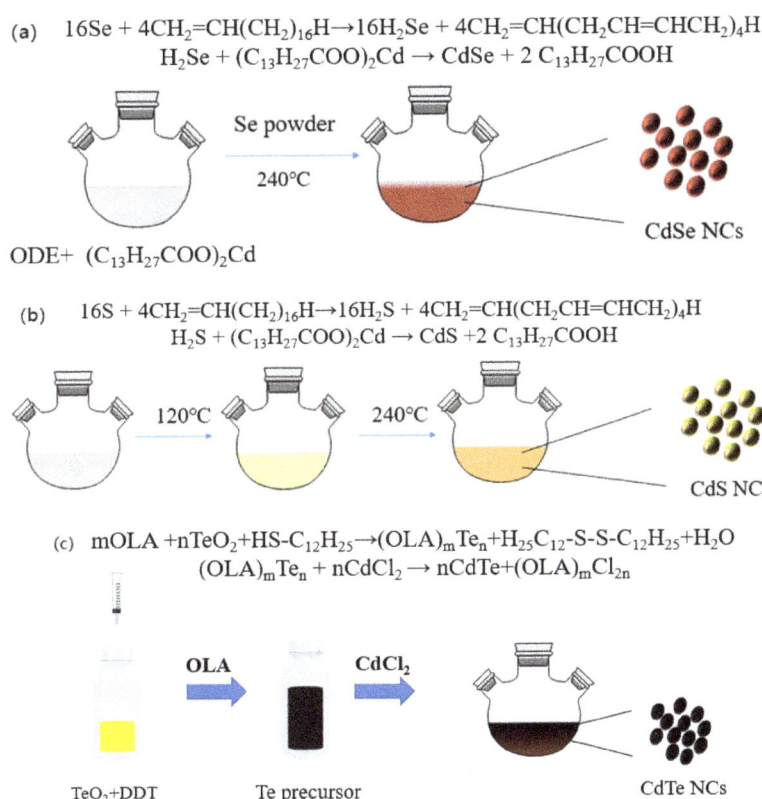

Figure 1. A schematic diagram of the NC fabricating process (**a**) CdSe, (**b**) CdS, (**c**) CdTe NCs.

To understand the properties of the NCs synthesized by our phosphine-free method, various characterizations were carried out. Figure 2 shows the transmission electron microscope (TEM) images of the as-prepared CdS (Figure 2a), CdSe (Figure 2b), and CdTe (Figure 2c) NCs. All the NCs exhibit a spherical shape. To investigate the uniformity of the size of NCs, a statistical number of 100 is selected for each NCs. As shown in Figure 2d–f, the size of all the NCs is homogeneous with mean size of 4.2 nm, 4.9 nm, and 5.9 nm for CdS, CdSe, and CdTe NCs, respectively. This result shows that our phosphine-free synthesis method allows effective control of the size and shape of the nanocrystals. We speculate that the origin of this uniformity is the in situ formation of carboxyl acid during the reaction, which has been confirmed in our previous work [39].

The crystal structures of the as-prepared CdS, CdSe, and CdTe NCs products are characterized by X-ray diffraction (XRD) (Figure 3). The XRD pattern of spherical CdS NCs exhibited diffraction peaks at about 25.1°, 41.8°, and 49.6° (Figure 3a), corresponding to the (111), (311), and (331) planes of the zinc-blende crystal structure of CdS. For CdSe NCs, diffraction patterns with peaks at about 25.3°, 42.0°, 49.7°, 67.1°, and 76.3° could be identified, corresponding to the (111), (220), (311), (331), and (422) facets of zinc–blende CdSe NCs [39]. A similar zinc–blende structure is revealed by the XRD pattern of CdTe NCs (Figure 3b). It is noted that the half peak width of CdTe (the main peak (111)) is narrower than that of CdSe or CdS NCs. As the CdS, CdSe, and CdTe NCs are collected by washing the NCs products with acetone, methanol, and toluene, no further treatment such as reflowing is carried out; the residual metal precursor or other chemical may affect the width of XRD peaks. Therefore, we speculate that the uniformity of the nanoparticles, or their size may not be related to the width of the XRD peaks.

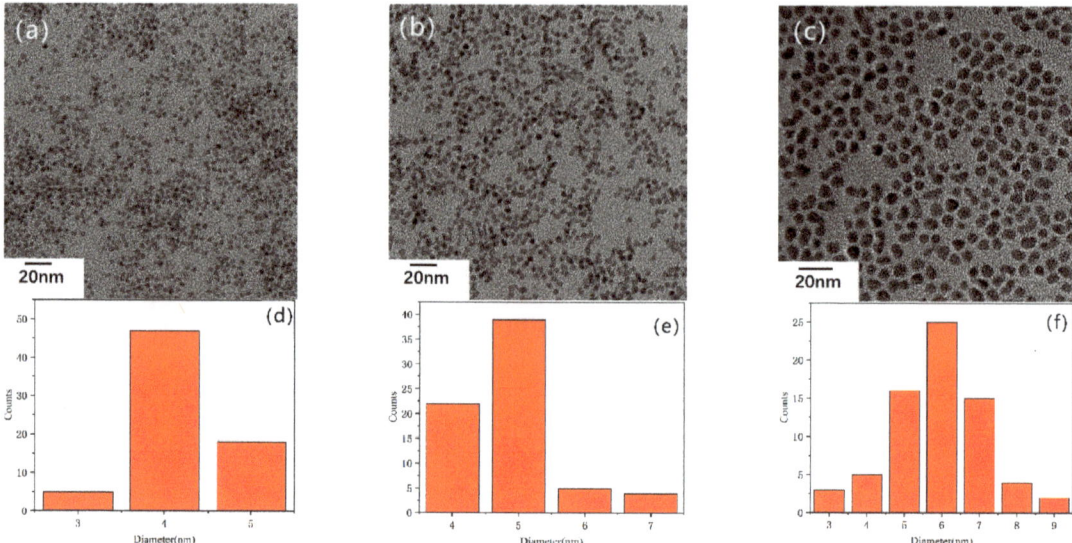

Figure 2. TEM images of (**a**) CdS (**b**) CdSe (**c**) CdTe NCs. (Scale bar 20 nm), particle size distribution of (**d**) CdS (**e**) CdSe (**f**) CdTe NCs.

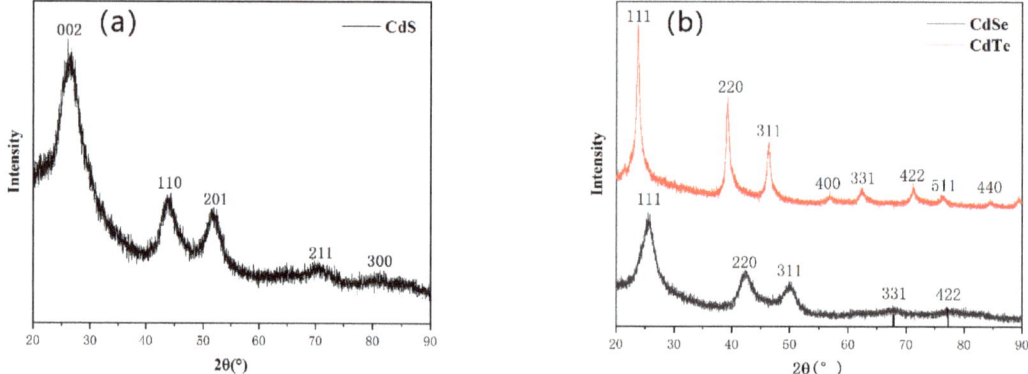

Figure 3. The XRD pattern of the (**a**) CdS (**b**) CdSe, and CdTe NCs.

The growing process is of great importance to control the optoelectronic properties of NCs as particle size has a significant influence on their electronic band structures. To monitor the growth kinetics of NCs, we characterized the UV–vis absorption and PL (Photoluminescence) spectra for different growth times after the formation of semiconductor NCs. For instance, Figure 4 showed the temporal evolutions of the UV–vis (Figure 4a) and PL spectra (Figure 4b) of CdSe NCs prepared by the phosphine-free non-injection method. As shown in Figure 2a, after the reaction temperature reached 220 °C, CdSe NCs are formed and the absorption peak is located at 543 nm, while the PL peak is at 572 nm. The reaction temperature is maintained at 230 °C and several drops of the NCs solution are taken out at different time intervals. After crystal nucleus formation, the absorption and PL peaks shift to 564 nm and 594 nm, respectively, after growing for 0.5 min. Although the absorption/PL peaks undergo a redshift to longer wavelengths as growing time increases, this trend slows down when the reaction progressed from 7 min to 30 min, which could be seen in the UV–vis (Figure 4a) and PL spectra (Figure 4b). In addition, from Figure 4b, one can see that the PL peaks are sharp and narrow with half-band width around 37 nm after the reaction time reaches 7 min, which implies a narrow NCs size distribution. Therefore, the phosphine-free method is promising for high-quality CdSe NCs fabrication; similar results can be found in the case of CdTe NCs (Figure S1). However, although a redshift is also found in the case of CdS NC (Figure S2) PL main peaks, the PL shows very broad emissions at high wavelengths and more mixed peaks appear in this case. We speculate that there are more defects in the CdS NCs and/or the effects of other chemicals such as 2,2′-Dithiobisbenzothiazole, tetraethylthiuram disulfide. The changes in fluorescent properties of NC with different reaction times are also monitored by using a portable UV light source. As shown in Figure 4c, the PL is yellow and green color at the beginning of reaction. Then the color slowly turns to yellow, light red, and to deep red, which implies that the NCs grow into larger sizes. After 20 min, there are almost no changes in the NC color, suggesting their low monomer concentration. The final size of CdS, CdSe, and CdTe NCs are calculated to be 1.92 nm, 4.50 nm, and 10.51 nm by using an experience formula [17]:

$$D = (1.6122 \times 10^{-9}) \lambda^4 - (2.6575 \times 10^{-6}) \lambda^3 + (1.6242 \times 10^{-3}) \lambda^2 - (0.4277) \lambda + (41.57)$$

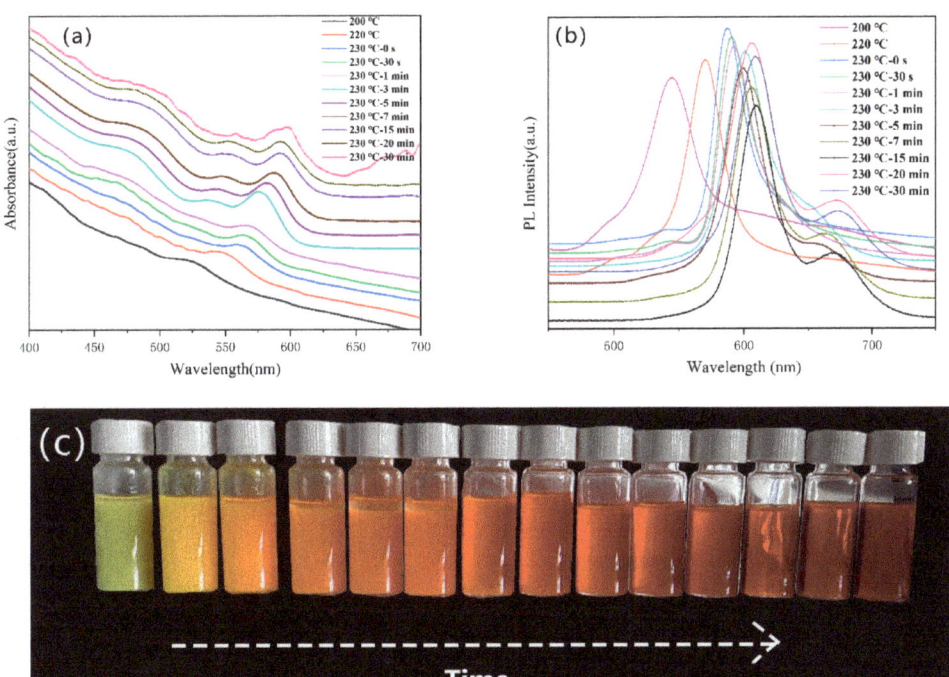

Figure 4. (**a**) UV absorbance of CdSe NCs with different growth time; (**b**) PL spectra of the CdSe NCs with different growth time; (**c**) PL images with excitation by 365 nm UV light.

Therefore, the calculated value for CdSe NC conforms to the value from the TEM measurement. However, the calculated value for CdS or CdTe NCs is quite different from the TEM results. We speculate that this formula is suitable for predicting the size of CdSe NCs but not for CdS or CdTe NCs. Moreover, a cadmium precursor or other chemical may affect the absorption of CdS or CdTe NCs; more work needs to be carried out to clarify this.

To exhibit the potential of our phosphine-free semiconductor NCs in the field of optoelectronic devices, photovoltaic devices were fabricated based on these materials. Devices with an inverted structure of ITO/ZnO (40 nm)/CdSe (60 nm)/CdTe (~400 nm)/Au (80 nm), and ITO/ZnO (40 nm)/CdS (60 nm)/CdTe (~400 nm)/Au (80 nm) were fabricated through a layer-by-layer spin-coating process, as reported before [38,39]. The device structure and band alignment with CdS NCs and CdSe NCs as n-type materials and CdTe as a p-type material are presented in Figure 5a,b. In this experiment, the CdTe NCs solution is dissolved into toluene at a concentration of 40 mg/mL, which is different from the works reported before [40]. For CdTe NCs solar cells, an appropriate annealing temperature is essential to increase the grain size and eliminate interface defects in the CdTe NCs thin film. From the AFM image presented in Figure S3, the grain size of CdTe NCs is up to several hundred nanometers, while the thickness of CdTe NCs is ~389.9 nm (characterized by using a step profiler). The J-V curves for the CdTe NC solar cells with CdSe and CdS NCs as the n-type partner under different annealing temperatures are exhibited in Figure 5c,d, while the photovoltaic parameters are as shown in Figure S4. The PCE of CdSe and CdS NC devices with different annealing temperatures is also summarized in Figure 6a,b. In the case of CdSe/CdTe junction solar cells, short circuit current (J_{sc}) increases almost linearly from 1.54 mA/cm^2 to 12.74 mA/cm^2 when the annealing temperature rises from 240 to 360 °C. When the annealing temperature further increases, the J_{sc} drops linearly. Similar laws are found for the changes in PCE of NCs solar cells. The V_{oc} of devices are around 0.4 V at low annealing temperatures (below 300 °C), while V_{oc} below 0.3 V is obtained for annealing temperatures higher than 300 °C. The champion device is obtained at an annealing temperature of 360 °C; it exhibited a short circuit current density (J_{sc}) of 12.74 mA/cm^2, an open circuit voltage (V_{oc}) of 0.26 V, a fill factor (FF) of 32.35%, and a PCE of 1.08%. This value is significantly lower than those ever reported (5.81%) with the same device structure (NCs are fabricated by using phosphine mixture) and similar annealing strategy [26,40]. The low device performance is mainly attributed to the low active layer (CdTe/CdSe) quality. We speculate that, in the case of CdTe NC solar cells with CdSe NC as n-type partner, the CdSe NCs is capped with oleic acid. During the heat-treatment of the CdSe NC film, the oleic acid cannot be removed effectively. The oleic acid is insulating and acts as a carrier recombination center, which will drastically decrease the device performance. When CdS NCs are used as n-type partners for CdTe NC solar cells, the trend of how devices performance changes with annealing temperature are similar to that of CdSe NCs devices. The V_{oc} of devices are kept stable at around 0.3–0.4 V when annealing temperature is below 360 °C. The PCE increases linearly when annealing temperature rises from 240 to 360 °C. At low annealing temperature (below 320 °C), the PCE is less than 1%. On the contrary, devices annealed at a moderate temperature of 340–360 °C show optimal performance. At a temperature of 360 °C, we obtain our champion device, which shows the following merits: a short circuit voltage (J_{sc}) of 18.01 mA/cm^2, an open circuit voltage (V_{oc}) of 0.33 V, a fill factor (FF) of 37.84% and a PCE of 2.28%. This value is two times higher than that for CdSe NC devices. It is noted that the defects of CdS NCs are higher than that of CdSe NCs (before purify). However, as no long chain alky acid ligands are used for the fabrication of CdS NCs, during the sintering process (the NCs thin film fabrication), the CdS NCs' thin films are more compact and low interface defects between CdS NCs is obtained. On the other hand, as OA cannot be removed completely during the sintering process (OA is insulated material), more defects are existed between CdSe NCs. Therefore, low carrier recombination is attained in the case of CdS NC solar cells, which will lead to higher J_{sc} and PCE. Both NC devices (with CdSe of CdS NCs as n-type partner) decay at higher annealing temperatures. This may be due to oxidation of the CdTe NC film at high temperatures, which has been confirmed before [38]. When comparing devices based on NCs fabricated by the traditional method [38,39] with phosphine mixture, the series resistance (R_s) obtained here (Tables 1 and 2) is several times higher than those reported before. The high R_s will result in a higher carrier recombination in the active layer and decrease the FF of the NC devices. Further investigation should be carried out to further increase the quality of the NC active layer. Figure 5e,f show the EQE spectra of the CdS NC and CdSe NC champion devices. Comparing to the CdSe NC device, the CdS NC device has a better EQE response at wavelengths from 400 nm to 800 nm, implying that the CdS NC device has a better capability to transfer photons to valence electrons and generate electron–hole pairs than that of the CdSe NC device. When the EQE curves are integrated, the calculated J_{sc} of 16.57 mA/cm^2 (for CdS NC device) and 14.45 mA/cm^2 (for CdSe NC device) are predicted, which are consistent with the J_{sc} value from the J-V curves under light (Figure 5e,f). From the dark J–V curves for CdSe NC and CdS NC devices (Figure 5g,h), it is evident that typical diode properties are obtained in both cases.

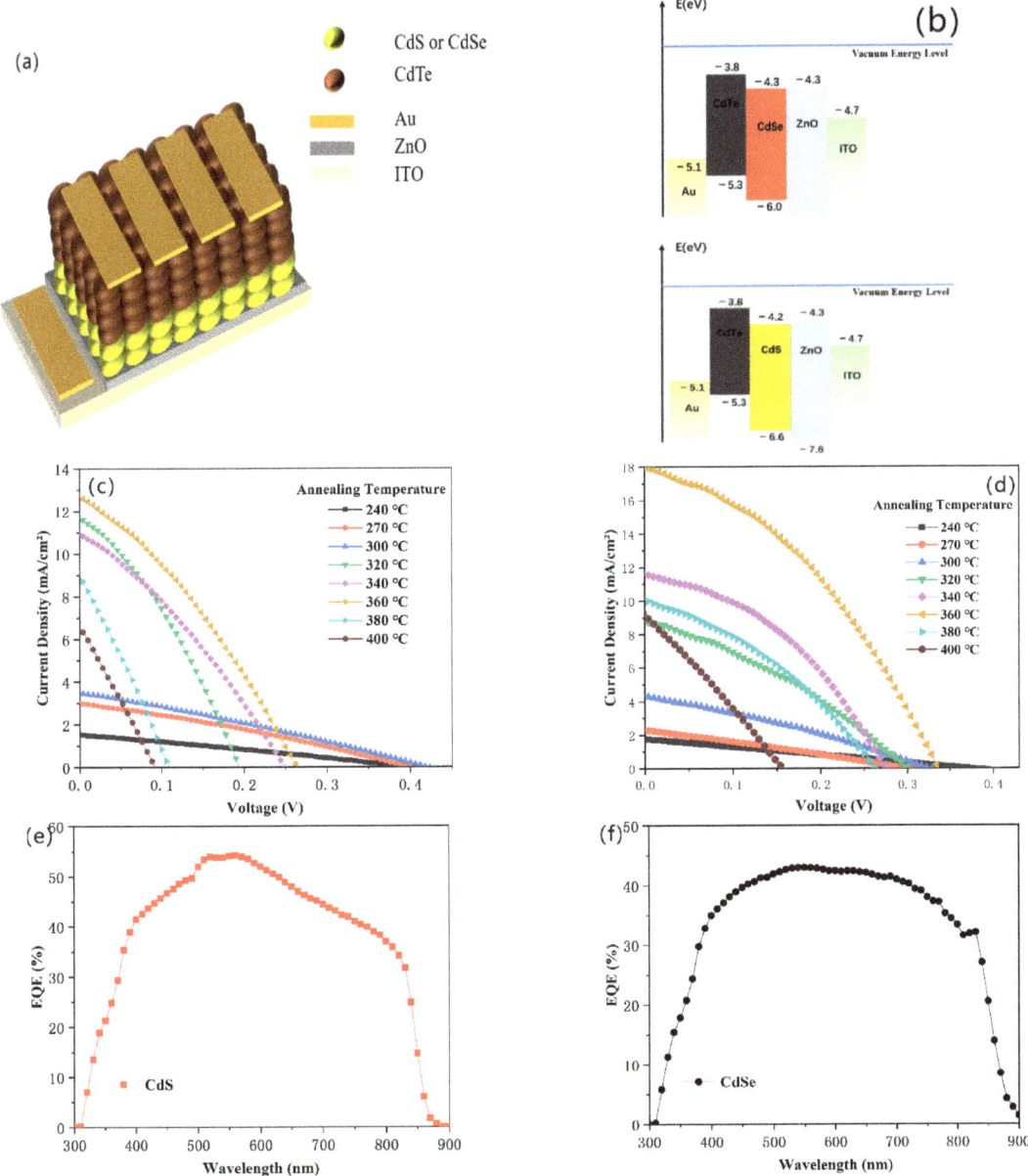

Figure 5. *Cont.*

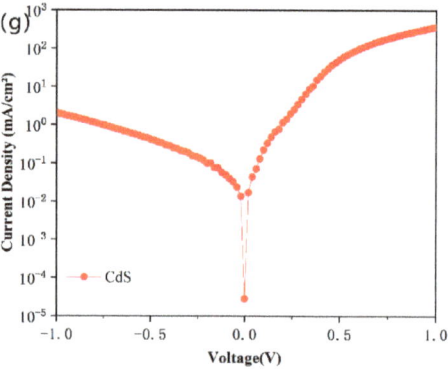
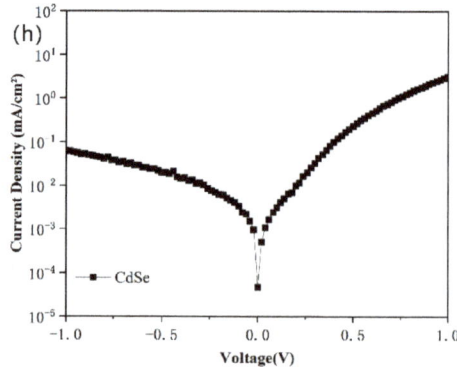

Figure 5. (a) The NC solar cell configuration; (b) Band alignment of the NC solar cell; *J-V* curves of (c) ITO/ZnO/CdSe/CdTe/Au and (d) ITO/ZnO/CdS/CdTe/Au with different annealing temperatures under light with different annealing temperatures under light. The corresponding EQE spectra of (e) ITO/ZnO/CdS/CdTe/Au (f) ITO/ZnO/CdSe/CdTe/Au; *J-V* curves of (g) ITO/ZnO/CdS/CdTe/Au and (h) ITO/ZnO/CdSe/CdTe/Au under dark.

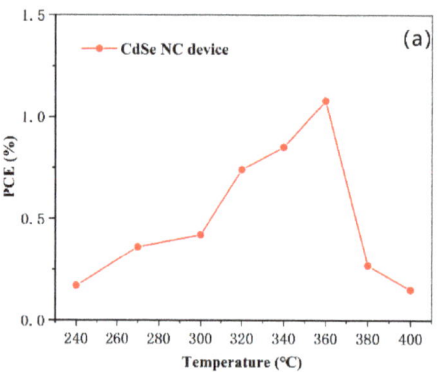
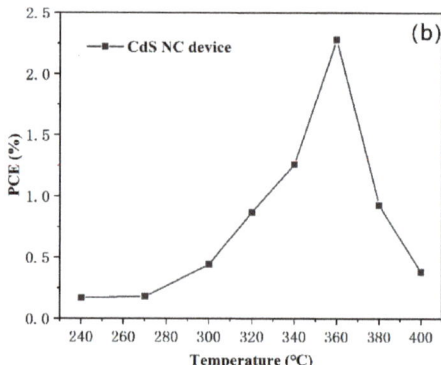

Figure 6. Evolution of parameters PCE for (a) CdTe/CdSe NC devices (b) CdTe/CdS NC devices under different annealing temperatures.

Table 1. Summarized performance of CdTe/CdSe NC solar cells with different annealing temperatures (Figure 5a).

Annealing Temperature (°C)	PCE (%)	J_{sc} (mA/cm^2)	FF (%)	V_{oc} (V)	R_s (Ω·cm^2)	R_{sh} (Ω·cm^2)
240	0.17 (± 0.02)	1.54 (± 0.2)	27.27 (± 2)	0.40 (± 0.02)	1330.42	1671.38
270	0.36 (± 0.05)	3.03 (± 0.5)	29.68 (± 4)	0.40 (± 0.03)	611.66	1458.51
300	0.42 (± 0.04)	3.52 (± 0.2)	28.35 (± 3)	0.42 (± 0.02)	649.64	904.81
320	0.74 (± 0.02)	12.04 (± 0.4)	30.32 (± 3)	0.20 (± 0.01)	74.16	176.07
350	0.85 (± 0.02)	11.01 (± 0.5)	30.93 (± 2)	0.25 (± 0.03)	98.84	234.84
360	1.08 (± 0.05)	12.74 (± 0.3)	32.35 (± 4)	0.26 (± 0.02)	93.84	334.76
380	0.27 (± 0.012)	9.16 (± 0.4)	25.93 (± 1)	0.11 (± 0.02)	60.89	84.61
400	0.15 (± 0.02)	6.80 (± 0.2)	24.39 (± 2)	0.09 (± 0.01)	87.54	86.20

Table 2. Summarized performance of CdTe/CdS NC solar cells with different annealing temperatures (Figure 5b).

Annealing Temperature (°C)	PCE (%)	J_{sc} (mA/cm^2)	FF (%)	V_{oc} (V)	R_s (Ω·cm^2)	R_{sh} (Ω·cm^2)
240	0.17 (± 0.02)	1.77 (± 0.3)	25.26 (± 2)	0.37 (± 0.02)	1306.76	1500.15
270	0.18 (± 0.01)	2.32 (± 0.2)	26.63 (± 3)	0.30 (± 0.03)	721.03	1103.55
300	0.44 (± 0.02)	4.40 (± 0.5)	30.36 (± 3)	0.33 (± 0.02)	337.50	781.55
320	0.87 (± 0.03)	8.81 (± 0.3)	33.12 (± 3)	0.30 (± 0.03)	145.38	670.95
340	1.26 (± 0.02)	11.64 (± 0.6)	39.61 (± 3)	0.27 (± 0.01)	77.86	611.93
360	2.28 (± 0.02)	18.01 (± 0.6)	37.84 (± 4)	0.33 (± 0.02)	70.40	600.15
380	0.93 (± 0.01)	10.13 (± 0.3)	34.91 (± 3)	0.26 (± 0.02)	97.51	315.32
400	0.38 (± 0.2)	9.29 (± 0.2)	26.84 (± 2)	0.15 (± 0.03)	105.35	113.92

In summary, CdS, CdSe, and CdTe NCs are fabricated successfully by a phosphine-free receipt. NCs with homogeneous morphology and size can be well controlled by this method. Based on CdS, CdSe, and CdTe NCs, CdTe NC solar cells with configuration of ITO/ZnO/CdSe/CdTe/Au and ITO/ZnO/CdS/CdTe/Au were fabricated and investigated. It was found that at optimal annealing temperature, we obtain champion devices with PCE of 1.08% and 2.28% by using CdSe and CdS NC as n-type partners, respectively. We believe that by optimizing device-fabricating techniques (such as using ligands exchange technology, designing active layer thickness, etc.), the PCE of these NC devices can be further improved. With a low-cost and environmentally-friendly fabricating process, these NC solar cells may pave the way for next-generation photovoltaic devices.

Supplementary Materials: The following are available online at https://www.mdpi.com/article/10.3390/nano11082071/s1, Table S1. The price of chemicals from the website. Table S2. The cost of materials for fabricating 1g CdS, CdSe, and CdTe NCs with different methods. Figure S1. (a) UV absorbance of CdTe NCs with different growth times; (b) PL emission spectra of the CdTe NCs with different growth times; (c) PL images with excitation by 365 nm UV light. Figure S2. (a) UV absorbance of CdS NCs with different growth times; (b) PL emission spectra of the CdS NCs with different growth times. Figure S3. Atomic for microscopy (AFM) images of devices. Figure S4. The microscopy image of a NC solar cell (one single substrate containing 5 devices with active areas of 0.16 cm^2).

Author Contributions: D.Q. and M.H. conceived and designed the experiments; M.H., Z.Z., A.X., K.X., and J.L. conducted the experiments; M.H. analyzed the data; D.Q., W.X., and L.H. contributed reagents/materials/analysis tools; and D.Q., and M.H. compiled the paper. All authors have read and agreed to the published version of the manuscript.

Funding: The authors are thankful for the financial support of the National Natural Science Foundation of China (No. 21875075, 61774077, 61274062, 61775061), Guangdong Natural Science Fund (No. 2018A0303130041), Guangzhou Science and Technology Plan Project (No. 201804010295, 2018A0303130211), National Undergraduate Innovative and Entrepreneurial Training Program (No. 202010561010), and the Fundamental Research Funds for the Central Universities for financial support.

Data Availability Statement: Not applicable.

Acknowledgments: We would like to thank Songwei Liu from the Department of Electronic Engineering, Chinese University of Hong Kong, for help in revising this manuscript.

Conflicts of Interest: The authors declare no conflict of interest.

References

De Giorgi, M.; Tarì, D.; Manna, L.; Krahne, R.; Cingolani, R. Optical properties of colloidal nanocrystal spheres and tetrapods. *Microelectron. J.* **2005**, *36*, 552–554. [CrossRef]

Dai, X.; Deng, Y.; Peng, X.; Jin, Y. Quantum-Dot Light-Emitting Diodes for Large-Area Displays: Towards the Dawn of Commercialization. *Adv. Mater.* **2017**, *29*, 1607022. [CrossRef] [PubMed]

Dai, X.; Zhang, Z.; Jin, Y.; Niu, Y.; Cao, H.; Liang, X.; Chen, L.; Wang, J.; Peng, X. Solution-processed, high-performance light-emitting diodes based on quantum dots. *Nature* **2014**, *515*, 96–99. [CrossRef]

4. Pu, C.; Qin, H.; Gao, Y.; Zhou, J.; Wang, P.; Peng, X. Synthetic Control of Exciton Behavior in Colloidal Quantum Dots. *J. Am. Chem. Soc.* **2017**, *139*, 9. [CrossRef]
5. Gur, I.; Fromer, N.A.; Geier, M.L.; Alivisatos, A.P. Air-Stable All-Inorganic Nanocrystal Solar Cells Processed from Solution. *Science* **2005**, *310*, 462–465. [CrossRef] [PubMed]
6. Yin, Y.; Yang, Q.; Liu, G. Ammonium Pyrrolidine Dithiocarbamate-Modified CdTe/CdS Quantum Dots as a Turn-on Fluorescent Sensor for Detection of Trace Cadmium Ions. *Sensors* **2020**, *20*, 312. [CrossRef] [PubMed]
7. Luo, K.; Wu, W.; Xie, S.; Jiang, Y.; Liao, S.; Qin, D. Building Solar Cells from Nanocrystal Inks. *Appl. Sci.* **2019**, *9*, 1885. [CrossRef]
8. Capizzi, G.; Lo Sciuto, G.; Napoli, C.; Shikler, R.; Woźniak, M. Optimizing the organic solar cell manufacturing process by means of AFM measurements and neural networks. *Energies* **2018**, *11*, 1221. [CrossRef]
9. Anderson, I.E.; Breeze, A.J.; Olson, J.D.; Yang, L.; Sahoo, Y.; Carter, S.A. All-inorganic spin-cast nanoparticlesolar cells with nonselective electrodes. *Appl. Phys. Lett.* **2009**, *94*, 063101. [CrossRef]
10. Ju, T.; Yang, L.; Carter, S. Thickness dependence study of inorganic CdTe/CdSe solar cells fabricated fromcolloidal nanoparticle solutions. *J. Appl. Phys.* **2010**, *10*, 104311. [CrossRef]
11. Olson, J.D.; Rodriguez, Y.W.; Yang, L.D.; Alers, G.B.; Carter, S.A. CdTe Schottky diodes from colloidalnanocrystals. *Appl. Phys. Lett.* **2010**, *96*, 242103. [CrossRef]
12. Girard, S.; He, J.; Zhou, X.; Shoemaker, D.; Jaworski, C.; Uher, C.; Heremans, J.; Kanatzidis, M. High performance Na-doped PbTe-PbS thermoelectric materials: Electronic density of states modification and shape-controlled nanostructures. *J. Am. Chem. Soc.* **2011**, *133*, 16588–16597. [CrossRef]
13. Reiss, P.; Protiere, M.; Li, L. Core/Shell semiconductor nanocrystals. *Small* **2009**, *5*, 154–168. [CrossRef]
14. Murray, C.; Norris, D.; Bawendi, M. Synthesis and characterization of nearly monodisperse CdE (E = sulfur, selenium, tellurium) semiconductor nanocrystallites. *J. Am. Chem. Soc.* **1993**, *115*, 8706–8715. [CrossRef]
15. Peng, X.; Wickham, J.; Alivisatos, A. Kinetics of II-VI and III-V colloidal semiconductor nanocrystal growth:"focusing" of size distributions. *J. Am. Chem. Soc.* **1998**, *120*, 5343–5344. [CrossRef]
16. Qu, L.; Peng, X. Control of photoluminescence properties of CdSe nanocrystals in growth. *J. Am. Chem. Soc.* **2002**, *124*, 2049–2055. [CrossRef]
17. Yu, W.W.; Qu, L.; Guo, W.; Peng, X. Experimental determination of the extinction coefficient of CdTe, CdSe, and CdS nanocrystals. *Chem. Mater.* **2003**, *15*, 2854–2860. [CrossRef]
18. Yang, Y.A.; Wu, H.; Williams, K.R.; Cao, Y.C. Synthesis of CdSe and CdTe Nanocrystals without Precursor Injection. *Angew. Chem. Int. Ed.* **2005**, *117*, 6870–6873. [CrossRef]
19. Yao, D.; Xin, W.; Liu, Z.; Wang, Z.; Feng, J.; Dong, C.; Liu, Y.; Yang, B.; Zhang, H. Phosphine-Free Synthesis of Metal Chalcogenide Quantum Dots by Directly Dissolving Chalcogen Dioxides in Alkylthiol as the Precursor. *ACS Appl. Mater. Interfaces* **2017**, *9*, 9840–9848. [CrossRef]
20. Galanakis, I.; Mavropoulos, P. Zinc-blende compounds of transition elements with N, P, As, Sb, S, Se, and Te as half-metallic systems. *Phys. Rev. B* **2003**, *67*, 104417. [CrossRef]
21. Zare, H.; Marandi, M.; Fardindoost, S.; Sharma, V.K.; Yeltik, A.; Akhavan, O.; Demir, H.V.; Nima Taghavinia, N. High efficiency CdTe/CdS core/shell nanocrystals in water enabled by photo-induced colloidal hetero-epitaxy of CdS shelling at room temperature. *Nano Res.* **2015**, *8*, 2317–2328. [CrossRef]
22. Kini, S.; Kulkarni, S.D.; Ganiga, V.; Nagarakshit, T.K.; Chidangil, S. Dual functionalized, stable and water dispersible CdTe Quantum Dots: Facile, one-pot aqueous synthesis, optical tuning and energy transfer applications. *Mater. Res. Bull.* **2018**, *110*, 57–66. [CrossRef]
23. Li, L.; Qian, H.; Fang, N.; Ren, J. Significant enhancement of the quantum yield of CdTe nanocrystals synthesized in aqueous phase by controlling the pH and concentrations of precursor solutions. *J. Lumin.* **2005**, *116*, 59–66. [CrossRef]
24. Zeng, Q.; Chen, Z.; Zhao, Y.; Du, X.; Liu, F.; Jin, G.; Dong, F.; Zhang, H.; Yang, B. Aqueous-Processed Inorganic Thin-Film Solar Cells Based on CdSe$_x$Te$_{1-x}$ Nanocrystals: The Impact of Composition on Photovoltaic Performance. *ACS Appl. Mater. Interfaces* **2015**, *7*, 23223–23230. [CrossRef] [PubMed]
25. Zeng, Q.; Hu, L.; Cui, J.; Feng, T.; Du, X.; Jin, G.; Liu, F.; Ji, T.; Li, F.; Zhang, H.; et al. High-Efficiency Aqueous-Processed Polymer/CdTe Nanocrystals Planar Heterojunction Solar Cells with Optimized Band Alignment and Reduced Interfacial Charge Recombination. *ACS Appl. Mater. Interfaces* **2017**, *9*, 31345–31351. [CrossRef]
26. Wen, S.; Li, M.; Yang, J.; Mei, X.; Wu, B.; Liu, X.; Heng, J.; Qin, D.; Hou, L.; Xu, W.; et al. Rationally Controlled Synthesis of CdSe$_x$Te$_{1-x}$ Alloy Nanocrystals and Their Application in Efficient Graded Bandgap Solar Cells. *Multidiscip. Digit. Publ. Inst.* **2017**, *7*, 380. [CrossRef]
27. Guo, X.; Rong, Z.; Wang, L.; Liu, S.; Liu, Z.; Luo, K.; Chen, B.; Qin, D.; Ma, Y.; Wu, H.; et al. Surface passivation via acid vapor etching enables efficient and stable solution-processed CdTe nanocrystal solar cells. *Sustain. Energy Fuels* **2020**, *4*, 399–406. [CrossRef]
28. Rong, Z.; Guo, X.; Lian, S.; Liu, S.; Qin, D.; Mo, Y.; Xu, W.; Wu, H.; Zhao, H.; Hou, L. Interface Engineering for Both Cathode and Anode Enables Low-Cost Highly Efficient Solution-Processed CdTe Nanocrystal Solar Cells. *Adv. Funct. Mater.* **2019**, *29*, 1904018. [CrossRef]
29. Jasieniak, J.; MacDonald, B.I.; Watkins, S.E.; Mulvaney, P. Solution-processed sintered nanocrystal solar cells via layer-by-layer assembly. *Nano Lett.* **2011**, *11*, 2856–2864. [CrossRef]

20. Guo, X.; Tan, Q.; Liu, S.; Qin, D.; Mo, Y.; Hou, L.; Liu, A.; Wu, H.; Ma, Y. High-efficiency solution-processed CdTe nanocrystal solar cells incorporating a novel crosslinkable conjugated polymer as the hole transport layer. *Nano Energy* **2018**, *46*, 150–157. [CrossRef]
21. Panthani, M.G.; Kurley, J.M.; Crisp, R.W.; Dietz, T.C.; Ezzyat, T.; Luther, J.M.; Talapin, D.V. High efficiency solution processed sintered CdTe nanocrystal solar cells: The role of interfaces. *Nano Lett.* **2014**, *14*, 670–675. [CrossRef]
22. MacDonald, B.I.; Gengenbach, T.R.; Watkins, S.E.; Mulvaney, P.; Jasieniak, J.J. Solution-processing of ultra-thin CdTe/ZnO nanocrystal solar cells. *Thin Solid Film.* **2014**, *558*, 365–373. [CrossRef]
23. MacDonald, B.I.; Martucci, A.; Rubanov, S.; Watkins, S.E.; Mulvaney, P.; Jasieniak, J.J. Layer-by-layer assembly of sintered CDSE x te1−x nanocrystal solar cells. *ACS Nano* **2012**, *6*, 5995–6004. [CrossRef]
24. MacDonald, B.I.; Della Gaspera, E.; Watkins, S.E.; Mulvaney, P.; Jasieniak, J.J. Enhanced photovoltaic performance of nanocrystalline CdTe/ZnO solar cells using sol-gel ZnO and positive bias treatment. *J. Appl. Phys.* **2014**, *115*, 184501. [CrossRef]
25. Webber, D.H.; Brutchey, R.L. Alkahest for V2VI3 chalcogenides: Dissolution of nine bulk semiconductors in a diamine-dithiol solvent mixture. *J. Am. Chem. Soc.* **2013**, *135*, 15722–15725. [CrossRef]
26. Wu, M.; Wang, Y.; Wang, H.; Wang, H.; Sui, Y.; Du, F.; Yang, X.; Zou, B. Phosphine-free engineering toward the synthesis of metal telluride nanocrystals: The role of a Te precursor coordinated at room temperature. *Nanoscale* **2018**, *10*, 21928–21935. [CrossRef] [PubMed]
27. Han, L.; Qin, D.; Jiang, X.; Liu, Y.; Wang, L.; Chen, J.; Cao, Y. Synthesis of high quality zinc-blende CdSe nanocrystals and their application in hybrid solar cells. *Nanotechnology* **2006**, *17*, 4736. [CrossRef]
28. Liu, S.; Liu, W.; Heng, J.; Zhou, W.; Chen, Y.; Wen, S.; Qin, D.; Hou, L.; Wang, D.; Xu, H. Solution-processed efficient nanocrystal solar cells based on CdTe and CdS nanocrystals. *Coatings* **2018**, *8*, 26. [CrossRef]
29. Liu, H.; Tao, H.; Yang, T.; Kong, L.; Qin, D.; Chen, J. A surfactant-free recipe for shape-controlled synthesis of CdSe nanocrystals. *Nanotechnology* **2010**, *22*, 045604. [CrossRef]
30. Liu, H.; Tian, Y.; Zhang, Y.; Gao, K.; Lu, K.; Wu, R.; Qin, D.; Wu, H.; Peng, Z.; Hou, L.; et al. Solution processed CdTe/CdSe nanocrystal solar cells with more than 5.5% efficiency by using an inverted device structure. *J. Mater. Chem. C* **2015**, *3*, 4227–4234. [CrossRef]

Construction of Chemically Bonded Interface of Organic/Inorganic g-C$_3$N$_4$/LDH Heterojunction for Z-Schematic Photocatalytic H$_2$ Generation

Yuzhou Xia [1], Ruowen Liang [1], Min-Quan Yang [2,*], Shuying Zhu [3,*] and Guiyang Yan [1,4,*]

1. Fujian Province University Key Laboratory of Green Energy and Environment Catalysis, Ningde Normal University, Ningde 352100, China; yzxia@ndnu.edu.cn (Y.X.); t1629@ndnu.edu.cn (R.L.)
2. Fujian Key Laboratory of Pollution Control & Resource Reuse, College of Environmental Science and Engineering, Fujian Normal University, Fuzhou 350007, China
3. College of Chemistry, Fuzhou University, Fuzhou 350116, China
4. Provincial Key Laboratory of Featured Materials in Biochemical Industry, Ningde Normal University, Ningde 352100, China
* Correspondence: yangmq@fjnu.edu.cn (M.-Q.Y.); syzhu@fzu.edu.cn (S.Z.); ygyfjnu@163.com (G.Y.)

Abstract: The design and synthesis of a Z-schematic photocatalytic heterostructure with an intimate interface is of great significance for the migration and separation of photogenerated charge carriers, but still remains a challenge. Here, we developed an efficient Z-scheme organic/inorganic g-C$_3$N$_4$/LDH heterojunction by in situ growing of inorganic CoAl-LDH firmly on organic g-C$_3$N$_4$ nanosheet (NS). Benefiting from the two-dimensional (2D) morphology and the surface exposed pyridine-like nitrogen atoms, the g-C$_3$N$_4$ NS offers efficient trap sits to capture transition metal ions. As such, CoAl-LDH NS can be tightly attached onto the g-C$_3$N$_4$ NS, forming a strong interaction between CoAl-LDH and g-C$_3$N$_4$ via nitrogen–metal bonds. Moreover, the 2D/2D interface provides a high-speed channel for the interfacial charge transfer. As a result, the prepared heterojunction composite exhibits a greatly improved photocatalytic H$_2$ evolution activity, as well as considerable stability. Under visible light irradiation of 4 h, the optimal H$_2$ evolution rate reaches 1952.9 µmol g^{-1}, which is 8.4 times of the bare g-C$_3$N$_4$ NS. The in situ construction of organic/inorganic heterojunction with a chemical-bonded interface may provide guidance for the designing of high-performance heterostructure photocatalysts.

Keywords: chemically bonded interface; heterojunction; Z-scheme; g-C$_3$N$_4$/LDH; photocatalytic

1. Introduction

The worsening environmental pollution and energy crisis caused by the large-scale consumption of fossil fuels has posed a great threat to the sustainable development of mankind. Solar-driven photocatalytic H$_2$ evolution is deemed to be a promising approach to meet the challenges, due to the easy accessibility and renewability of solar energy [1–7]. Since the first report of photoelectrochemical water splitting over a TiO$_2$ electrode by Fujishima and Honda, diverse catalysts have been developed for photocatalytic H$_2$ evolution, including TiO$_2$ [8–12], g-C$_3$N$_4$ [13–15], ZnIn$_2$S$_4$ [16,17], COFs [18], MOFs [19–21] and so on. However, despite much progress being made in the area, the overall efficiency is still much less than satisfactory, owing to some stubborn issues, such as insufficient light harvesting and fast recombination of photogenerated electron–hole pairs.

Over the past decades, the construction of heterostructure, especially for Z-scheme heterostructure, is an effective strategy to mediate these problems [22–31]. In such a system, the photogenerated electrons and holes are spatially separated to the component with a more negative conduction band (CB) position and the counterpart with more positive valence band (VB) position, respectively. This not only benefits the light utilization and boosts the separation of photogenerated charge carriers, but also maintains the high redox ability of the charge carriers. In order to construct an efficient Z-scheme photocatalyst, it is necessary to design a high-quality junction with matched band structures and Fermi energy levels (E_f) between two components to drive the migration of electrons–holes, as well as a compactly bonded interface to facilitate the charge transfer across the boundary. Currently, great efforts have been devoted to the design of band-aligned heterojunction systems. Relative mature guiding principles are formed. However, less attention has been paid to the interfacial engineering [32–34]. Most of the reported heterojunctions are mixtures of two components with weak interaction through Van der Waals forces. As such, photogenerated electrons and holes are easily accumulated and recombined at the interface.

Very recently, the construction of strong interacted interfaces connected via chemical bonds have provided an insight for optimizing the charge migration in terms of efficiency and accuracy, thereby improving the photocatalytic H_2 evolution performance. For example, Li et al. reported the synthesis of a Mo-S bonded Z-scheme heterojunction by in situ growth of $MoSe_2$ on $ZnIn_2S_4$ nanosheets with an S defect [35]. The $MoSe_2$–$ZnIn_2S_4$ demonstrates a greatly improved H_2 generation rate than pristine $ZnIn_2S_4$. Nevertheless, most of the developed chemical-bond linked heterostructures are all-inorganic systems, which suffer from limitations such as the difficulty in creating anchoring sites for growing the other semiconductor component, and the instability of the surface coordinative unsaturated atoms [36–38]. In this context, inspired by the efficient electron transfer of bio-enzyme systems in nature, which are composed of an inorganic metal center and organic coenzyme through the coordination of metal and protein, the construction of organic/inorganic hybrids may be a more convenient and universal approach to obtain strong interacted heterojunctions.

g-C_3N_4 is expected to be a promising organic support due to a good coordination ability, as well as appropriate band-edge positions and high visible light absorption. Especially, an ultrathin 2D g-C_3N_4 NS is rich in pyridine-like nitrogen atoms, which can coordinate with metal precursors via strong nitrogen–metal interaction, yielding a chemical-bonded interface [39–41]. As for the inorganic counterpart, we focus on the layered double hydroxides (CoAl-LDHs) due to the large exposure of transition metal atoms. As such, herein we purposely design and synthesize an organic/inorganic g-C_3N_4/CoAl-LDH heterojunction with a chemical-bond-connected interface for efficient photocatalytic H_2 generation. The structure, morphology and photoabsorption properties of the prepared samples are characterized in detail. Photocatalytic activity test results reveal that the heterojunction composites show a much higher H_2 evolution rate than pure g-C_3N_4 NS under visible light irradiation. Collective photoelectrochemical characterizations disclose that the excellent photocatalytic behavior can be attributed to the intimate interface with a sufficient contact area, which endows the heterojunction with more high-speed channels for the migration of charge carriers.

2. Materials and Methods

2.1. Materilas

All reagents were analytical grade and used without further purification. Urea, $Co(NO_3)_2$, $Al(NO_3)_3$, and NH_4F were purchased from Sinopharm Chemical Reagent Co. Ltd. (Beijing, China).

2.2. Preparation of Catalyst

Fabrication of g-C_3N_4 NS: Bulk g-C_3N_4 was firstly synthesized by annealing urea (10 g) at 550 °C for 4 h at the rate of 2.3 °C min^{-1} in air. The obtained sample was ground into power. Then, 1 g bulk g-C_3N_4 power was added into 50 mL H_2O and ethyl alcohol with a volume ratio of 1:1 and sonicated for 2 h. The resultant suspension was centrifuged at 3500 rpm for 10 min to remove the residual layered precursor. Consequently, a g-C_3N_4 NS suspension with a concentration of ~2 mg mL^{-1} was obtained, denoted as CN.

Organic/inorganic g-C_3N_4/CoAl-LDH heterojunction was synthesized by in situ growth of CoAl-LDH onto g-C_3N_4 NS. Typically, 0.25 mmol $Co(NO_3)_2$, 0.125 mmol

Al(NO$_3$)$_3$, 4 mmol urea and 8 mmol NH$_4$F were added into the g-C$_3$N$_4$ NS suspension (50 mL) under vigorous stirring for 30 min. The mixture was heated at 120 °C in a Teflon-lined autoclave for 12 h, then cooled to room temperature naturally. The obtained precipitation was centrifuged and washed with deionized water several times and dried in a vacuum oven overnight at 60 °C, denoted as CN-CoAl$_{0.25}$. A series of g-C$_3$N$_4$/CoAl-LDH heterojunctions were prepared by varying the amounts of CoAl-LDH precursors via the same synthesis method, and were named as CN-CoAl$_x$. For comparison, pure CoAl-LDH was synthesized via the same method without the addition of g-C$_3$N$_4$ NS.

2.3. Characterization

The as-prepared samples were characterized by powder X-ray diffraction (XRD) on a Bruker D8 Advance X-ray diffractometer (Bruker AXS GmbH, Karlsruhe, Germany), operated at 40 kV and 40 mA with Ni-filtered Cu K irradiation (λ = 1.5406 Å). The Fourier-transform infrared (FTIR) spectra were carried out on a Nicolet 670 FTIR spectrometric analyzer (Thermo Electron, Waltham, MA, USA). UV-vis diffuse reflectance spectra (UV-vis DRS) were obtained by using a UV-vis spectrophotometer (Varian Cary 500, Varian, CA, America). The morphologies of the products were observed by scanning electron microscopy (FEI Nova NANO-SEM 230 spectrophotometer, Hillsboro, OR, USA). Transmission electron microscopy (TEM) images were obtained using a JEOL model JEM 2010 EX instrument (JEOL, Tokyo, Japan) at an accelerating voltage of 200 kV. X-ray photoelectron spectroscopy (XPS) measurements were carried out by using a VG Scientific ESCA Lab Mark II spectrometer (VG Scientific Ltd., Manchester, UK), equipped with two ultra-high vacuum 6 (UHV) chambers. The binding energies of all tested samples were calibrated by C 1 s at 284.6 eV. BET surface area tests were performed on an ASAP2020M apparatus (Micromeritics Instrument Corp., Atlanta, GA, USA). Electron paramagnetic resonance (EPR) spectroscopic measurements were tested via Bruker A300 EPR spectrometer (Bruker AXS GmbH, Karlsruhe, Germany). Raman spectra were recorded on a Renishaw Raman spectrometer (Renishaw InVia, Gloucestershire, UK) with a laser beam of λ = 325 nm. PL was measured by a fluorophotometer (Edinburgh FLS1000, Edinburgh Instruments, Livingston, UK) with an excitation wavelength of 375 nm.

2.4. Electrochemistry Measurement

The working electrode was prepared on fluorine-doped tin oxide (FTO) glass, which was cleaned by sonication in acetone and ethanol for 30 min, and 5 mg of the as-prepared samples were dispersed in 0.5 mL N, N-dimethylformamide under sonication for 2 h. Additionally, 10 µL of slurry was dip coated onto the FTO side with exposed areas of 0.25 cm^2. The uncoated parts of the FTO electrodes were sealed with epoxy resin. The electrochemical measurements were performed in a conventional three-electrode cell, using a Pt plate and an Ag/AgCl electrode as a counter electrode and reference electrode, respectively. The working electrode was immersed in a 0.2 M Na$_2$SO$_4$ aqueous solution for 40 s before measurement. The photocurrent measurement was conducted with a CHI650E electrochemical workstation (Chenhua Instruments, Shanghai, China). Electrochemical impedance spectroscopy (EIS) was recorded on a ZENNIUM IM6 electrochemical workstation (Zahner, Germany). A 300 W Xe lamp (PLS-SXE300C, Perfectlight Co., Beijing, China) was used as a light source.

2.5. Evaluation of Photocatalytic Activity

The photocatalytic H$_2$ evolution activity was evaluated in a Pyrex top-irradiation-type reaction vessel connected to a glass-closed gas circulation system. In the typical photocatalytic experiment, 40 mg photocatalyst with 40 µL of 10 mg mL^{-1} H$_2$PtCl$_6$·6H$_2$O were added into 50 mL solution with 10% triethanolamine (TEOA), which acted as sacrificial agent to trap the photogenerated holes. The suspension was vacuum treated for 30 min to eliminate the air. A 300 W Xenon lamp (PLS-SXE300C, Perfectlight Co., Beijing, China) equipped with a 420 nm cut-off filter was used as the light source. H_2 was detected using an online gas chromatograph (Tianmei, TCD, Ar carrier, Shanghai, China).

3. Results and Discussion

Figure 1a schematically illustrates the synthesis of the organic/inorganic g-C_3N_4/CoAl-LDH heterojunction. Bulk g-C_3N_4 was ultrasonically exfoliated into 2D nanosheets, firstly. Then, the nanosheets were employed as an organic support to mix with the LDH precursor for hydrothermal treatment, which enabled the in situ growth of CoAl-LDH onto the CN forming 2D/2D heterojunction. Figure 1b depicts the XRD patterns of the as-prepared samples. Two typical diffraction peaks at 13° and 27° were observed in CN, associated with the trigonal N linkage of tri-s-triazine motifs (100) and periodic stacking of layers for conjugated aromatic systems (001), respectively. For pure CoAl-LDH, all of the peaks can be well indexed to a hexagonal CoAl-LDH phase (JCPDS NO. 51-0045). The diffraction peaks at 11.5°, 23.3°, 34.4°, 38.9° and 46.5° corresponded to the (003), (006), (012), (015) and (018) lattice planes of CoAl-LDH, respectively. In case of CN-CoAl$_x$ samples, both characteristic peaks of CN and CoAl-LDH were observed. The peak intensity of LDH increased gradually with the increment in CoAl-LDH content, demonstrating the successful integration of the two components in the heterojunction.

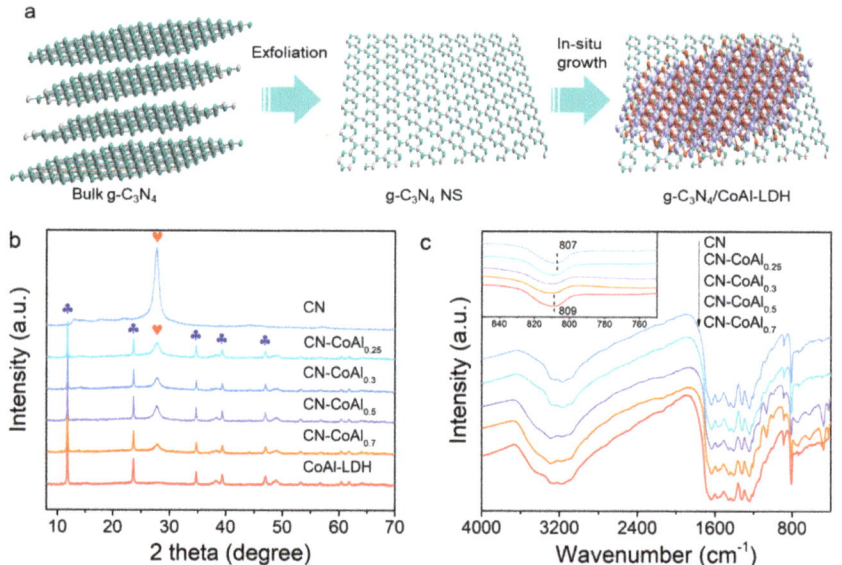

Figure 1. Schematic illustration of preparation of g-C_3N_4/CoAl-LDH heterojunction (**a**), XRD patterns (**b**) and FTIR spectra (**c**) of the as-prepared samples.

The chemical structure of the nanocomposites was investigated by FTIR analysis, as shown in Figure 1c. In comparison with CN, the FTIR spectra of CN-CoAl$_x$ showed similar characteristic peaks in the range of 900–1700 cm^{-1} assigned to the stretching vibrations of tri-s-triazine heterocyclic rings [42], validating the preservation of the major chemical structure of g-C_3N_4 in the heterojunction. As can be seen from the enlarged inset in Figure 1c, the peak at 807 cm^{-1} ascribing to the breathing vibration of tri-s-triazine showed a gradual shift towards a higher wavenumber, with the increasing amount of CoAl-LDH in the composites. The corresponding peak shifted to 809 cm^{-1} for CN-CoAl$_{0.7}$ suggested a strong chemical interaction between CN and CoAl-LDH in the heterojunction.

Figure 2 shows the SEM and TEM images of the CN, CoAl-LDH and CN-CoAl$_{0.5}$, which were employed to analyze the micromorphology and structure information of the samples. As shown in Figure 2a, pure CN exhibited a typical 2D layered structure composed of ultrathin nanosheets. CoAl-LDH (Figure 2b) displayed a nanoflower assembly structure of nanosheets. As for CN-CoAl$_{0.5}$, the characteristic nanosheet structure of CN was observed (Figure 2c). However, no obvious LDH nanoflower could be detected. This might have been caused by the CN nanosheets that assisted the in situ growth of the CoAl-LDH, which promoted the dispersion of the LDH and inhibited its aggregation. When it further increased the loading amount of CoAl-LDH onto g-C$_3$N$_4$, both the characteristic structure of CN and CoAl-LDH were observed (Supplementary Materials, Figure S1). The excessive loading of the CoAl-LDH caused the self-aggregation. Figure 2d–f displays the TEM images of the CN-CoAl$_{0.5}$. It is obvious that the heterojunction displayed a sheet-on-sheet structure with planar interface, which is conductive for the high flow and fast transference of charges, due to the large contact interface and excellent electron mobility [43]. HRTEM analysis revealed obvious lattice fringes of 0.26 nm that are assigned to the (012) facets of CoAl-LDH [44]. Meanwhile, an amorphous CN was observed to attach closely to the CoAl-LDH, suggesting a good interfacial contact between the two components. EDS elemental mappings of CN-CoAl$_{0.5}$ presented the co-existence of C, N, Co, Al and O, further validating the formation of a heterojunction structure.

Figure 2. SEM images of CN (**a**), CoAl-LDH (**b**), CN-CoAl$_{0.5}$ (**c**), TEM of CN-CoAl$_{0.5}$ (**d–f**) and the corresponding EDS element mappings (**g**).

Moreover, the composition and chemical state of the as-prepared samples were measured by XPS. Both elements of CN and CoAl-LDH existed in the CN-CoAl$_{0.5}$ (Figure 3a), verifying the integration of CN with CoAl-LDH. For the C 1 s spectrum of CN (Figure 3b), two peaks located at the binding energies of 284.6 and 288.0 eV were detected, which corresponded to the sp^2 C-C and N-C=N units, respectively. In comparison, the C 1 s of CN-CoAl$_{0.5}$ could be fitted into three peaks at 284.6, 287.9 and 289.5 eV. The new peak at 289.5 eV could be attributed to C=O, which was generated from the hydrolysis of urea in the synthesis of CoAl-LDH [44]. The N 1 s spectrum of CN-CoAl$_{0.5}$ could be fitted into three main peaks at 398.5, 399.9 and 401.0 eV (Figure 3c), which were assigned to sp^2-hybridized nitrogen (C-N=C), tertiary nitrogen N-(C)$_3$ and free amino units (C-N-H), respectively [45]. The N 1s of CN was similar with that of CN-CoAl$_{0.5}$, excepting for a shift of the peak responding to C-N=C at 398.4 eV. In Figure 3d, three pairs of peaks were detected for Co 2p spectra in both CN-CoAl$_{0.5}$ and CoAl-LDH. The main peaks at 781.1

and 783.9 eV of CoAl-LDH were assigned to Co^{3+} and Co^{2+}, respectively [46]. A slight shift towards low binding energy of Co^{2+} (783.8 eV) was observed for CN-CoAl$_{0.5}$. In addition, the Al 2p spectra of both CN-CoAl$_{0.5}$ and CN were located at 74.0 eV (Figure 3e) confirming the Al^{3+} in the samples. The O 1 s spectra of CN-CoAl$_{0.5}$ and CN also showed no difference (Figure 3f). The predominant peak at 531.4 eV was attributed to the lattice oxygen, while the peak at 533.4 eV was assigned to the chemisorbed oxygen [47]. Thus, it is notable that there was an increase in the binding energy of N and decrease in Co^{2+} in the CN-CoAl$_{0.5}$ sample, as compared to that in CN and CoAl-LDH, while other elements showed analogous chemical states. This result suggests that a strong interaction between CoAl-LDH and CN was formed through the coordination of Co with N in the heterojunction.

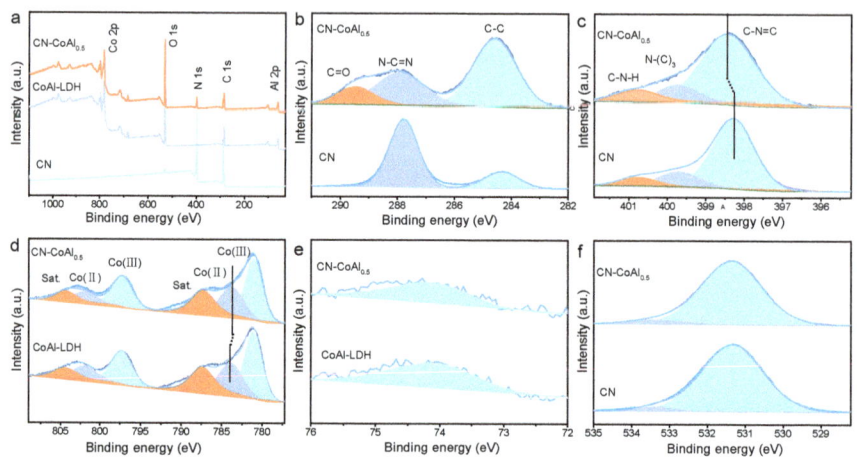

Figure 3. The survey spectra (**a**) and high-resolution XPS spectra of C (**b**), N (**c**), Co (**d**), Al (**e**) O (**f**) of CN-CoAl$_{0.5}$, CN and CoAl-LDH.

The chemical structure of the CN-CoAl heterojunction and the interaction between the CoAl-LDH and CN components was further studied by an EPR and Raman spectra. As shown in Figure S2 (Supplementary Materials), bulk g-C$_3$N$_4$ showed no EPR signal in the g range of 1.92–2.08. When bulk g-C$_3$N$_4$ was exfoliated into 2D nanosheets, an obvious signal at g = 2.003 for CN was observed (Figure 4a), corresponding to the unpaired electrons in π-bonded aromatic rings caused by C defects [48]. Notably, with the integration with CoAl-LDH, the peak was gradually intensified with the increasing amount of CoAl-LDH, which may be ascribed to the redistribution of π-electrons caused by the strong coordination of N with metal Co. The defect could serve as effective electron "traps" to accelerate the separation of photocarriers [49], thus benefiting the photocatalytic performance. Raman spectra of the as-prepared samples were recorded and are displayed in Figure 4b. The characteristic peaks of pure CN at 481, 592, 766, 874, 978 and 1119 cm^{-1} were assigned to the C-N extended network, consistent with those obtained from pristine CN in the literature [50]. The Raman peak at 707 cm^{-1} arose from the breathing mode of the s-triazine ring in g-C$_3$N$_4$, while the peak at 664 cm^{-1} was associated with the heptazine ring structure of CN. The CN-CoAl heterojunctions displayed similar spectra as those of pure CN, suggesting the chemical structure of CN was preserved. However, the peak ascribed to the heptazine ring structure at 664 cm^{-1} of CN-CoAl composites exhibited a slightly negative shift when compared to the spectrum of the bare CN, which may have been caused by the formation of a new chemical bond at the interface between g-C$_3$N$_4$ and CoAl-LDH. The result well matched the FTIR and XPS analyses, consolidating the strong chemical interaction between CN and CoAl-LDH.

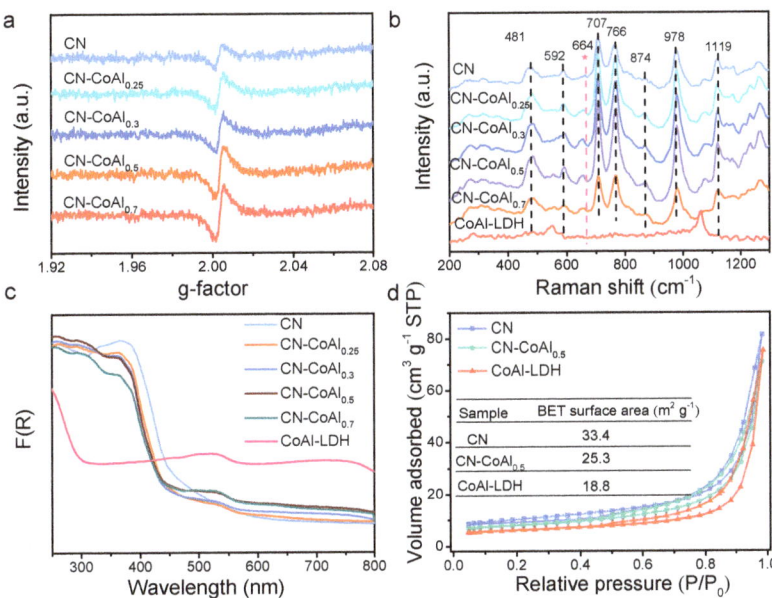

Figure 4. EPR (**a**), Raman (**b**), DRS (**c**) and BET (**d**) analyses of the prepared samples.

Furthermore, the optical absorption properties of the bare CN, CoAl-LDH and CN-CoAl$_x$ composites were measured by DRS. As shown in Figure 4c, the absorption edge of CN was around 480 nm, revealing its visible-light response characteristic. CoAl-LDH showed two distinct absorption peaks at 280 nm and in the range of 450–550 nm. The absorption edge at 280 nm was assigned to the ligand-to-metal charge transfer of CoAl-LDH, while the absorption at 530 nm was generated from d-d transitions of Co^{2+} in an octahedral geometry [44,51]. As for the heterojunction composites, both characteristic peaks of the CN and CoAl-LDH were observed, suggesting the good integration of the two components in CN-CoAl$_x$. Based on the transformed Kubelka–Munk function plots (Supplementary Materials, Figure S3), the band gaps (E_g) of the CN and CoAl-LDH were measured to be 2.6 eV and 2.1 eV, respectively. Nitrogen (N$_2$) adsorption–desorption measurements were measured to investigate the surface properties of the obtained catalysts. As presented in Figure 4d, all of the measured samples showed a type IV adsorption isotherm, revealing their mesoporous structure. The BET surface area of the CN, CN-CoAl$_{0.5}$ and CoAl-LDH were determined to be 33.4, 25.3 and 18.8 m^2 g^{-1}, respectively. The CN-CoAl$_{0.5}$ showed a moderate surface area, which likely resulted from the hybridization of CN with CoAl-LDH.

The photocatalytic H$_2$ evolution activities of the as-prepared samples are presented in Figure 5a. No H$_2$ was detected after 4 h irradiation for pristine CoAl-LDH. In the case of bare CN, a relatively low H$_2$ production with the value of 233.2 μmol g^{-1} was detected under visible light irradiation of 4 h, which should have been restricted by the fast recombination of the photogenerated electrons and holes. After coupling with CoAl-LDH, the photocatalytic H$_2$ production activities of the composites significantly increased. The optimal CN-CoAl$_{0.5}$ showed the highest H$_2$ evolution amount of 1952.9 μmol g^{-1}, which is 8.4 times of CN. Moreover, the loading amount of the CoAl-LDH on g-C$_3$N$_4$ played a significant role in the photocatalytic H$_2$ generation performance. With the increase in the CoAl-LDH content from 0.25 mmol to 0.5 mmol in the hybrids, the H$_2$ evolution rate enhanced gradually. The boosted photocatalytic activity may be due to the increased interacted interface and surface active sites in the heterojunction, which facilitated the separation of photogenerated carriers and promoted the surface reaction. However, a further increase in CoAl-LDH to 0.7 mmol depressed the H$_2$ evolution activity. This may have been caused by the large amount of CoAl-LDH that shielded the light absorption of CN. The excessive loading of CoAl-LDH caused self-aggregation, which decreased the exposure of surface active sites, thus leading to a decline in the photocatalytic performance. XPS measurement of the as-prepared samples was performed to investigate the effect of the Co^{2+}/Co^{3+} ratio on photocatalytic activity. As shown in Figure S4 (Supplementary Materials), the Co species were in the form of a mixed state of +2 and +3 for all of the hybrid samples. By normalizing the peak areas, the atom ratios of Co^{2+}/Co^{3+} were calculated to be 2.11, 2.14, 2.07 and 2.01 for the CN-CoAl$_{0.25}$, CN-CoAl$_{0.3}$, CN-CoAl$_{0.5}$ and CN-CoAl$_{0.7}$, respectively. The Co^{2+}/Co^{3+} ratios were almost the same for the different CN-CoAl samples, while their photocatalytic activities varied a lot. The result suggests that the Co^{2+}/Co^{3+} ratio was not a main factor in affecting the H_2-generation activity of the CN-CoAl composites.

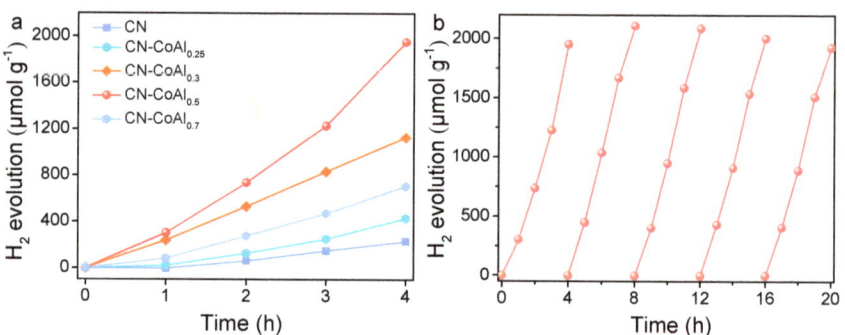

Figure 5. Photocatalytic hydrogen evolution over CN and CN-CoAl$_x$ (**a**), stability test of hydrogen evolution over CN-CoAl$_{0.5}$ (**b**).

For comparison, a reference catalyst of a CN/CoAl-Mix was also prepared by physical mixing of CN with CoAl-LDH. As shown in Figure S5 (Supplementary Materials), the H_2 evolution rate of the CN/CoAl-Mix was slightly higher than the pure CN, while much lower than that of CN-CoAl$_{0.5}$. The result directly proves that the construction of strong chemical bond interacted 2D/2D heterojunction is more effective for boosting the photocatalytic H_2 production activity. The stability of photocatalytic H_2 production over the optimized CN-CoAl$_{0.5}$ was also evaluated. As presented in Figure 5b, no obvious decrease in catalytic activity was observed during the five recycle tests. Moreover, the surface area and morphology of the CN-CoAl$_{0.5}$ after five cycles of stability text were investigated and are displayed in Figure S6 (Supplementary Materials). The BET surface area was measured to be 28.1 $m^2\ g^{-1}$, which was similar to the value before the reaction. The SEM and TEM images reveal that the 2D/2D sheet-to-sheet structure barely changed. These results verify that the 2D/2D CN-CoAl heterojunction had a satisfactory stability and reusability for photocatalytic H_2 generation.

To gain mechanistic insight into the enhanced photoactivity of the CN-CoAl$_x$, photo electric responses of the heterojunction composites were carried out. Figure 6a shows the transient photocurrent test of the samples. The CoAl-LDH-modified CN samples exhibited much higher photocurrent strength compared to the CN, suggesting the constructed 2D/2D heterojunction with a large contact area could efficiently reduce the recombination rate of photogenerated carriers. The electrochemical impedance spectrum (EIS) results reveal that CN-CoAl0.5 exhibited a decreased semicircle compared to the bare CN and CoAl-LDH (Figure 6b), illustrating that the 2D/2D interface strongly favored the migration of photogenerated charge carriers, and thus enhancing the photocatalytic H_2 evolution activity. The behavior of photogenerated carriers was also monitored via PL spectra. As shown in Figure 6c, pure CN displayed an emission peak at around 470 nm. When the CoAl-LDH was introduced, the heterojunction showed a dramatically depressed PL inten sity, suggesting the inhibited recombination of photogenerated electron–hole pairs over the strong interacted heterojunction structure.

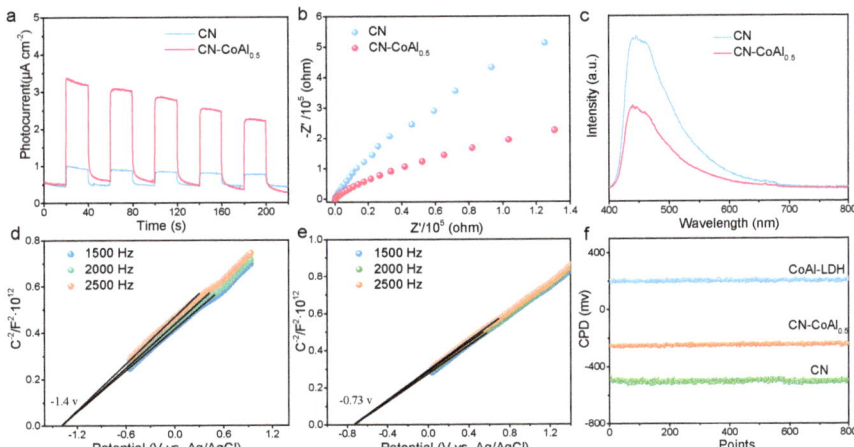

Figure 6. Transient photocurrent response (**a**), EIS (**b**), PL (**c**), Mott–Schottky plots of CN (**d**) and CoAl-LDH (**e**), CPDs of the as-prepared samples surface related to Au reference (**f**).

Moreover, the specific band structures of the CN and CoAl-LDH were determined by the Mott–Schottky test. As presented in Figure 6d,e, the spectra of both samples under three different constant frequencies showed a positive slope of the line segment, suggesting that the prepared CN and CoAl-LDH were typical n-type semiconductors. The derived values of the flat-band potentials (E_f) of CN and CoAl-LDH were -1.4 V and -0.73 V (vs Ag/AgCl) at pH 7, respectively. According to the conversion equation of $E_{NHE} = E_{Ag/AgCl} + E^0_{Ag/AgCl}$, ($E^0_{Ag/AgCl}$ is about 0.197 V at 25 °C, pH = 7), E_f of CN and CoAl-LDH were determined to be -1.20 V and -0.53 V versus the normal hydrogen electrode (NHE, pH = 7). Thereby, the valence band (VB) of CN and CoAl-LDH could be calculated to be 1.40 V and 1.57 V. Furthermore, the Kelvin probe test was measured to study the interfacial electronic structure. Figure 6f depicts the measured contact potential difference (CPDs) of the as-prepared samples related to the Au (5.1 eV) reference. The work function of the CN CoAl-LDH and CN-CoAl$_{0.5}$ were calculated to be 4.62 eV, 5.4 eV and 4.85 eV, respectively.

Based on the above analysis, a plausible photogenerated charge carrier transfer mechanism over the CN-CoAl heterojunction could be proposed. As shown in Figure 7a, due to the lower Femi level of CoAl-LDH than that of CN, free electrons transferred from CN to CoAl-LDH through the N–metal bond until an equilibrium state formed (Figure 7b). As such, an interfacial electric field oriented in the direction from the CN to CoAl-LDH emerged. Under light irradiation, both CN and CoAl-LDH could be photoexcited to generate electron–hole pairs. With the driving force of the built-in electric field, photogenerated electrons from the CB of CoAl-LDH transferred to the VB of CN and recombined with the holes (Figure 7c), thus realizing the Z-scheme charge transfer. Concurrently, the accumulated electrons on the CB of CN reacted with H_2O for H_2 production, while the holes of CoAl-LDH were trapped by TEOA.

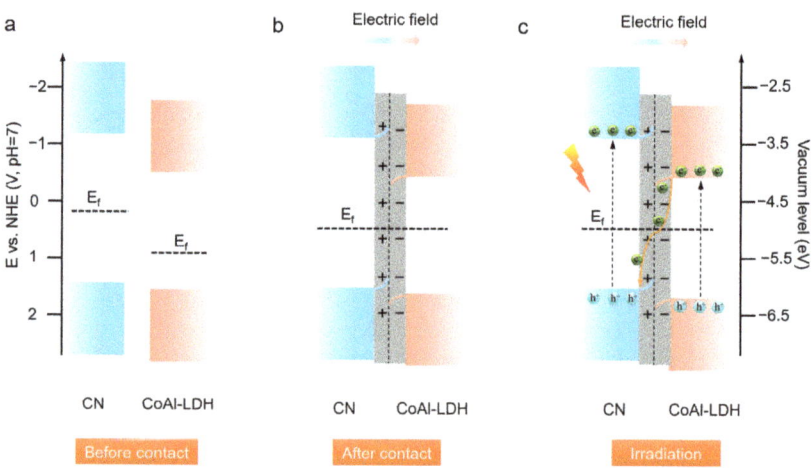

Figure 7. Schematic diagram for band structure of CN and CoAl-LDH before contact (**a**), after contact (**b**) and the light-induced charge transfer from CN to CoAl-LDH toward H_2 evolution (**c**).

4. Conclusions

In summary, a 2D/2D organic/inorganic g-C_3N_4/CoAl-LDH heterojunction with a strong interacted interface was synthesized via an in situ hydrothermal method. The resulting composites showed a markedly improved photocatalytic H_2 production activity. The optimal CN-CoAl composite displayed a H_2 evolution of 1952.9 μmol g^{-1} for 4 h visible light irradiation, which is 8.4 times as that of pure CN. The enhanced photocatalytic activity crucially relied on the well-matched band positions of g-C_3N_4 and CoAl-LDH components, as well as the sufficient interfacial contact between them, which greatly benefited the migration and separation of the photogenerated charge carriers.

Supplementary Materials: The following are available online at https://www.mdpi.com/article/10.3390/nano11102762/s1. Figure S1: SEM image of the prepared CN-CoAl$_{0.7}$. Figure S2: EPR spectra of bulk g-C_3N_4, CN and CoAl-LDH. Figure S3: Tauc plots of CN and CoAl-LDH. Figure S4: XPS spectra of Co in CN-CoAlx samples. Figure S5: Photocatalytic H_2 evolution rates of CoAl-LDH, CN, CN/CoAl-Mix and CN-CoAl$_{0.5}$. Figure S6: BET (a), SEM (b) and TEM (c) analyses of CN-CoAl$_{0.5}$ after 5 cycles of H_2-generation stability text.

Author Contributions: Y.X.: Conceptualization, Methodology, Data Curation, Writing—Reviewing R.L.: Investigation, Data curation. M.-Q.Y.: Resources, Writing—Reviewing and Editing. S.Z.: Conceptualization, Writing—Reviewing and Editing. G.Y.: Resources, Supervision. All authors have read and agreed to the published version of the manuscript.

Funding: This work was financially supported by the National Natural Science Foundation of China (22108129, 21806085, 21902030), Natural Science Foundation of Fujian Province (2020J05223, 2019J01016), Natural Science Foundation of Ningde Normal University (2019Y07, 2019Q101, 2020T01), and Middle-Aged Teacher Education Project of Fujian Provincial Department of Education (JAT190808).

Conflicts of Interest: The authors declare no conflict of interest.

References

1. Yang, M.; Shen, L.; Lu, Y.; Chee, S.W.; Lu, X.; Chi, X.; Chen, Z.; Xu, Q.; Mirsaidov, U.; Ho, G.W. Disorder engineering in monolayer nanosheets enabling photothermic catalysis for full solar spectrum (250–2500 nm) harvesting. *Angew. Chem. Int. Ed.* **2019**, *58*, 3077–3081. [CrossRef]
2. Zhang, Y.C.; Afzal, N.; Pan, L.; Zhang, X.; Zou, J.J. Structure-activity relationship of defective metal-based photocatalysts for water splitting: Experimental and theoretical perspectives. *Adv. Sci.* **2019**, *6*, 1900053. [CrossRef] [PubMed]
3. Li, C.; Xu, Y.; Tu, W.; Chen, G.; Xu, R. Metal-free photocatalysts for various applications in energy conversion and environmental purification. *Green Chem.* **2017**, *19*, 882–899. [CrossRef]

8. Wang, Z.; Li, C.; Domen, K. Recent developments in heterogeneous photocatalysts for solar-driven overall water splitting. *Chem. Soc. Rev.* **2019**, *48*, 2109–2125. [CrossRef]
9. Lu, Q.; Yu, Y.; Ma, Q.; Chen, B.; Zhang, H. 2D transition-metal-dichalcogenide-nanosheet-based composites for photocatalytic and electrocatalytic hydrogen evolution reactions. *Adv. Mater.* **2016**, *28*, 1917–1933. [CrossRef] [PubMed]
10. Yin, S.; Liu, S.; Yuan, Y.; Guo, S.; Ren, Z. Octahedral shaped $PbTiO_3$-TiO_2 nanocomposites for high-efficiency photocatalytic hydrogen production. *Nanomaterials* **2021**, *11*, 2295. [CrossRef]
11. Martino, M.; Ruocco, C.; Meloni, E.; Pullumbi, P.; Palma, V. Main hydrogen production processes: An overview. *Catalyst* **2021**, *11*, 547. [CrossRef]
12. Zhou, W.; Li, W.; Wang, J.Q.; Qu, Y.; Yang, Y.; Xie, Y.; Zhang, K.; Wang, L.; Fu, H.; Zhao, D. Ordered mesoporous black TiO_2 as highly efficient hydrogen evolution photocatalyst. *J. Am. Chem. Soc.* **2014**, *136*, 9280–9283. [CrossRef]
13. Geng, R.; Yin, J.; Zhou, J.; Jiao, T.; Feng, Y.; Zhang, L.; Chen, Y.; Bai, Z.; Peng, Q. In situ construction of $Ag/TiO_2/g$-C_3N_4 heterojunction nanocomposite based on hierarchical co-assembly with sustainable hydrogen Evolution. *Nanomaterials* **2020**, *10*, 1. [CrossRef]
14. Sajan, C.P.; Wageh, S.; Al-Ghamdi, A.A.; Yu, J.; Cao, S. TiO_2 nanosheets with exposed {001} facets for photocatalytic applications. *Nano Res.* **2015**, *9*, 3–27. [CrossRef]
15. Zhang, P.; Wang, T.; Gong, J. Mechanistic understanding of the plasmonic enhancement for solar water splitting. *Adv. Mater.* **2015**, *27*, 5328–5342. [CrossRef]
16. Meng, A.; Wu, S.; Cheng, B.; Yu, J.; Xu, J. Hierarchical $TiO_2/Ni(OH)_2$ composite fibers with enhanced photocatalytic CO_2 reduction performance. *J. Mater. Chem. A* **2018**, *6*, 4729–4736. [CrossRef]
17. Jia, J.; Sun, W.; Zhang, Q.; Zhang, X.; Hu, X.; Liu, E.; Fan, J. Inter-plane heterojunctions within 2D/2D $FeSe_2/g$-C_3N_4 nanosheet semiconductors for photocatalytic hydrogen generation. *Appl. Catal. B* **2020**, *261*, 118249. [CrossRef]
18. Che, W.; Cheng, W.; Yao, T.; Tang, F.; Liu, W.; Su, H.; Huang, Y.; Liu, Q.; Liu, J.; Hu, F.; et al. Fast photoelectron transfer in (cring)-C_3N_4 plane heterostructural nanosheets for overall water splitting. *J. Am. Chem. Soc.* **2017**, *139*, 3021–3026. [CrossRef]
19. Ran, J.; Guo, W.; Wang, H.; Zhu, B.; Yu, J.; Qiao, S.Z. Metal-free 2D/2D Phosphorene/g-C_3N_4 Van der Waals heterojunction for highly enhanced visible-light photocatalytic H_2 production. *Adv. Mater.* **2018**, *30*, e1800128. [CrossRef] [PubMed]
20. Wang, S.; Wang, Y.; Zhang, S.L.; Zang, S.Q.; Lou, X.W.D. Supporting ultrathin $ZnIn_2S_4$ nanosheets on Co/N-doped graphitic carbon nanocages for efficient photocatalytic H_2 generation. *Adv. Mater.* **2019**, *31*, 1903404. [CrossRef]
21. Yang, M.; Xu, Y.; Lu, W.; Zeng, K.; Zhu, H.; Xu, Q.; Ho, W. Self-surface charge exfoliation and electrostatically coordinated 2D hetero-layered hybrids. *Nat. Commun.* **2017**, *8*, 1–9. [CrossRef] [PubMed]
22. Lian, J.; Li, D.; Qi, Y.; Yang, N.; Zhang, R.; Xie, T.; Guan, N.; Li, L.; Zhang, F. Metal-seed assistant photodeposition of platinum over Ta_3N_5 photocatalyst for promoted solar hydrogen production under visible light. *J. Energy Chem.* **2021**, *55*, 444–448. [CrossRef]
23. Zuo, Q.; Liu, T.; Chen, C.; Ji, Y.; Gong, X.; Mai, Y.; Zhou, Y. Ultrathin metal-organic framework nanosheets with ultrahigh loading of single Pt atoms for efficient visible-light-driven photocatalytic H_2 evolution. *Angew. Chem. Int. Ed.* **2019**, *58*, 10198–10203. [CrossRef]
24. Liang, R.; Huang, R.; Ying, S.; Wang, X.; Yan, G.; Wu, L. Facile in situ growth of highly dispersed palladium on phosphotungstic-acid-encapsulated MIL-100(Fe) for the degradation of pharmaceuticals and personal care products under visible light. *Nano Res.* **2017**, *11*, 1109–1123. [CrossRef]
25. Zhang, F.M.; Sheng, J.L.; Yang, Z.D.; Sun, X.J.; Tang, H.L.; Lu, M.; Dong, H.; Shen, F.C.; Liu, J.; Lan, Y.Q. Rational design of MOF/COF hybrid materials for photocatalytic H_2 evolution in the Presence of Sacrificial Electron Donors. *Angew. Chem. Int. Ed.* **2018**, *57*, 12106–12110. [CrossRef]
26. Cheng, C.; He, B.; Fan, J.; Cheng, B.; Cao, S.; Yu, J. An inorganic/organic s-scheme heterojunction H_2-production photocatalyst and its charge transfer mechanism. *Adv. Mater.* **2021**, *33*, 2100317. [CrossRef] [PubMed]
27. Wang, Z.; Lin, Z.; Shen, S.; Zhong, W.; Cao, S. Advances in designing heterojunction photocatalytic materials. *Chin. J. Catal.* **2021**, *42*, 710–730. [CrossRef]
28. Xue, C.; An, H.; Shao, G.; Yang, G. Accelerating directional charge separation via built-in interfacial electric fields originating from work-function differences. *Chin. J. Catal.* **2021**, *42*, 583–594. [CrossRef]
29. Xia, Y.; Chen, W.; Liang, S.; Bi, J.; Wu, L.; Wang, X. Engineering a highly dispersed co-catalyst on a few-layered catalyst for efficient photocatalytic H_2 evolution: A case study of $Ni(OH)_2/HNb_3O_8$ nanocomposites. *Catal. Sci. Technol.* **2017**, *7*, 5662–5669. [CrossRef]
30. Chen, J.; Zhan, J.; Zhang, Y.; Tang, Y. Construction of a novel $ZnCo_2O_4/Bi_2O_3$ heterojunction photocatalyst with enhanced visible light photocatalytic activity. *Chin. Chem. Lett.* **2019**, *30*, 735–738. [CrossRef]
31. Zheng, Y.; Fan, M.; Li, K.; Zhang, R.; Li, X.; Zhang, L.; Qiao, Z.-A. Ultraviolet-induced Ostwald ripening strategy towards a mesoporous Ga_2O_3/GaOOH heterojunction composite with a controllable structure for enhanced photocatalytic hydrogen evolution. *Catal. Sci. Technol.* **2020**, *10*, 2882–2892. [CrossRef]
32. Wu, S.; Yu, X.; Zhang, J.; Zhang, Y.; Zhu, Y.; Zhu, M. Construction of $BiOCl/CuBi_2O_4$ S-scheme heterojunction with oxygen vacancy for enhanced photocatalytic diclofenac degradation and nitric oxide removal. *Chem. Eng. J.* **2021**, *411*, 128555. [CrossRef]
33. Geng, Y.; Chen, D.; Li, N.; Xu, Q.; Li, H.; He, J.; Lu, J. Z-Scheme 2D/2D α-Fe_2O_3/g-C_3N_4 heterojunction for photocatalytic oxidation of nitric oxide. *Appl. Catal. B* **2021**, *280*, 119409. [CrossRef]
34. He, F.; Meng, A.; Ho, W.; Yu, J. Enhanced photocatalytic H_2-production activity of WO_3/TiO_2 step-scheme heterojunction by graphene modification. *Chin. J. Catal.* **2020**, *41*, 9–20. [CrossRef]

31. Xia, Y.; Liang, S.; Wu, L.; Wang, X. Ultrasmall NiS decorated HNb_3O_8 nanosheeets as highly efficient photocatalyst for H evolution reaction. *Catal. Today* **2019**, *330*, 195–202. [CrossRef]
32. Feng, N.; Lin, H.; Deng, F.; Ye, J. Interfacial-bonding Ti-N-C boosts efficient photocatalytic H_2 evolution in close coupling g-C_3N_4/TiO_2. *J. Phys. Chem. C* **2021**, *125*, 12012–12018. [CrossRef]
33. Xia, M.; Yan, X.; Li, H.; Wells, N.; Yang, G. Well-designed efficient charge separation in 2D/2D N doped $La_2Ti_2O_7$/$ZnIn_2S$ heterojunction through band structure/morphology regulation synergistic effect. *Nano Energy* **2020**, *78*, 105401. [CrossRef]
34. Wang, X.; Wang, X.; Huang, J.; Li, S.; Meng, A.; Li, Z. Interfacial chemical bond and internal electric field modulated Z-scheme Sv-$ZnIn_2S_4$/$MoSe_2$ photocatalyst for efficient hydrogen evolution. *Nat. Commun.* **2021**, *12*, 4112. [CrossRef] [PubMed]
35. Xia, Y.; Zhu, S.; Liang, R.; Huang, R.; Yan, G.; Liang, S. Interfacial reconstruction of 2D/2D $ZnIn_2S_4$/HNb_3O_8 through Nb-S bonds for efficient photocatalytic H_2 evolution performance. *Mater. Des.* **2021**, *209*, 110007. [CrossRef]
36. Qi, Y.; Jiang, J.; Liang, X.; Ouyang, S.; Mi, W.; Ning, S.; Zhao, L.; Ye, J. Fabrication of black In_2O_3 with dense oxygen vacancy through dual functional carbon doping for enhancing photothermal CO_2 hydrogenation. *Adv. Funct. Mater.* **2021**, *31*, 2100908. [CrossRef]
37. Liang, X.; Wang, P.; Gao, Y.; Huang, H.; Tong, F.; Zhang, Q.; Wang, Z.; Liu, Y.; Zheng, Z.; Dai, Y.; et al. Design and synthesis of porous M-ZnO/CeO_2 microspheres as efficient plasmonic photocatalysts for nonpolar gaseous molecules oxidation: Insight into the role of oxygen vacancy defects and M=Ag, Au nanoparticles. *Appl. Catal. B* **2020**, *260*, 118151. [CrossRef]
38. Chen, X.; Li, J.-Y.; Tang, Z.-R.; Xu, Y.-J. Surface-defect-engineered photocatalyst for nitrogen fixation into value-added chemical feedstocks. *Catal. Sci. Technol.* **2020**, *10*, 6098–6110. [CrossRef]
39. Chen, L.; Zhu, D.; Li, J.; Wang, X.; Zhu, J.; Francis, P.S.; Zheng, Y. Sulfur and potassium co-doped graphitic carbon nitride for highly enhanced photocatalytic hydrogen evolution. *Appl. Catal. B* **2020**, *273*, 119050. [CrossRef]
40. Wang, D.; Zeng, H.; Xiong, X.; Wu, M.-F.; Xia, M.; Xie, M.; Zou, J.-P.; Luo, S.-L. Highly efficient charge transfer in CdS-covalent organic framework nanocomposites for stable photocatalytic hydrogen evolution under visible light. *Sci. Bull.* **2020**, *65*, 113–122. [CrossRef]
41. Wang, X.; Maeda, K.; Thomas, A.; Takanabe, K.; Xin, G.; Carlsson, J.M.; Domen, K.; Antonietti, M. A metal-free polymeric photocatalyst for hydrogen production from water under visible light. *Nat. Mater.* **2009**, *8*, 76–80. [CrossRef] [PubMed]
42. Lin, Q.; Li, L.; Liang, S.; Liu, M.; Bi, J.; Wu, L. Efficient synthesis of monolayer carbon nitride 2D nanosheet with tunable concentration and enhanced visible-light photocatalytic activities. *Appl. Catal. B* **2015**, *163*, 135–142. [CrossRef]
43. Zhang, S.; Si, Y.; Li, B.; Yang, L.; Dai, W.; Luo, S. Atomic-level and modulated interfaces of photocatalyst heterostructure constructed by external defect-induced strategy: A critical review. *Small* **2021**, *17*, e2004980. [CrossRef]
44. Jo, W.K.; Tonda, S. Novel CoAl-LDH/g-C_3N_4/RGO ternary heterojunction with notable 2D/2D/2D configuration for highly efficient visible-light-induced photocatalytic elimination of dye and antibiotic pollutants. *J. Hazard. Mater.* **2019**, *368*, 778–787. [CrossRef] [PubMed]
45. Li, C.; Du, Y.; Wang, D.; Yin, S.; Tu, W.; Chen, Z.; Kraft, M.; Chen, G.; Xu, R. Unique P-Co-N surface bonding states constructed on g-C_3N_4 nanosheets for drastically enhanced photocatalytic activity of H_2 evolution. *Adv. Funct. Mater.* **2017**, *27*, 1604328. [CrossRef]
46. Ping, J.; Wang, Y.; Lu, Q.; Chen, B.; Chen, J.; Huang, Y.; Ma, Q.; Tan, C.; Yang, J.; Cao, X.; et al. Self-assembly of single-layer CoAl-layered double hydroxide nanosheets on 3D graphene network used as highly efficient electrocatalyst for oxygen evolution reaction. *Adv. Mater.* **2016**, *28*, 7640–7645. [CrossRef] [PubMed]
47. Wu, Y.; Wang, H.; Sun, Y.; Xiao, T.; Tu, W.; Yuan, X.; Zeng, G.; Li, S.; Chew, J.W. Photogenerated charge transfer via interfacial internal electric field for significantly improved photocatalysis in direct Z-scheme oxygen-doped carbon nitrogen/CoAl-layered double hydroxide heterojunction. *Appl. Catal. B* **2018**, *227*, 530–540. [CrossRef]
48. Cao, S.; Fan, B.; Feng, Y.; Chen, H.; Jiang, F.; Wang, X. Sulfur-doped g-C_3N_4 nanosheets with carbon vacancies: General synthesis and improved activity for simulated solar-light photocatalytic nitrogen fixation. *Chem. Eng. J.* **2018**, *353*, 147–156. [CrossRef]
49. Duan, L.; Li, G.; Zhang, S.; Wang, H.; Zhao, Y.; Zhang, Y. Preparation of S-doped g-C_3N_4 with C vacancies using the desulfurized waste liquid extracting salt and its application for NO_x removal. *Chem. Eng. J.* **2021**, *411*, 128551. [CrossRef]
50. Hu, S.; Yang, L.; Tian, Y.; Wei, X.; Ding, J.; Zhong, J.; Chu, P. Simultaneous nanostructure and heterojunction engineering of graphitic carbon nitride via in situ Ag doping for enhanced photoelectrochemical activity. *Appl. Catal. B* **2015**, *163*, 611–622. [CrossRef]
51. Dou, Y.; Zhang, S.; Pan, T.; Xu, S.; Zhou, A.; Pu, M.; Yan, H.; Han, J.; Wei, M.; Evans, D.G.; et al. TiO_2@layered double hydroxide core-shell nanospheres with largely enhanced photocatalytic activity toward O_2 generation. *Adv. Funct. Mater.* **2015**, *25*, 2243–2249. [CrossRef]

Review

Surveying the Synthesis, Optical Properties and Photocatalytic Activity of Cu₃N Nanomaterials

Patricio Paredes, Erwan Rauwel and Protima Rauwel *

Institute of Forestry and Engineering Sciences, Estonian University of Life Sciences, Kreutzwaldi 56/1, 51014 Tartu, Estonia; patricio.paredes@emu.ee (P.P.); erwan.rauwel@emu.ee (E.R.)
* Correspondence: protima.rauwel@emu.ee

Abstract: This review addresses the most recent advances in the synthesis approaches, fundamental properties and photocatalytic activity of Cu₃N nanostructures. Herein, the effect of synthesis conditions, such as solvent, temperature, time and precursor on the precipitation of Cu₃N and the formation of secondary phases of Cu and Cu₂O are surveyed, with emphasis on shape and size control. Furthermore, Cu₃N nanostructures possess excellent optical properties, including a narrow bandgap in the range of 0.2 eV–2 eV for visible light absorption. In that regard, understanding the effect of the electronic structure on the bandgap and on the optical properties of Cu₃N is therefore of interest. In fact, the density of states in the d-band of Cu has an influence on the band gap of Cu₃N. Moreover, the potential of Cu₃N nanomaterials for photocatalytic dye-degradation originates from the presence of active sites, i.e., Cu and N vacancies on the surface of the nanoparticles. Plasmonic nanoparticles tend to enhance the efficiency of photocatalytic dye degradation of Cu₃N. Nevertheless, combining them with other potent photocatalysts, such as TiO₂ and MoS₂, augments the efficiency to 99%. Finally, the review concludes with perspectives and future research opportunities for Cu₃N-based nanostructures.

Keywords: Cu₃N; nanostructures; synthesis; optical properties; photocatalysis

1. Introduction

The production of nanoscale structures has attracted interest as they exhibit size and shape-dependent physical, electrical and chemical properties that are absent in their bulk counterparts [1–3]. Transition metal nitride nanostructures such as Co₂N [4], TiN [5], OsN₂ [6], Zn₃N₂ [7] and IrN₂ [8] are rising as strong candidates for a variety of industrial applications, as they display excellent catalytic, electrochemical and optoelectronic properties. Transition metal nitrides simultaneously behave as ionic crystals, transition metals and covalent compounds [9]. They are therefore being investigated as potential contenders for electrochemical energy conversion, heterogeneous catalysis, fuel cells, batteries and supercapacitors [10–14]. Additionally, their optical absorption, electrical properties and tolerance to structural defects make them a new class of semiconductors of interest [15]. The most popular metal nitrides in the semiconductor family are GaN, AlGaN and InGaN, which produce blue light-emitting diodes (LEDs) and lasers [16–18]. Furthermore, semiconductors, such as Ta₃N₅ [19], InN [20,21] and InGaN [22,23], are also potential candidates for photocatalysis due to their simple chemical composition, tunable narrow band gap and high stability. However, the metals, i.e., In, Ta and Ga, are relatively rare in the Earth's crust, and the price of indium has shot up by 900% in the last 10 years. Therefore, more sustainable metallic raw materials are required for industrial applications.

In that regard, Cu₃N, composed of earth-abundant elements, is a potential candidate for industrial applications. Moreover, owing to its cost-effectiveness of production and environmental friendliness, it has gained importance. In addition, due to its physicochemical characteristics and tunable optical and electrical properties, Cu₃N is the focus of several research investigations [24,25]. The shape and size-controlled synthesis of Cu_3N nanoparticles is one of the main challenges today. The current synthesis methods usually involve the use of reactive nitrogen precursors, high pressure and temperature [26] Synthesis techniques, such as thermal decomposition [27], electroplating [28], solvothermal [26], chemical vapor deposition (CVD) [29], radio-frequency (RF) and direct current (DC) magnetron reactive sputtering [30–32] have been employed for the synthesis of Cu_3N thin films. For nanoparticle synthesis, several reports claim that solution-based synthesis approaches tend to offer a better way of controlling the morphology and properties of Cu_3N nanoparticles [24,33] by carefully varying the synthesis conditions, i.e., time, temperature and precursors. Although Cu_3N itself has been extensively studied during the last decade, there is still a significant disparity between the theoretical and experimental optical band gaps. Nevertheless, all these studies concede that Cu_3N has a narrow bandgap with values ranging from 0.2 to 2.0 eV, exhibiting either metallic or semiconductor behavior [30,34,35]. In fact, the physical and chemical properties of the material not only depend on the synthesis conditions but also on the presence of dopants in the Cu_3N lattice [36,37].

Cu_3N has been introduced as a new class of materials for the next generation of photovoltaic and photocatalytic applications due to its high optical absorption, good electrical properties and p-type defect-tolerant semiconductor characteristics [38–40]. Photocatalysis is an advanced oxidation process in which illumination is used to separate excitons that are then transferred to the surface of the nanoparticles [41]. The oxidation–reduction reactions produce reactive oxygen species (ROS) that oxidize organic matter on the surface of the nanomaterial leading to its degradation. The mechanism is explained in detail in Section 4 of this review. According to Cheng et al., transition metal nitrides are capable of reducing the overpotential or activation energy for photocatalytic reactions by providing additional active sites that promote electron–hole separation [39]. Furthermore, improvement in the catalytic yield of Cu_3N has been achieved by doping with transition metal ions, such as Au, Ni, Cr, Fe and Co, at the interstitial sites of the cubic structure. In addition, some metallic nanoparticles such as Ag, Au and Cu display localized surface plasmon resonance (LSPR) that enhances their optical and electrical properties. In general, metal–semiconductor junctions tend to enhance the physicochemical performances of the heterostructures [38,42].

This review attempts to provide the first comprehensive compilation of the current research on synthesis approaches, electronic structure, optical properties and photocatalytic activity of Cu_3N nanostructures. Herein, we analyze the effect of synthesis conditions on the precipitation of Cu_3N and the formation of secondary phases of Cu and Cu_2O, with emphasis on shape and size control. In fact, synthesis conditions determine the shape, size and surface defects of the nanoparticle. These have a direct on influence the electronic structure and subsequently on the band gap, optical and photocatalytic properties of the materials. Furthermore, it summarizes the emerging applications of Cu_3N in photocatalysis, and for the first time, to the best of our knowledge, a possible photocatalytic reaction mechanism of Cu_3N is proposed. The outlook of the material in commercial applications and future research is described in the summary and outlook section.

2. Synthesis of Cu_3N Nanomaterials

During the last decade, significant progress has been made in the nitride chemistry synthesis field. The main synthesis routes consist of solid–gas state synthesis, solvothermal, sol-gel, non-thermal plasma, electrospinning, electrodeposition, atomic layer deposition, chemical vapor deposition and laser ablation methods among others [43–45]. These new strategies for nanomaterial synthesis must consider the reaction conditions, along with the most relevant precursors in order to obtain reliable and reproducible properties of the synthesized materials. Compared to metal oxides, carbides, sulphides and oxynitrides, the synthesis of metal nitrides is relatively challenging due to the requirements of inert conditions during the growth process in order to suppress the formation of copper oxide. In addition, metal nitride nanostructures also tend to oxidize in ambient conditions, making their stability a challenging issue.

In particular, synthesis routes for Cu_3N nanomaterials can be divided into chemical and physical. The chemical methods have been the focus of recent research for the synthesis of Cu_3N nanoparticles, as they offer better control of the morphology and particle size distribution during the synthesis process. However, maintaining a controlled inert atmosphere in order to impede the formation of secondary phases (Cu, CuO and Cu_2O) remains a concern. Alternatively, physical methods mostly focus on thin-film growth. In both methods, the growth of Cu_3N nanomaterials requires different types of N sources, including metal-nitride, ammonia, alkylamines and nitrogen-based reagents. Solid–gas-state synthesis consists of Cu precursors reacting with nitrogen gas. On the other hand, solution-based syntheses are usually carried out in non-aqueous media in order to avoid oxidation. The synthesis methods for Cu_3N nanostructures are summarized and discussed in detail below (Table 1).

Table 1. Summary of synthesis conditions and experimental band gaps for Cu_3N nanostructures.

Cu_3N Morphology	Synthesis Method	Experimental Details Precursors and Substrates	Conditions	Band Gap (eV) Direct	Band Gap (eV) Indirect	Second Phases	Ref
Nanocubes	Single source precursor method	$Cu(NO_3)_2\ 3H_2O$, ODA	1. 110 °C, 1 h 2. 260 °C, 5 min	1.89	-	CuO	[24]
	Thermal decomposition	$Cu(NO_3)_2·3H_2O$, OAm, ODE	1. 110 °C, 1 h 2. 210 °C, 15 min	-	1.6 eV	-	[38]
	Ammonolysis reaction	$Cu(OAc)_2$, urea	1. 300 °C, 2 h	-	-	-	[46]
	One-phase process	$Cu(NO_3)_2·3H_2O$, ODA (or HAD or OAm) + ODE	1. 150 °C, 3 h 2. 250 °C, 30 min	1.5	1.04	-	[47]
	One-step synthesis	$Cu(NO_3)_2·3H_2O$, ODA	1. 115 °C, 1 h 2. 240 °C, 5 min	2.41	-	-	[48]
	Thermal decomposition	$Cu(NO_3)_2·3H_2O$, ODA	1. 240 °C, 10 min	-	-	Cu/Cu_2O	[49]
	Solvothermal synthesis method	$Cu(NO_3)_2·3H_2O$, ODA, OAm	1. 110 °C, 1 h 2. 240 °C, 40 min	-	-	-	[50]
Spherical nanoparticles	Single source precursor method	PPC, ODA	1. 110 °C, 1 h 2. 260 °C, 5 min	2.21	-	Cu	[24]
	Pyridine-based synthesis	CuI, pyridine, NH_3aq, KNH_2	1. −35 °C 2. 130 °C, 30 min	2.0	-	-	[51]
	Thermal decomposition	$Cu(NO_3)_2·3H_2O$, HAD	1. 110 °C, 1h 2. 230 °C, 5 min	2.92	-	Cu/CuO	[52]
	Ammonolysis reaction	$Cu(CO_2CH_3)_2\ H_2O$, 1-nonanol, NH_3 gas	1. 190 °C, 1h 2. 170 °C 3. 185 °C	-	-	CuO	[53]
	Surfactant-free Solution-phase approach	$Cu(OMe)_2$, BZA	140 °C, 15 min	-	-	-	[54]
	Ammonolysis reaction	CuF_2, NH_3 gas	1. 140 °C, 6 h 2. 300 °C, 8 h	-	-	-	[55]
	Ammonolysis reaction	$CuCO_3$, pivalic acid, NH_3 gas	1. 70 °C, 30 min 2. 250 °C, 10 h	-	-	Cu	[55]
	Ammonolysis reaction	CuC_{10}, 1-nonanol, NH_3 gas	1. 190 °C, 40 min	-	-	-	[56]
Powders	Solid state reaction	CuO, $NaNH_2$	1. 170 °C, 60 h	-	-	Cu/CuO	[57]
	Wet processing and ammonolysis	$Cu(CF_3COO)_2$, NH_3 gas	250–350 °C, 45 min–5 h	1.48	-	Cu	[58]
	Solvothermal synthetic method	$CuCl_2$, NaN_3, Toluene/ solvent	1. ~50 °C, 4 h 2. ~100 °C, 10–12 h 3. T ↑ for several days: ~40 °C/day	-	-	CuO	[59]
	Ammonolysis reaction	CuF_2, NH_3 gas	250–350 °C, 6 h–18 h	-	-	-	[60]

Table 1. Cont.

Cu$_3$N Morphology	Synthesis Method	Experimental Details Precursors and Substrates	Conditions	Band Gap (eV) Direct	Band Gap (eV) Indirect	Second Phases	Ref
Nanocrystals	Single-step solvothermal approach	Cu(NO$_3$)$_2$·5H$_2$O, hexamethylenetetramine (HMT)	200 °C, 1 h	1.6	-	-	[26]
	PEALD	(Cu(hfac)$_2$), NH$_3$ plasma gas	ALD cycle: 2 s for Cu(hfac)$_2$ (80 °C) in 0.5 Torr, 5 s for NH$_3$ plasma, and 5 s of N$_2$ purge at 1 Torr.	1.92	-	-	[34]
	Solution-phase synthesis	Cu(NO$_3$)$_2$·3H$_2$O, ODE, OAm	1. 120 °C, 10 min 2. 240 °C, 15 min	-	-	-	[61]
	Ammonolysis reaction	Cu$_2$O, NH$_3$ gas	250 °C, 21 h	0.95	-	-	[62]
	Solvothermal process	Cu(NO$_3$)$_2$·3H$_2$O, ODE, OAm or HDA	1. 150 °C, 3 h 2. 220 °C, 10 min 3. 250 °C, 10 min	-	-	Cu/Cu$_2$O	[63]
Thin films	Thermal evaporation and ammonolysis reaction	Silicon substrate and ammonia solution	1. 120 °C, 1 h 2. 180 °C, 2 h 3. 310 °C, 4 h	2.0	-	Cu/Cu$_2$O	[28]
	RF and magnetron sputtering	On silicon slice and quartz plate substrate	P = 1.0 Pa, N$_2$ gas flow 40 sccm, RF power 300 W	1.23–1.91	-	-	[25]
	DC magnetron sputtering	Quartz glass substrates	P = 1.0 Pa, N$_2$ gas flow 3.5–4.0 sccm, RF power 100–130 W.	-	1.44	-	[64]
	Magnetron sputtering	Single-crystal silicon and quartz substrates	P = 1.0 Pa, N$_2$ gas flow 40 sccm, RF power 250 W.	2.0	-	Cu/Cu$_2$O	[65]
	Modified activated reactive evaporation	Borosilicate glass substrate	P = 10 mTorr, N$_2$ gas flow 40 sccm, RF power 50 W	2.15	1.60	-	[66]
	Thermal evaporation method	Glass substrate	1. P = 10^{-2} Torr, 1000 °C 2. N$_2$ gas flow 100 sccm, 300 °C	-	-	Cu	[67]
	RF reactive sputtering	Glass substrate	P = 10 mTorr nitrogen balanced by 10 mTorr of argon, 140–280 °C	-	-	-	[68]
Doped Cu$_3$N and their nanocomposites							
Cu$_3$N:Pd films	RF and DC magnetron sputtering	Single-crystal silicon substrate	P = 2 × 10^{-3} Pa, Ar gas flow 10 sccm, N$_2$ gas flow 30 sccm, RF power 200 W, DC power 0–7 W for Pb	-	-	-	[30]
Cu$_3$N:Ag Thin film	RF and DC magnetron sputtering	Monocrystalline silicon and glass substrate	P = 10 × 10^{-3} Pa, N$_2$ gas flow 40 sccm, RF power 200 W	-	1.59	-	[69]
Cu$_3$N nanocrystals on CNTs	PEALD	(Cu(hfac)$_2$), NH$_3$ gas	P = 1 torr, plasma power 100–400 W, NH$_3$ gas 250 °C	1.9	-	-	[70]
Cu$_3$N@SiO$_2$ spheres	Ammonolysis	CuSiO$_3$, NH$_3$ gas	350 °C, 1 h	-	-	-	[71]

Acronyms: Octadecene (ODE), 1-octadecylamine (ODA), hexadecylamine (HAD), oleylamine (OAm) and benzylamine (BZA), PPC = pyrrole-2-carbaldpropyliminato Cu(II).

2.1. Gas-state Synthesis and the Ammonia Source

In this method, the corresponding Cu-based metal precursor undergoes nitration under gas flow (i.e., NH$_3$ or a mixture of N$_2$:H$_2$) to produce Cu$_3$N nanostructures [72]. A variety of N-rich complexes, including hydrazine, urea, ammonium salt, carbon nitride and their mixtures, act as nitrogen sources in order to tailor the nanoparticle size and shape. Nevertheless, ammonia gas for ammonolysis has been widely used as it creates a reductive atmosphere and simultaneously acts as a powerful nitrogen source [73]. In ammonolysis, the reaction time, temperatures under 200 °C, heating and cooling rates, as well as the amount of ammonia are often adjusted in order to optimize the reaction conditions and the nanoparticle features. In addition, the properties of the produced nitride depend on the precursor and preparation conditions, such as heating rate and final temperature. For example, Panda et al. reported a method for the synthesis of Cu_3N nanocubes by nitration of copper acetate at 300 °C inside a quartz tube, based on the in situ decomposition of urea leading to the release of ammonia, as shown in Figure 1a. Their TEM results show that the Cu_3N nanoparticles are cube-shaped with particle sizes ranging from 60–100 nm presented in Figure 1b,c [46]. On the other hand, Nakamura et al. used cooper (II) acetate monohydrate in an alcoholic solution of 1-nonanol along with bubbling ammonia for 1 h at 190 °C to produce Cu_3N nanoflowers (Figure 1d–f) [53]. In their study, the nanoparticles had granular shapes with a diameter of less than 200 nm. The differences in morphology and size arise from a combination of copper precursors, synthesis temperature and nitrogen sources. The utilization of urea as a nitrogen source, as opposed to NH_3, is very advantageous owing to facile handling and controllability of flux, implying that the risk of over-reduction of copper acetate to form metallic copper is mitigated [46]. The mechanism of nanoparticle formation was elucidated by Nakamura et al. and followed certain reaction steps: (1) formation of copper (II) amine complex, (2) reduction, by the long-chain alcohol, of Cu^{2+} to Cu^+ and (3) the reaction between ammonia and Cu^+ [53].

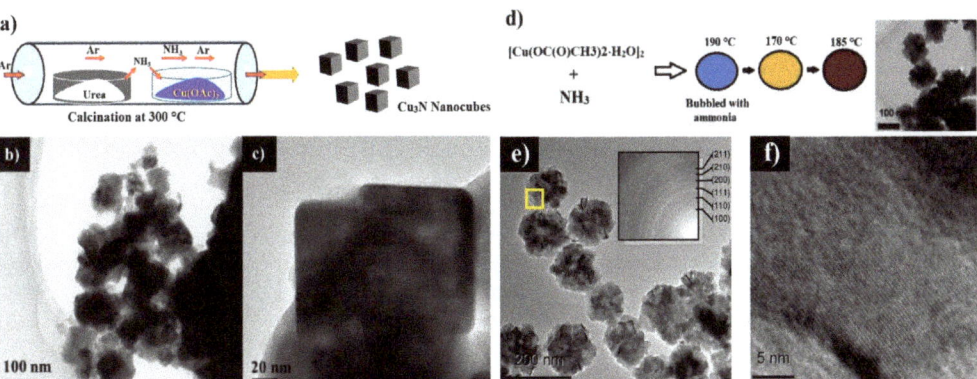

Figure 1. Schematic illustration of the synthesis and TEM micrographs of: Cu_3N nanocubes using urea (**a–c**). Reprinted with permission from Ref [46], ACS publications, 2019. Schematic illustration of the synthesis of Cu_3N nanoflowers using ammonia gas in long-chain alcohol solvent and corresponding TEM micrographs (**d–f**), adapted with permission from Ref [53], ACS publications, 2014.

Multistep synthesis approaches have also been applied to the synthesis of Cu_3N. These approaches are used to reduce copper oxide nanoparticles produced in the first step of the synthesis, followed by nitration. For example, Szczesny et al. synthesized Cu_3N nanoparticles with diverse morphologies from oxygen-containing precursors using a two-step process that combines solvothermal and solid–gas ammonolysis stages [74]. In the first step, copper (II) chloride dihydrate was used as a precursor for the fabrication of copper (II) oxide and copper (II) hydroxide nanoarchitectures by solvothermal methods. In the second step, Cu_3N was obtained through ammonolysis of $Cu(OH)_2$ and CuO nanoparticles [74]. Deshmukh et al. synthesized $Cu_3N@SiO_2$ composites using a multistep approach [71]. First, CuO nanoparticles were synthesized within hollow mesoporous silica spheres by binding or adsorbing Cu^{2+} ions onto the surface of carbon spheres, followed by the formation of a mesoporous silica shell by a sol-gel method. Cu_3N nanoparticles (size < 30 nm) were formed within silica spheres through the nitration of copper (II) oxide composite

(CuO@SiO$_2$) with ammonia gas at 300 °C for 10 h [71]. In the two-step approach, chemical nitration can only occur after the copper oxide is reduced to copper ions, which then react with ammonia to form Cu$_3$N. During the synthesis, a change in the color of the solution from yellow to brown indicates the reaction of copper ions with ammonia leading to the precipitation of nitride nanoparticles [62]. Furthermore, Nakamura et al. suggested that for successful nitration, ammonia must come into contact with the copper ions at an optimum temperature as soon as the reduction of copper oxide starts [53].

The ammonolysis reactions provide a versatile route to synthesize Cu$_3$N nanoparticles. Based on reports, ammonolysis is performed under specific conditions in solid-state chemistry in order to obtain a complete exclusion of oxygen and water during synthesis and handling. In addition, this approach can be optimized by varying the solid precursor, ammonia flow rate, heating and cooling rates, temperature and reaction time. However, the control of the nanoparticle shape and size in these processes remains a challenge. Moreover, this pathway involves the use of NH$_3$, which is toxic and corrosive at high temperatures.

2.2. Solution-Based Synthesis

In addition to ammonia gas, nitrogen-based compounds can also be used as nitrogen sources for the synthesis of Cu$_3$N nanostructures. In consequence, the use of non-aqueous solutions in the synthesis of metal nitrides has gained interest, as it suppresses oxidation and allows controlling the nanoparticle size and shape. The technique of non-aqueous synthesis requires water or hydrate-free solvents, as well as regulated temperatures and pressures in a controlled environment, as it relies primarily on the solubility and reactivity of precursors at relatively high temperatures and pressures [75]. Primary amines are commonly used as capping agents to synthesize metal nitrides and semiconductor nanomaterials. In general, solvothermal methods offer good control of the shape, size and crystallinity of nanostructured materials by controlling synthesis parameters, such as reaction time, temperature, precursor, surfactant and solvent, in order to vary the physical and chemical properties of the nanomaterials [76].

2.2.1. Effect of Solvent

Presently, the thermal decomposition of copper salt-based precursors in long-chain amines or alcohols is one of the most popular synthesis techniques for Cu$_3$N nanoparticles. It not only serves as a nitrogen source but can also performs simultaneous functions as a solvent, surface stabilizer and reducing reagent. Therefore, special attention is being paid to the synthesis of Cu$_3$N nanoparticles using solvents, such as 1-octadecene (ODE) and amine sources, including 1-octadecylamine (ODA), hexadecylamine (HAD), oleylamine (OAm) and benzylamine (BZA) [33,38,47,49,54,63,77]. Synthesis of Cu$_3$N nanocubes with tunable sizes was reported by Wu et al. through the decomposition of Cu(NO$_3$)$_2$ in ODE with different capping agents by a facile one-step process [47]. The nanoparticle size depended on the amine source used and ranged from 10 to 30 nm. In fact, the size was reduced significantly to 26 nm with ODA, to 18.6 nm with HAD and even further to 10.8 nm with OAm [47]. A similar synthesis approach for Cu$_3$N nanocubes using ODA and ODE was also reported by Wang et al. [49]. They obtained Cu$_3$N nanocubes (size = 15 nm) in Figure 2a after the thermal decomposition of Cu(NO$_3$)$_2$·3H$_2$O at 240 °C in ODA solvent [49]. Similarly, Barman et al. reported the synthesis of Cu$_3$N nanocubes with an average particle size of ~10 nm using a 1:1 volume ratio of ODE and OAm solvents [38]. The TEM image consists of smaller nanoparticles that coalesce into a nanocube, as visible in Figure 2b.

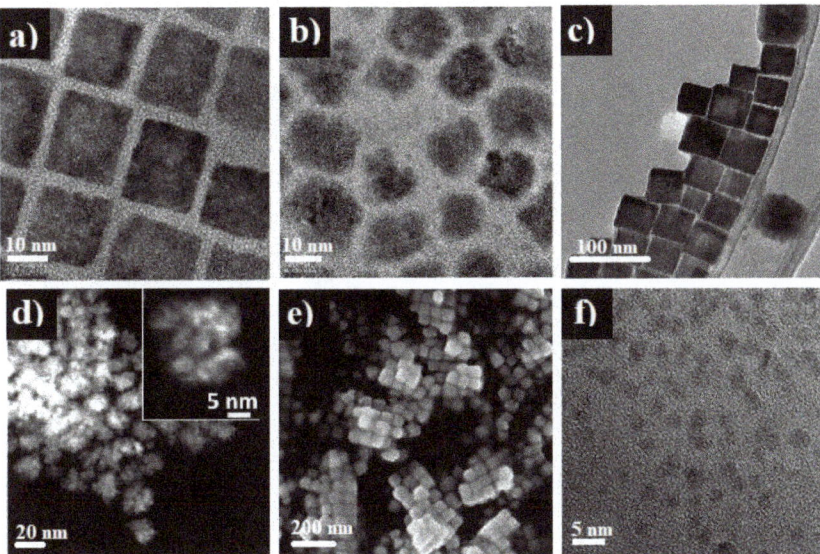

Figure 2. TEM micrographs of Cu$_3$N nanostructures synthesized using different amines: (**a**) Cu$_3$N nanocubes from Cu(NO$_3$)$_2$·3H$_2$O in ODA. Reprinted with permission from Ref [49], RSC publications, 2011. (**b**) cubic-like Cu$_3$N nanoparticles from Cu(NO$_3$)$_2$·3H$_2$O in OAm. Reprinted with permission from Ref [38], ACS publications, 2019. (**c**) Cu$_3$N nanocubes from Cu(NO$_3$)$_2$·3H$_2$O in ODA. Reprinted with permission from Ref [48], Elsevier publications, 2019. (**d**) HAADF-STEM micrograph of Cu$_3$N mesocrystals from PPC in ODA. Reprinted with permission from Ref [63], RSC publications, 2021. (**e**) FE-SEM of Cu$_3$N cubic-like nanoparticles from Cu(NO$_3$)$_2$·3H$_2$O in ODA. Reprinted with permission from Reprinted with permission from Ref [63] RSC publications, 2021. (**f**) ultrasmall nanoparticles from Cu(OMe)$_2$ in BZA solvent. Reprinted with permission from Ref [54], RSC publications, 2015.

In addition, different solvents and reaction parameters influence the size and morphology of the nanoparticles. For example, Princ et al. synthesized two types of Cu$_3$N nanoparticles with different morphologies using two different capping agents dissolved in ODE [63]. The report revealed that when OAm is used as a capping agent, spherical nanoparticles of 8 nm were produced (Figure 2d). In contrast, HAD produces cube-like nanoparticles of 50 nm (Figure 2e) [63]. Sithole et al., in two different reports, synthesized Cu$_3$N nanoparticles and Cu$_3$N nanocubes using amines as solvents [24,48]. Cu$_3$N nanoparticles with an average particle size of 2.8 nm were synthesized using pyrrole-2-carbaldpropyliminato Cu(II) (PPC) as a single-source precursor with ODA [24]. On the other hand, Cu$_3$N nanocubes with an average size of 41 nm, Figure 2c, were synthesized using Cu(NO$_3$)$_2$·3H$_2$O at 240 °C in ODA [48]. Finally, Deshmukh et al. reported the synthesis of ultrasmall Cu$_3$N nanoparticles of ~2 nm, Figure 2f, via a one-step reaction of copper (II) methoxide (Cu(OMe)$_2$) precursor with BZA at lower temperatures for short reaction periods (e.g., 140 °C for 15 min) [54].

2.2.2. Effect of Capping Agent

The above results suggest that the nanoparticle size and shape vary as a function of capping agents or surfactants. In fact, the long-chained amine surfactant plays an important role in shaping the nanoparticles at the nucleation and growth stages. They modify the decomposition routes of the copper nitride (II) reagent contrarily to other solvents or solid-state processes. According to Princ et al., the reasons for differences in the overall morphologies of nanoparticles using different capping agents may be related to the different mesocrystal subunits or the subsequent crystallographic orientations of subunits [63]. From the TEM images of Figure 2b–d, the coalescence of small spherical nanoparticles into nanocubes could also be another mechanism. The formation of nanocubes via such a mechanism appears to be more favorable in the case of long-chained amines. In fact, the subunits of nanocrystals are subject to various competitive forces, such as van der Waals, dipole–dipole and dissolution. Nonetheless, it is important to mention that so far, there is no report that elucidates the effect of the amine chain length on the size and morphology of the nanoparticles. Nevertheless, a complex combination of the synthesis environment, reaction time, temperature and amount of reagents influence the final shape and size of the nanoparticles.

2.2.3. Effect of Reaction Time

The reaction time related to the formation mechanism of Cu_3N nanoparticles (nucleation, growth, aggregation and breakage) is a primary factor to be considered. It has been reported that the nanoparticle size, shape and stability are strongly dependent on the reaction time [78]. For example, Xi et al. reported the solvothermal synthesis of magnetic Cu_3N nanocubes with high electrocatalytic reduction properties [50]. The cube-shaped Cu_3N nanoparticles were synthesized by dissolving copper (II) nitrate in organic solvents of ODA and OAm at a high temperature of 240 °C. The detailed growth process of Cu_3N nanocrystals is shown in Figure 3a,b, consisting of three different stages: (1) nucleation stage to form nanoparticles, (2) growth stage in which the nanoparticles adopt the cubic shape, owing to the surface tension exerted by the solvent and (3) molding stage in which nanocubes are formed [50]. In addition, Sithole et al. (shown in Figure 3c) reported a fourth step, i.e., the transformation of Cu_3N into Cu nanoparticles [33], which is due to a spontaneous decomposition of Cu_3N nanocubes depending on the reaction time. In fact, a cloud of nuclei with a few developed Cu_3N nanocubes was obtained after 5 min, followed by well-defined nanocubes after 15 min. Finally, after 20 min of heating, degradation of the nanocubes occurs, and the copper nitride decomposes into metallic Cu within 60 min.

2.2.4. Formation of Secondary Phases

Both routes have their share of advantages and drawbacks. One common disadvantage is the precipitation of secondary phases during Cu_3N synthesis. For instance, it is reported that by prolonging the reaction time at a particular temperature, a partial phase transformation of Cu_3N into Cu and Cu_2O occurs [63]. Kieda et al. applied a spray pyrolysis technique using a copper–amine complex solution. However, obtaining a pure phase of Cu_3N using only $Cu_2CO_3(OH)_2$ precursor [79] is unlikely, as the co-precipitation of Cu, Cu_2O and CuO secondary phases is inevitable. On the other hand, changing the molar ratio of the reactants ($Cu(NO_3)_2 \cdot 3H_2O$:ODA) results in the formation of Cu and Cu_2O nanocrystals with or without Cu_3N [33]. The formation of CuO is suppressed most likely due to a more reductive synthesis environment. Therefore, the development of a scalable, reliable and reproducible method for the synthesis of high-quality Cu_3N nanostructures with a high density of active sites is the need of the hour. Additionally, the degassing process, which removes water molecules from the solution, is an important step in suppressing the formation of secondary phases. The most commonly used gases are nitrogen or argon, which eliminate water molecules from the solution while heating the solution between 100 °C to 150 °C. Wet chemical routes tend to be more reliable and reproducible in controlling nanoparticle shape and size compared to gas-state methods, thus making them viable for larger-scale production. Major limitations in upscaling these methods include high reaction temperatures and the use of surfactants, which increase their costs and complexity.

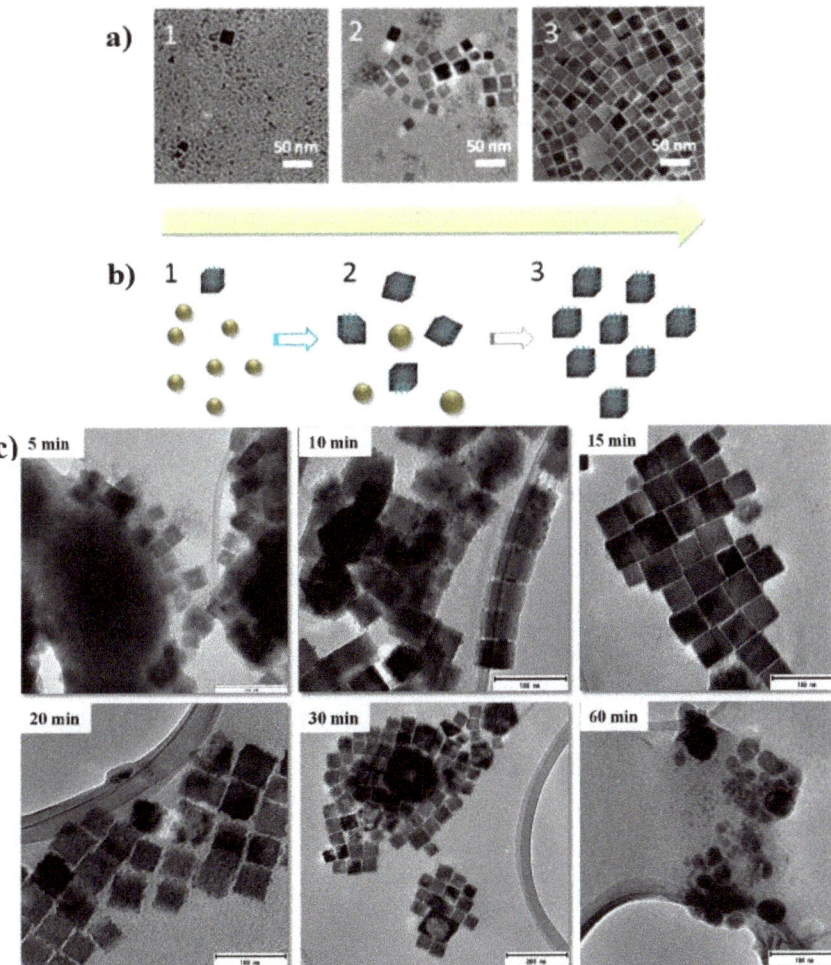

Figure 3. TEM micrographs of Cu_3N nanocubes synthesized at different reaction times: (**a,b**) 2, 5, 10 min of reaction. Reprinted with permission from Ref [50], RSC publications, 2014. (**c**) 5, 10, 15, 20, 30 and 60 min of reaction time. Reprinted from Ref [33], Creative Commons agreement from RSC publications, 2019.

2.3. Thin Film Deposition Techniques

Several studies have focused on the growth of Cu_3N thin films. A number of techniques, such as DC and RF reactive magnetron sputtering, thermal evaporation, pulse laser deposition (PLD) and CVD, are commonly used to grow Cu_3N thin films. These deposition techniques produce nitride thin films with controlled stoichiometry, thickness and composition [39]. In fact, the efficient thin film deposition of Cu_3N has mainly been achieved via physical methods for applications in optical data storage [80], solar energy conversion [15], BioMEMs [81] and photovoltaics [82]. In general, physical methods are useful for growing single-phase thin films but have the drawback of producing only a limited variety of compounds [76]. Furthermore, the doping of Cu_3N with other compounds has marked a breakthrough in nitride thin film research. In that regard, thin-film deposition techniques provide a simple pathway to introduce foreign atoms while retaining the Cu_3N lattice stability.

Most of the previous work on Cu_3N nanostructures has been on thin film deposition using magnetron sputtering. The method is often used to produce transition metal nitride thin film electrodes, which exhibit excellent adhesion, controllable composition and thickness of nanostructured films with excellent electrochemical properties [83]. During the reactive sputtering of Cu_3N, a reductive gas such as N_2 or Ar reacts with a Cu metal or a Cu_3N target, sputtering off the compound layer. Then, the sputtered Cu and N species are deposited on substrates forming Cu_3N compound layers [68]. For instance, Jiang et al deposited Cu_3N:Pb thin films using different Pb concentrations on monocrystalline silicon through DC and RF reactive magnetron sputtering [30]. In fact, Pb doping leads to a high adhesion of the films to the substrate, as it promotes nucleation in the early stages of thin-film growth [30]. Figure 4a shows the cross-sectional view of dense and well-adhered films. Similarly, Xiao et al. investigated the growth of Cu_3N:Ag thin films using RF and DC magnetron sputtering (Figure 4b) [69]. In this research, Cu_3N:Ag thin films were prepared with variable sputtering power of Ag. They observed that an increase in the Ag content brings about an increase in the grain size. Additionally, the optical band gap of the films also increases owing to the plasmonic-induced Burstein–Moss effect leading to an increase in the number of carriers in the conduction band of Cu_3N, which in turn displaces the Fermi level towards the conduction band [69]. In both Ag and Pb doped thin films, the growth mode is columnar, typical of the sputtering deposition process. However, the roughness of the Pb doped films is higher than the Ag-doped films, most probably due to the Pb atom being slightly larger than the Ag atom. Since the lattice mismatch for Pb doped films is higher, the lattice relaxation process enhances the thin film roughness. Similarly, nanocomposites of Cu_3N tend to have enhanced optical properties. For example, TiO_2-Cu_3N nanocomposites manifest an improved bandgap compared to Cu_3N alone. Zhu et al. also reported the growth of TiO_2-Cu_3N and Cu_3N-MoS_2 nanocomposite films by magnetron sputtering that demonstrate enhanced photocatalytic properties [35,84]. Both TiO_2 and MoS_2 are n-type semiconductors that, when combined with p-type Cu_3N, create p-n junctions. For catalytic activity, large specific surfaces of active materials are required. Moreover, when the roughness increases, the number of active sites on the surface also augments, which leads to an enhancement of photocatalytic activity [35,84]. Industrially viable thin film growth techniques, such as sputtering, allow thickness control with the possibility of depositing on large surface areas in order to obtain a high specific surface. The deposition parameters of Cu_3N thin films include sputtering power, N_2 pressure r ($r = N_2/[N_2^+ Ar]$), substrate temperature and deposition pressure, which have a significant influence on Cu_3N thin film growth and their properties. Additionally, deposition parameters also affect the crystal orientation and grain size of Cu_3N thin films [65,83].

Cu_3N thin films can also be deposited by thermal evaporation using nitrogen as a gas source. In this method, the Cu powder is evaporated onto a glass substrate and then exposed to N_2 gas in order to fabricate Cu_3N thin films. Ali et al. thermally evaporated Cu powder on a glass substrate in a horizontal glass tube furnace at 1000 °C in a high vacuum for 25 min [67]. Then, the temperature of the tube furnace was set to 300 °C, and an N_2 gas flow rate of 100 sccm was maintained for different reaction times. Figure 5a–d show the SEM images of Cu_3N thin films deposited for a duration of 2, 4, 6 and 8 h under nitrogen gas flow at 300 °C. When the growth time is prolonged, an increase in the grain size is observed [67]. Cu_3N thin films with a rough surface were deposited by Lindahl et al. using a gas pulsed CVD [29], which is a gas phase chemical method by which thin films are grown on a suitable substrate through precursor decomposition at elevated temperatures. In that case, Cu-Ni-N thin films were prepared using CVD processes with intermittent NH_3 gas flow. Ni-doped Cu_3N can be described as a solid solution of Ni in Cu_3N formed in a pseudo-binary system composed of the two metastable phases Cu_3N and Ni_3N (Figure 5b). According to Lindahl et al., the texture of the deposited films changes with the increase of Ni content, which was attributed to changes in the deposition mechanism through the probable occupancy of the interstitial sites of Cu_3N in the crystal structure [29].

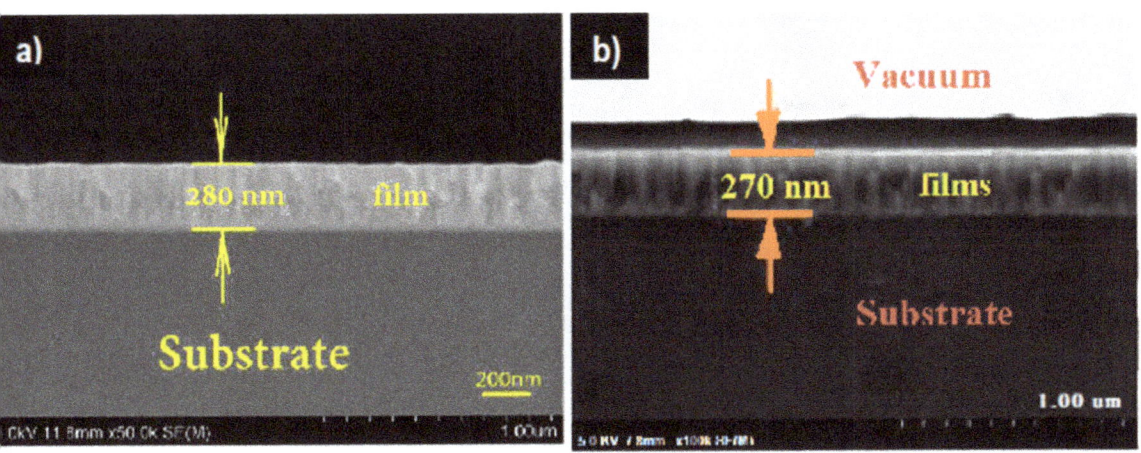

Figure 4. Cross-section view of thin films: (**a**) Cu$_3$N:Pb thin film. Reprinted with permission from Ref [30], Elsevier publications, 2019. (**b**) Cu$_3$N:Ag thin film. Reprinted with permission from Ref [69], IOP publishing, 2018.

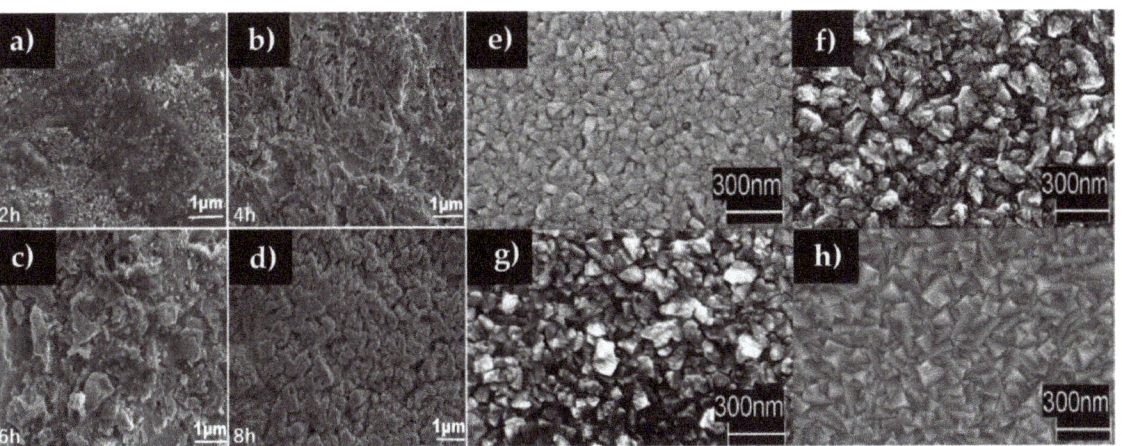

Figure 5. (**a**–**d**) SEM images of Cu$_3$N thin films grown by thermal evaporation technique at different reaction times. Reprinted with permission from Ref [67], Elsevier publications, 2021. (**e**–**h**) SEM images of Cu$_3$N films with different Ni metal content: (**e**) pure Cu$_3$N, (**f**) 21% Ni, (**g**) 46% Ni and (**h**) 76% Ni. Reprinted with permission from Ref [29], Elsevier publications, 2018.

3. Electronic Structure and Band Gap

Cu$_3$N has a primitive open anti-rhenium trioxide cubic structure (space group Pm3m, lattice constant a = 3.82 Å), presented in Figure 6 [85]. In the structure, N atoms occupy the cube corners of the unitary cell in (0, 0, 0) and Cu atoms are located between two consecutive nitrogen atoms in ($\frac{1}{2}$,0,0), (0, $\frac{1}{2}$, 0) and (0, 0, $\frac{1}{2}$) [26,51,86,87]. Each N atom in the cubic structure of Cu$_3$N is shared by eight cells, while each Cu atom is shared by four cells. In pure Cu$_3$N, the face centers and the body of the cube are empty. These void interstitial sites are, therefore, available to host Cu, N or other foreign atoms under certain conditions [88]. However, the insertion of additional metallic elements in the body of the Cu$_3$N lattice alters the chemical interactions between Cu and N, thereby modifying or changing the electrical and optical properties. Experimental and computational studies reveal that metal-doped Cu$_3$N adopts metallic characteristics due to the reduction of the metallic inclusions (M^0) [88]. For instance, Hahn et al. demonstrated that Cu_3N is a semiconductor, while Cu_3N-Pd exhibits semi-metallic conductivity due to Pd atoms causing modifications in the energy bands [89]. Additionally, Cu_3N:Pd presents higher mass activities and stability in comparison with commercial Pd alone. In addition, Moreno et al. established that the addition of an extra Cu atom into the body of the unitary cell of Cu_3N endows it with metallic properties [90]. In turn, the unit cell of Cu_3N expands, owing to the progressive insertion of Cu atoms. Furthermore, their computational studies show that the lattice parameter of Cu_3N-Cu expands by 0.06 Å. Moreno et al. inferred that Cu_3N with lattice constants higher than 3.868 Å behave as conductors [90]. Finally, these types of doped-compounds, i.e., Cu_3M_xN, present an anti-perovskite structure, owing to the metal dopant occupation of ($\frac{1}{2}, \frac{1}{2}, \frac{1}{2}$), i.e., the body-center position [88]. Anti-perovskite structures consist of interchanged A and B site atoms. They are defect tolerant, present excellent ionic conductivities and are ideal candidates for electrodes of batteries [91]. This suggests that they possess interesting redox capabilities, enabling their application as photocatalysts.

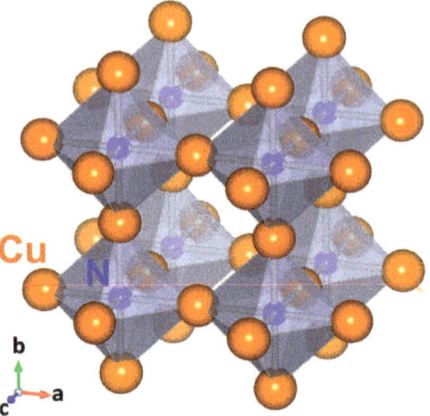

Figure 6. Schematic view of the anti-ReO_3 crystal structure of Cu_3N. Reprinted with permission from Ref [85], Wiley publications, 2018.

Furthermore, the crystal chemistry of transition metal nitrides is largely dictated by the bond coordination of N and M atoms. In these compounds, the M atoms are bonded with nitrogen via covalent or ionic bonds. According to Rao et al., the addition of N atoms to a transition metal converts the metal–metal bond to a covalent bond or a mixture of covalent-metal bonds [44]. Moreover, the addition of N increases the bond length and, consequently the lattice parameters, resulting in a contraction of the metal d-band, further suggesting that bonds in metal nitrides are ionic. This d-band contraction causes an increased density of states (DOS) near the fermi level, giving rise to novel catalytic properties, which are different from the parent metal but rather similar to noble metals [14]. Moreover, recent studies suggest that both covalent and ionic bonds connect Cu and N atoms in undoped Cu_3N [30]. On the other hand, for M-doped Cu_3N, the bond coordination has a covalent character as the interactions between the inserted M and Cu atoms dominate [42,92].

Undoped Cu_3N tends to be insulating due to defect states created by metallic and non-metallic vacancies of Cu and N, respectively [93]. However, due to the varying nature of the vacancies giving rise to shallow defect states, the bandgap of Cu_3N shows variations and ranges from 0.2 to 2.0 eV. Another study demonstrated that doping with iodine gives rise to states that are located very close to the band edges and therefore resulting in the absence of band gap states [94]. Furthermore, Cu_3N being a p-type semiconductor, displays a strong hybridization of the Cu 3d and N 2p orbitals close to the valence band owing to their anti-bonding states, which causes their high tolerance to defects. Theoretical calculations confirm that Cu 3d electrons are a major contributor to the DOS and have an important influence on the Cu_3N band gap.

Since nanostructured defect-tolerant transition metal nitrides possess semiconducting properties with band gaps corresponding to visible wavelengths, they are therefore considered suitable materials for photocatalysis and photovoltaic applications [39,95]. In the literature, there are several discrepancies in the band gaps of Cu_3N arising from both theoretical and experimental studies. Theoretical calculations predict an indirect band gap ranging from 0.23 to 1.0 eV depending on the calculation method [58,88]. Density Functional Theory (DFT) tends to underestimate the band gap mainly due to self-interaction errors and changes in potential upon changing the number of electrons in the conventional framework of DFT [96]. Co-workers also calculated the band structure and DOS before and after the addition of a foreign atom into the cubic structure of Cu_3N. In fact, the electronic structure changes in the presence of foreign atoms, accompanied by an increase in the Fermi energy, along with a consequential change in the band gap. For instance, computational calculations reveal that pure Cu_3N is a semiconductor with a small indirect band gap equal to 0.38 eV. In contrast, for M-doped Cu_3N (M = Sc, Ti, V, Cr, Mn, Fe, Co and Ni), the band gap becomes negligible, giving rise to the metallic character of Cu_3N [42]. Other studies confirm a semiconducting behavior but suggest that pure Cu_3N presents an indirect band gap of 0.32 eV and direct gaps of 1.09 and 0.87 eV, whereas for Cu_3N-N rich compounds, a partially filled spin-resolved narrow band of new electronic bandgap states at the Fermi energy is observed [97]. This new band modifies the optical properties of Cu_3N and makes the material susceptible to infrared absorption.

On the other hand, experimental results ascertain both direct and indirect band gaps ranging from 1.17 to 2.38 eV that depend on the experimental conditions and N content. Table 1 summarize the synthesis parameters and band gaps of Cu_3N and Cu_3N-M nanostructures. Band gaps determined experimentally vary, for example, between 1.91 to 2.15 eV for Cu_3N thin films deposited by sputtering, whereas a maximum band gap of 2.92 eV was reported for nanoparticles of Cu_3N [52]. Furthermore, band gaps of 1.04 eV (indirect) and 1.5 eV (direct) were obtained from cyclic voltammetry (CV) measurements on Cu_3N nanocubes based on onset Redox potentials [47]. In general, the experimental band gap depends on the synthesis conditions, which in turn determines the morphology and chemical composition of the nanomaterial.

4. Photocatalytic Activity of Cu_3N Nanoparticles

Photocatalysis is a redox process involving three main steps: (i) the generation of electron-hole pairs (e^-/h^+) by photoexcitation, (ii) transportation of the excitons to the semiconductor surface and (iii) the utilization of charge for surface oxidation–reduction reactions [39,41]. The total efficiency in these processes is strongly determined by the properties of the semiconductor photocatalyst, such as the electronic structure, band gap, surface properties along with kinetic and thermodynamic processes [98]. An ideal photocatalytic material should primarily have a bandgap that is optimal for absorbing in the entire range of the solar spectrum in order to dissociate water molecules while retaining its stability during the reaction process. In addition, it must be cost-effective, easy to process, readily available and non-toxic. Moreover, in order to efficiently harvest visible light and accelerate reactions, the band gap should preferably be under 3.2 eV. However, since redox reactions require a certain amount of energy, narrow band gap semiconductors may therefore not be the most suitable for catalytic reactions. In addition, the probability of excitonic recombination increases for lower bandgaps. Therefore, an optimum balance between the band gap and recombination lifetime is required. One method for efficient charge separation is the combination of plasmonic nanoparticles with photocatalytic semiconductors.

Current reports demonstrate a keen interest in the photocatalytic activity of metal nitride nanomaterials. Their metal-like properties, such as conductivity, optical band gap and visible light activation, are beneficial for highly efficient separation and delivery of photogenerated carriers. Unlike traditional semiconductors (oxides, sulfides and carbides), metal nitrides can lower the overpotential or activation energy for photocatalytic reactions on the surface of semiconductors. They also provide additional active sites and promote electron-hole separation at the interface of the co-catalyst [39,99]. Furthermore, their low-cost production, high thermal stability and tolerance against acids and bases are advantages in photocatalysis [39]. Inspired by these properties, the design and construction of semiconductor-based metal-nitride nanomaterials have been carried out and applied to photocatalytic processes for water splitting, organic dye degradation, CO_2 reduction and decomposition. For example, CoN, [100] Ta_2N, [100] Ta_3N_5, [101], Ni_3N, [102,103] and InGaN [104] have been tested as potential semiconductors for water splitting. In photocatalytic water splitting reactions, the photo-induced process enables the production of H_2 and O_2 [44,105]. Additionally, metal nitrides are sturdy with good refractory properties at elevated temperatures of 2000 °C owing to their metallic character. The latter also dotes them with properties similar to plasmonic metal nanoparticles. However, compared to noble metal nanoparticles, nitrides display plasmonic wavelengths in the visible and infrared regions. For ex., TiN has been used as plasmonic interconnectors, further suggesting that nitrides can replace noble metals in such applications. Nitrides also display tunable optical responses and improved light-harvesting, equivalent to noble metals. They can, in general, significantly enhance the local electromagnetic field when the incident light interacts with the surface plasmons [44,106]. In other words, localized surface plasmon resonance (LSPR) enhances the electric field, facilitating light absorption and charge transfer processes of semiconductors [38,107]. In such cases, the generated excitons have higher energy than the Fermi level of the photocatalyst, which is ideal for driving photocatalytic reactions [108]. For example, ZrN [109], HfN [110] and TiN [111] are reported as plasmonic metal nitride nanostructures. However, Cu_3N has not yet been specifically reported as a plasmonic nanomaterial. Nevertheless, synthesis routes tend to produce secondary phases of Cu metal, which is a known plasmonic material that enhances the overall catalytic performance of the primary phase [33]. In all cases, metal nitrides need to be combined with a metal or themselves possess a metallic character in order to display surface plasmon resonance. This is likely if they are synthesized with an excess of Cu in the case of Cu_3N. In soft chemical synthesis, excess Cu could favor the co-precipitation of Cu metal and simultaneously incorporate Cu in the Cu_3N lattice. For thin-film synthesis, the Cu metal phase separation can be suppressed by controlling the growth conditions in order to ensure the incorporation of Cu in the Cu_3N lattice.

With further regard to Cu_3N-based nanostructures, several studies have reported photocatalytic degradation of dyes. A schematic representation of the Cu_3N-based photocatalytic process is represented in Figure 7. The process starts when Cu_3N absorbs energy equivalent to its bandgap, whereupon generating a wide range of electron-hole pairs (e^--h^+) under the photoelectric effect. The e^--h^+ pairs migrate to different positions on the surface of the Cu_3N under the action of the generated electric field. The electrons on the surface of Cu_3N are then transferred to the adsorbed molecular oxygen, generating reactive oxygen species such as superoxides ($\bullet O_2-$). The holes react with water and hydroxyl ions to generate hydroxyl radicals ($\bullet OH$) that can oxidize organic matter adsorbed on the surface of Cu_3N. The oxidizing ability of $\bullet OH$ radicals is the most potent among reactive oxygen species in aqueous media. It is capable of oxidizing most of the inorganic contaminants, as well as organic matter in water and degrading them to smaller inorganic molecules such as carbon dioxide, water and other harmless substances. The possible chemical reactions of the photocatalytic activity of Cu_3N are as follows:

$$Cu_3N + h\nu \rightarrow e^- + h^+$$

$$h^+ + H_2O \rightarrow \bullet OH + H^+ \tag{1}$$

$$e^- + O^2 \rightarrow O_2^-$$

$$OH + H^+ \, dye \rightarrow \ldots \rightarrow CO_2 + H_2O.$$

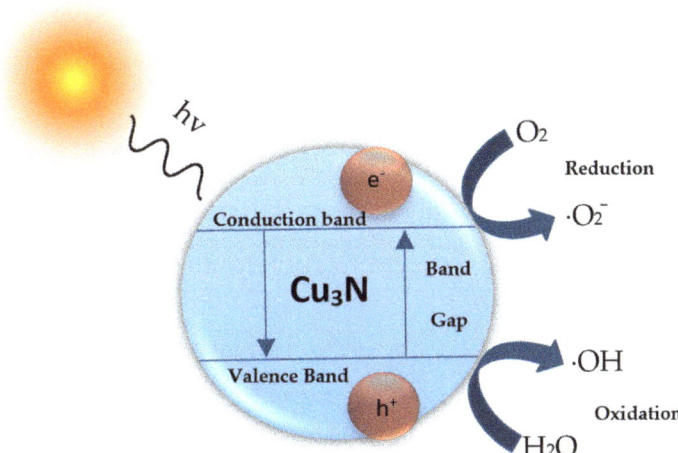

Figure 7. Schematic representation of photocatalytic mechanism for Cu_3N photocatalyst.

In general, Cu_3N demonstrates excellent electrochemical performance. Both photo- and electrocatalysis have the same mechanism of charge creation. The difference lies in the type of activation energy provided to create these charges. Therefore, electrochemistry can elucidate the redox reactions of Cu_3N in order to evaluate its potential for hydrogen and oxygen production. Sajeev et al. evaluated the potential of Cu_3N in hydrogen evolution reaction (HER), where they claim a lower overpotential than its oxide counterpart Cu_2O [112]. In general, the shift from oxides to nitrides for photo- or electrocatalysis has been facilitated owing to the higher potential of N 2p than O 2p orbitals. This signifies that metal nitrides have a lower bandgap than metal oxides, whereby they are more adapted to photocatalysis. They can therefore be also employed as working photoelectrodes for photocatalytic water splitting. Nevertheless, Cu_3N tends to be unstable in aqueous media and high and low pH conditions due to its tendency to oxidize. However, these concerns are surmounted by synthesizing nanocomposites of Cu_3N [98], which are particularly useful for dye degradation in aqueous media. Cu_3N is also used as an electrocatalyst in gas-phase reactions for the production of hydrocarbons via CO_2 reduction reactions [113]. Presently, there are no accounts of photoelectrocatalytic production of hydrocarbons with Cu_3N or its nanocomposites, even though the potential exists.

In recent years, several studies of Cu_3N nanomaterials for photocatalytic applications have been reported (Table 2). Jiang et al. reported the photocatalytic degradation of methyl orange (MO) using Cu_3N thin films with an optical band gap of 2.0 eV [65]. In this work, a MO solution of 20 mg/L in 50 mL water was used as the target degradation product and Cu_3N thin films with dimensions of 2.5 cm × 1.0 cm × 120 nm as a photocatalyst under a high-pressure mercury lamp (500 W). A degradation yield of 95.5% in 30 min was achieved (Figure 8a). They infer that the effective photogenerated electron-hole pairs mainly originate from Cu vacancies and interstitials in the film [65]. Although the Cu_3N thin-film alone presents excellent photocatalytic properties for degrading dyes, it has a few shortcomings. Therefore, Cu_3N nanostructures are combined with other semiconductors. For instance, Zhu et al. demonstrated that the degradation of MO using $TiO_2@Cu_3N$ thin films (effective band gap = 3.0 eV) is higher than in pure TiO_2 thin films, as seen in Figure 8b [35]. A degradation rate of 99.2% of MO solution with an initial concentration of 20 g/mL using a 500 W mercury lamp for 30 min was obtained [84]. Other studies using $Cu_3N@MoS_2$ thin films (band gap = 2.05 eV) as photocatalysts obtained a 98.3% degradation rate of MO with an initial concentration of 10 mg/mL in 30 min. In this case, controlled quantities of MoS_2 reduce the overall band gap of the nanocomposite to values lower than the band gap of Cu_3N [35]. On the other hand, during the past few decades researchers have also studied the photocatalytic activity of Cu_3N nanoparticles in aqueous media. Sithole et al. reported the photocatalytic degradation of methylene blue (MB) and MO using Cu_3N nanocubes with a band gap of 2.41 eV [48]. In that study, 0.1 g of Cu_3N nanocubes were added into the solution of MB and MO, with an initial concentration of 20 ppm under a solar simulator (AM 1.5 G 100 mW/cm^2). Degradation efficiencies of 89% of MO after 180 min and 61% of MB after 240 min were obtained. They argue that the difference in degradation efficiencies was due to differences in the chemical structure of the dyes. They affirm that catalytic reactions are more efficient in anionic than cationic dyes. Similarly, Barman et al. evaluated the photocatalytic degradation of MB and MO using Cu_3N-Au heterojunction as a photocatalyst (Figure 8c) [38]. The Au-Cu_3N heterostructures ACN1 (Au nanoparticle size is ~5 nm) and ACN2 (Au nanoparticle size is ~10 nm) exhibit enhanced photocatalytic activity in comparison with pure Cu_3N nanocubes, mostly due to the LSPR effect of Au nanoparticles that enhances charge carrier separation, Figure 8c. In addition, the ACN2 sample with an Au particle size ~10 nm presents much better photocatalytic behavior in MB under solar radiation. In this case, the enhancement of the photocatalytic activity requires coupling between the conduction band of Cu_3N and the Fermi level of plasmonic metals. Subsequently, charge carriers in semiconductors undergo plasmon-induced resonance energy transfer [38], which can improve the photocatalytic activity towards dye degradation.

Table 2. Summary of photocatalytic behavior of Cu_3N-based nanomaterials.

Semiconductor Composite	Band Gap (eV)	Dye Concentration	Condition	Degradation Time	Efficiency	Particle Size	Ref
Cu_3N@MoS_2 thin films	2.05	MO–10 mg/mL	500 W Hg–lamp	30 min	98.3%	-	[35]
Au-decorated Cu_3N nanocubes	1.67	MO/MB 0.0008 mM	250 W Universal arc lamp	25 min for MB and MO	84% for MO 93% for MB	NC: 10 ± 5 nm Au: 10 and 5 nm	[38]
Cu_3N nanocubes	2.41	MO/MB 20 ppm	Solar simulator source 100 mW/cm^2	180 min for MO 240 min for MB	89% for MO 61% for MB	NC: 41 ± 7 nm	[48]
Cu_3N thin films	2.0	MO–20 mg/L	500 W Hg–lamp	30 min	95.5%	-	[65]
TiO_2@Cu_3N thin films	3.0	MO–20 mg/L	500 W Hg–lamp	30 min	99.2%	-	[84]

Since photocatalysis is a surface phenomenon, the size of the nanocatalyst is a crucial parameter in evaluating the efficiency of its photocatalytic activity [114]. The decreasing size of the nanoparticle increases its specific surface, i.e., the number of active sites needed for the production of ROS on photoexcitation. Active sites should be able to chemisorb the dye molecule and simultaneously transfer the photogenerated charge carriers to them. These active sites could be defects such as Cu or N vacancies at the surface of the nanoparticle [112]. These surface defects are dominant in nanoparticles and augment with decreasing size of the nanoparticle. In addition, the morphology of the nanoparticles also plays an important role. Different morphologies such as rod, flower and cube tend to have different photocatalytic efficiencies. For example, cubic, octahedral and spherical platinum (Pt) nanoparticles present different photocatalytic activities under visible light [115]. Even though Cu_3N nanoparticles exhibit high photocatalytic efficiencies of 80%, combining them with plasmonic nanoparticles of Au further increases these efficiencies to 90%. Compared to nanoparticles, thin films of Cu_3N demonstrate extremely high efficiencies of 95%, most likely due to better control of the stoichiometry during thin film growth. Furthermore, the formation of secondary phases is easier to suppress during thin film growth by varying the growth parameters. In contrast, chemical routes for nanoparticle synthesis inadvertently generate secondary phases of Cu and Cu_2O that influence the overall photocatalytic activity of Cu_3N. Moreover, nanocomposites of Cu_3N consisting of TiO_2 and MoS_2 show a higher photocatalytic dye degradation efficiency [35,84]. In such heterostructures, the Fermi level alignment reduces the overall bandgap of the material, making the nanocomposite more susceptible to visible light absorption and thereby increasing the absorption cross-section.

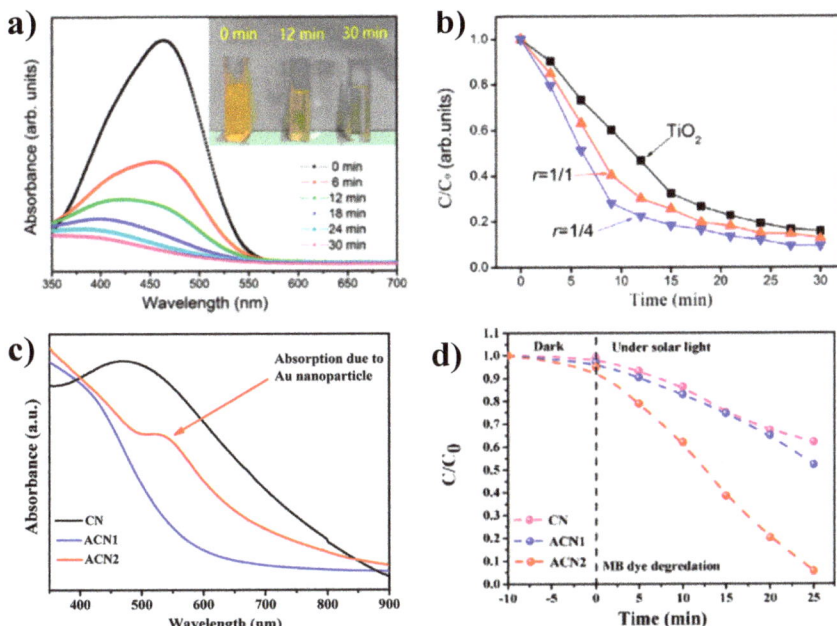

Figure 8. (**a**) Methyl orange degradation by UV-Vis analysis using Cu_3N thin films. Reprinted with permission from Ref [65], Creative Commons agreement from MDPI, 2020. (**b**) Degradation of methyl orange by TiO_2 films and TiO_2@Cu_3N composite films prepared by different gas flow radio (r = [N_2]/[Ar + N_2]), at r = 1/4 the degradation rate of MO is 99.2%. Reprinted with permission from Ref [84], Creative Commons agreement from Elsevier publications, 2021. (**c**) UV-Vis spectra of Cu_3N nanocubes and Au-decorated Cu_3N nanocubes (ACN1 Au nanoparticles is ~5 nm and ACN2 Au nanoparticles is ~10 nm) and (**d**) Solar light driven photo degradation of MB using Au-decorated Cu_3N nanocubes. Reprinted with permission from Ref [38], ACS publications, 2019.

Cu_3N-based hybrid nanomaterials are considered stable and can be reused in environmental remediation applications. Other than photocatalytic dye degradation in wastewaters, other environmental remediation applications of Cu_3N are also available. Li et al. synthesized 3D Hierarchical Ni_4N/Cu_3N nanotube arrays for the electrolysis of urea, which is a common effluent in wastewater [116]. Furthermore, Lee et al. demonstrate that the catalytic activity of copper-based nanomaterials shows high stability and reusability. They examined the reuse of the Cu_3N/Fe_3N@SiO_2 magnetic catalyst. The recycled catalysts were separated by a magnet and presented high stability and identical yield to those of the virgin catalyst after five cycles [92]. Yin et al. also support the stability and reusability of Cu_3N nanoparticles. They report a new perovskite-type copper (I) nitride (Cu_3N) nanocubes catalyst for selective carbon dioxide reduction reaction (CO_2RR), and they demonstrate the reproducibility of their results after several cycles [113].

5. Other Applications of Cu_3N

Not only does Cu_3N exhibit high photocatalytic activity, but it also has appreciable catalytic activity for a range of important reactions. For example, Cu_3N is considered a novel catalyst in the fields of sustainability, energy and the environment. Studies of the electrocatalytic activity of oxygen and hydrogen evolution reactions (OER and HER, respectively) and in the oxygen reduction reaction (ORR) have already shown that Cu_3N holds considerable promise for energy production systems [47,50,113]. Furthermore, recent reports also indicate that Cu_3N can be very effectively used as a catalyst in the CO_2RR. Yin and co-workers demonstrated that Cu_3N nanocubes were stable and selective to ethylene production at −1.6 V, with a Faradaic efficiency of 60% [113]. The electrocatalytic activity of Cu_3N is, therefore, also suitable for electrodes of Li- and Na-batteries, owing to lithium-copper nitride reversible processes for a large number of cycles at elevated temperatures [55,117,118]. The material displays high optical absorption and has therefore been applied to photovoltaics [119]. In the electronics industry, Cu_3N has successfully shown optical data storage capabilities when coupled with an Al_2O_3 protective layer [74]. The capacity to transform Cu_3N under laser pulses of different wavelengths to Cu is the basis of optical data storage, measured in terms of changes in the reflectivity of the nanomaterial [80]. However, the thermal stability of Cu_3N under laser irradiation is not suitable for spintronic applications. Therefore, FeN was combined with Cu_3N, and the nanocomposite subsequently demonstrated potential for spintronic applications [120]. Nevertheless, the ability of Cu_3N to decompose is an asset for other applications, such as Si-based large-scale integrated circuits. These Cu-metal lines on Si wafers offer higher signal speed than traditionally used Al-metal lines [121]. Furthermore, patterning substrates by exploiting the phase separation of Cu from Cu_3N allows applying the plasmonic effect of the segregated Cu nanoparticles to Surface-Enhanced Raman Scattering (SERS) applications [122,123]. In fact, SERS is already being applied in several biomedical fields, such as bio-imaging, cancer diagnostics and plasmonic sensors [124,125]. For antimicrobial applications, Cu is considered a potent nanomaterial. However, in high doses, Cu presents cytotoxicity also. Therefore, careful control of Cu quantities is required. Furthermore, the antimicrobial properties of NiTi were enhanced when Cu_3N-Cu catalyst was deposited on its surface, owing to a micro-galvanic effect [126]. Antibacterial activity against two strains of bacteria, i.e., *E. coli* and *S. aureus* were evaluated. In fact, the HER of the Cu_3N-Cu catalyst induces oxidative stress on the bacteria through the production of ROS.

6. Summary and Outlook

The global market for nanomaterials has grown considerably in the last few years. Recent reports suggest an increase in the production of nanomaterials with a compound annual growth rate (CAGR) of 14.1% between 2021 and 2028. Among the various types of nanomaterials, nanoparticles are likely to account for the largest market share. Recent promising results on their applications in the health, energy, electronics and water treatment sectors have been a major driving force towards their commercialization. However, the synthesis of nanoparticles with specific properties for different applications is one of the greatest challenges for researchers, especially in water treatment applications, including photocatalysis. There are several photocatalytic nanomaterials in the market, with TiO_2 and ZnO being the most effective. However, these nanomaterials suffer from certain drawbacks, such as inefficient absorption of visible light, large band gaps, the low adsorption capacity of hydrophobic contaminants, non-uniform distribution in aqueous suspension and reclamation methods after treatment. In addition, recent reports mention that there are regulatory restrictions in Europe on the use of TiO_2 due to concerns about its genotoxicity [127]. Therefore, novel and industrially viable photocatalysts are necessary in order to surpass these limitations.

Environmental pollution, including water contamination, is a global concern that deteriorates ecosystems and negatively affects the climate. Today, climate change is the focus of the hour. Wastewater effluents generated daily by chemical, electroplating and metallurgical industries contain various organic dyes, e.g., methyl orange, rhodamine B and methylene blue. These pollutants cause damage to water resources, in addition to causing harm to humans and aquatic life owing to their toxicity and carcinogenicity. In the literature, there are strategies to deal with dye-contaminated waters involving physical, chemical and biological methods [128]. However, their disadvantages, such as low efficiency and incomplete contaminant removal, need to be further addressed. In this regard, photodegradation using nanoparticles has gained enormous popularity in recent years. Photobleaching of dyes is considered a favorable technology for industrial wastewater treatment due to its environmental friendliness, low cost and absence of secondary pollution. Therefore, photocatalysis research has become a niche topic in water treatment.

In the present scenario, metal nitrides appear as a versatile class of materials of growing interest. Cu_3N specifically has attracted attention as an inexpensive, non-toxic material with potential applications in solar cells, high-density optical data storage, lithium-ion batteries, photocatalysis, electrocatalysis and photoelectrocatalysis. Chemical synthesis routes lead to the formation of nanocrystals with characteristics that can be controlled by varying the synthesis parameters. In fact, the size and shape of the nanoparticles are sensitive to the reaction time, temperature and chemical reagents. Although there are a good number of reports on the synthesis of pure phase Cu_3N, there is still progress to be made, as the routes are often multistep, may involve the use of hazardous gases such as ammonia and provoke the co-precipitation of different phases, such as Cu_2O and Cu during the synthesis process. Cu_2O is already a well-established photocatalyst, and Cu has shown a synergistic effect on the latter. Therefore, Cu_3N could be an excellent co-catalyst as it widens the visible absorption range of Cu_2O. The large discrepancy in the bandgap of Cu_3N depends on the synthesis conditions, including doping, which can be considered an advantage when tuning the bandgap of the material, especially when used as a co-catalyst with Cu_2O. Similarly, thin films based on Cu_3N have been extensively studied. Sputtering deposition of Cu_3N thin films being industrially viable, can be used to coat large surface areas. Substrate dimensions on industrial sputtering lines can be as large as $1\text{ m} \times 1.5\text{ m}$. Since photocatalysis is a surface phenomenon, a large surface area coverage of Cu_3N as obtained by such deposition techniques is necessary to ensure efficient photoabsorption and enhanced surface reactions. Therefore, Cu_3N nanostructures could be good candidates as catalysts or co-catalysts in the photodegradation of polluted waters.

Cu_3N nanomaterials present enormous potential for applications in several fields. However, the low stability of Cu_3N remains a shortcoming. The synthesis of pure Cu_3N without the presence of secondary phases will be the target for the coming decade. The field of application of Cu_3N nanomaterials and nanocomposites is very broad and includes energy and environmental applications, such as catalysis, photocatalysis and electrocatalysis (OER, HER and ORR), as mentioned in this review. Furthermore, the electrocatalytic activity of Cu_3N makes it a suitable electrode material for Li- and Na-ion batteries. Recent investigations have shown that Cu_3N can also be integrated into microelectronic applications, i.e., for optical data storage when coupled with an Al_2O_3 protective layer and in spintronics when combined with FeN. In the thin film form, Cu_3N presents promising attributes for the field of photovoltaics owing to its versatile band gap modified via the growth conditions in order to obtain higher optical absorption. Therefore, thin-film and nanoparticle growth of Cu_3N offer promising routes for the next generation of applications. Additionally, given its versatility, combining Cu_3N with other materials could augment its electronic, optical and photocatalytic properties. Extensively studied materials, such as carbon nanomaterials (graphene or CNTs), could enhance the adsorption of dyes and, in turn, photocatalytic properties of Cu_3N. Carbon-based nanomaterials themselves serve as efficient dye adsorbents and, at the same time, enhance the charge conductivity of the inorganic material, thereby facilitating charge separation and redox reactions. Furthermore, they also tend to protect the inorganic material against oxidation, whereupon providing them structural and chemical stability. Therefore, by combining the above strategies, highly efficient, stable, low-cost and advanced semiconductors for large-scale photocatalysis could be fabricated. Developing a stable, high efficiency and cost-effective photocatalytic system will be the central theme of future research. Cu_3N-based nanostructure research is clearly challenging and still young, but adding these nitrides to the portfolio of photocatalytic materials with excellent optical properties would open up many exciting possibilities for environmental remediation.

Author Contributions: P.P. is the main contributor. He conceptualized, compiled and wrote the manuscript. P.R. provided feedback, contributed to writing and revised the manuscript. E.R. provided feedback, participated in the writing and provided his expertise. All authors have read and agreed to the published version of the manuscript.

Funding: This research was funded by the Center of Excellence TK134 project, Archimedes foundation

Informed Consent Statement: Not Applicable.

Conflicts of Interest: The authors declare no conflict of interest.

References

1. Biswas, P.; Wu, C.Y. Nanoparticles and the environment. *J. Air Waste Manag. Assoc.* **2005**, *55*, 708–746. [CrossRef] [PubMed]
2. Jamkhande, P.G.; Ghule, N.W.; Bamer, A.H.; Kalaskar, M.G. Metal nanoparticles synthesis: An overview on methods of preparation, advantages and disadvantages, and applications. *J. Drug Deliv. Sci. Technol.* **2019**, *53*, 101174. [CrossRef]
3. Panigrahi, S.; Kundu, S.; Ghosh, S.; Nath, S.; Pal, T. General method of synthesis for metal nanoparticles. *J. Nanopart. Res.* **2004**, *6*, 411–414. [CrossRef]
4. Chen, P.; Xu, K.; Tong, Y.; Li, X.; Tao, S.; Fang, Z.; Chu, W.; Wu, X.; Wu, C. Cobalt nitrides as a class of metallic electrocatalysts for the oxygen evolution reaction. *Inorg. Chem. Front.* **2016**, *3*, 236–242. [CrossRef]
5. Braic, M.; Balaceanu, M.; Braic, V.; Vladescu, A.; Pavelescu, G.; Albulescu, M. Synthesis and characterization of TiN, TiAlN and TiN/TiAlN biocompatible coatings. *Surf. Coat. Technol.* **2005**, *200*, 1014–1017. [CrossRef]
6. Yu, R.; Zhan, Q.; De Jonghe, L.C. Crystal Structures of and Displacive Transitions in OsN_2, IrN_2, RuN_2, and RhN_2. *Angew. Chem. Int. Ed.* **2007**, *46*, 1136–1140. [CrossRef]
7. Prabha, S.; Durgalakshmi, D.; Subramani, K.; Aruna, P.; Ganesan, S. Enhanced Emission of Zinc Nitride Colloidal Nanoparticles with Organic Dyes for Optical Sensors and Imaging Application. *ACS Appl. Mater. Interfaces* **2020**, *12*, 19245–19257. [CrossRef]
8. Young, A.F.; Sanloup, C.; Gregoryanz, E.; Scandolo, S.; Hemley, R.J.; Mao, H.K. Synthesis of novel transition metal nitrides IrN_2 and OsN_2. *Phys. Rev. Lett.* **2006**, *96*, 155501. [CrossRef]
9. Wang, H.; Li, J.; Li, K.; Lin, Y.; Chen, J.; Gao, L.; Nicolosi, V.; Xiao, X.; Lee, J.-M. Transition metal nitrides for electrochemical energy applications. *Chem. Soc. Rev.* **2021**, *50*, 1354–1390. [CrossRef]
10. Zhu, C.; Yang, P.; Chao, D.; Wang, X.; Zhang, X.; Chen, S.; Tay, B.K.; Huang, H.; Zhang, H.; Mai, W.; et al. All Metal Nitrides Solid-State Asymmetric Supercapacitors. *Adv. Mater.* **2015**, *27*, 4566–4571. [CrossRef]
11. Xie, J.; Xie, Y. Transition Metal Nitrides for Electrocatalytic Energy Conversion: Opportunities and Challenges. *Chem. A Eur. J.* **2016**, *22*, 3588–3598. [CrossRef] [PubMed]
12. Mosavati, N.; Salley, S.O.; Ng, K.Y.S. Characterization and electrochemical activities of nanostructured transition metal nitrides as cathode materials for lithium sulfur batteries. *J. Power Sources* **2017**, *340*, 210–216. [CrossRef]
13. Hargreaves, J.S.J. Heterogeneous catalysis with metal nitrides. *Coord. Chem. Rev.* **2013**, *257*, 2015–2031. [CrossRef]
14. Dong, S.; Chen, X.; Zhang, X.; Cui, G. Nanostructured transition metal nitrides for energy storage and fuel cells. *Coord. Chem. Rev.* **2013**, *257*, 1946–1956. [CrossRef]
15. Zakutayev, A. Design of nitride semiconductors for solar energy conversion. *J. Mater. Chem. A* **2016**, *4*, 6742–6754. [CrossRef]
16. Dylewicz, R.; Patela, S.; Paszkiewicz, R. *Applications of GaN-Based Materials in Modern Optoelectronics*; SPIE: Bellingham, DC, USA, 2004; Volume 5484. [CrossRef]
17. Bochkareva, N.I.; Gorbunov, R.I.; Klochkov, A.V.; Lelikov, Y.S.; Martynov, I.A.; Rebane, Y.T.; Belov, A.S.; Shreter, Y.G. Optical properties of blue light-emitting diodes in the InGaN/GaN system at high current densities. *Semiconductors* **2008**, *42*, 1355–1361. [CrossRef]
18. Pankove, J.I. GaN: From fundamentals to applications. *Mater. Sci. Eng. B* **1999**, *61–62*, 305–309. [CrossRef]
19. Lian, J.; Li, D.; Qi, Y.; Yang, N.; Zhang, R.; Xie, T.; Guan, N.; Li, L.; Zhang, F. Metal-seed assistant photodeposition of platinum over Ta_3N_5 photocatalyst for promoted solar hydrogen production under visible light. *J. Energy Chem.* **2021**, *55*, 444–448. [CrossRef]
20. Menon, S.S.; Hafeez, H.Y.; Gupta, B.; Baskar, K.; Bhalerao, G.; Hussain, S.; Neppolian, B.; Singh, S. ZnO:InN oxynitride: A novel and unconventional photocatalyst for efficient UV–visible light driven hydrogen evolution from water. *Renew. Energy* **2019**, *141*, 760–769. [CrossRef]
21. Singh, P.; Ruterana, P.; Morales, M.; Goubilleau, F.; Wojdak, M.; Carlin, J.F.; Ilegems, M.; Chateigner, D. Structural and optical characterisation of InN layers grown by MOCVD. *Superlattices Microstruct.* **2004**, *36*, 537–545. [CrossRef]
22. Wang, L.; Zhao, W.; Hao, Z.-B.; Luo, Y. Photocatalysis of InGaN Nanodots Responsive to Visible Light. *Chin. Phys. Lett.* **2011**, *28*, 057301. [CrossRef]
23. Sánchez, A.M.; Gass, M.; Papworth, A.J.; Goodhew, P.J.; Singh, P.; Ruterana, P.; Cho, H.K.; Choi, R.J.; Lee, H.J. V-defects and dislocations in InGaN/GaN heterostructures. *Thin Solid Film.* **2005**, *479*, 316–320. [CrossRef]

4. Sithole, R.K.; Machogo, L.F.E.; Airo, M.A.; Gqoba, S.S.; Moloto, M.J.; Shumbula, P.; Van Wyk, J.; Moloto, N. Synthesis and characterization of Cu_3N nanoparticles using pyrrole-2-carbaldpropyliminato Cu(ii) complex and $Cu(NO_3)_2$ as single-source precursors: The search for an ideal precursor. *New J. Chem.* **2018**, *42*, 3042–3049. [CrossRef]
5. Xiao, J.; Qi, M.; Cheng, Y.; Jiang, A.; Zeng, Y.; Ma, J. Influences of nitrogen partial pressure on the optical properties of copper nitride films. *RSC Adv.* **2016**, *6*, 40895–40899. [CrossRef]
6. Mondal, S.; Raj, C.R. Copper Nitride Nanostructure for the Electrocatalytic Reduction of Oxygen: Kinetics and Reaction Pathway. *J. Phys. Chem. C* **2018**, *122*, 18468–18475. [CrossRef]
7. Nakamura, T.; Cheong, H.J.; Takamura, M.; Yoshida, M.; Uemura, S. Suitability of Copper Nitride as a Wiring Ink Sintered by Low-Energy Intense Pulsed Light Irradiation. *Nanomaterials* **2018**, *8*, 617. [CrossRef]
8. Scigala, A.; Szłyk, E.; Rerek, T.; Wiśniewski, M.; Skowronski, L.; Trzcinski, M.; Szczesny, R. Copper Nitride Nanowire Arrays—Comparison of Synthetic Approaches. *Materials* **2021**, *14*, 603. [CrossRef]
9. Lindahl, E.; Ottosson, M.; Carlsson, J.-O. Doping of metastable Cu_3N at different Ni concentrations: Growth, crystallographic sites and resistivity. *Thin Solid Film.* **2018**, *647*, 1–8. [CrossRef]
10. Jiang, A.; Xiao, J.; Gong, C.; Wang, Z.; Ma, S. Structure and electrical transport properties of Pb-doped copper nitride (Cu_3N:Pb) films. *Vacuum* **2019**, *164*, 53–57. [CrossRef]
11. Joo Kim, K.; Hyuk Kim, J.; Hoon Kang, J. Structural and optical characterization of Cu_3N films prepared by reactive RF magnetron sputtering. *J. Cryst. Growth* **2001**, *222*, 767–772. [CrossRef]
12. Zhu, W.; Zhang, X.; Fu, X.; Zhou, Y.; Luo, S.; Wu, X. Resistive-switching behavior and mechanism in copper-nitride thin films prepared by DC magnetron sputtering. *Phys. Status Solidi* **2012**, *209*, 1996–2001. [CrossRef]
13. Kadzutu-Sithole, R.; Machogo-Phao, L.F.E.; Kolokoto, T.; Zimuwandeyi, M.; Gqoba, S.S.; Mubiayi, K.P.; Moloto, M.J.; Van Wyk, J.; Moloto, N. Elucidating the effect of precursor decomposition time on the structural and optical properties of copper(i) nitride nanocubes. *RSC Adv.* **2020**, *10*, 34231–34246. [CrossRef] [PubMed]
14. Wang, L.-C.; Liu, B.-H.; Su, C.-Y.; Liu, W.-S.; Kei, C.-C.; Wang, K.-W.; Perng, T.-P. Electronic Band Structure and Electrocatalytic Performance of Cu_3N Nanocrystals. *ACS Appl. Nano Mater.* **2018**, *1*, 3673–3681. [CrossRef]
15. Zhu, L.; Gong, C.; Xiao, J.; Wang, Z. Photocatalytic Properties of Copper Nitride/Molybdenum Disulfide Composite Films Prepared by Magnetron Sputtering. *Coatings* **2020**, *10*, 79. [CrossRef]
16. Rahmati, A. Ti-containing Cu_3N nanostructure thin films: Experiment and simulation on reactive magnetron sputter-assisted nitridation. *IEEE Trans. Plasma Sci.* **2015**, *43*, 1969–1973. [CrossRef]
17. Zhao, Y.; Zhang, Q.; Huang, S.; Zhang, J.; Ren, S.; Wang, H.; Wang, L.; Yang, T.; Yang, J.; Li, X. Effect of Magnetic Transition Metal (TM = V, Cr, and Mn) Dopant on Characteristics of Copper Nitride. *J. Supercond. Nov. Magn.* **2016**, *29*, 2351–2357. [CrossRef]
18. Barman, D.; Paul, S.; Ghosh, S.; De, S.K. Cu_3N Nanocrystals Decorated with Au Nanoparticles for Photocatalytic Degradation of Organic Dyes. *ACS Appl. Nano Mater.* **2019**, *2*, 5009–5019. [CrossRef]
19. Cheng, Z.; Qi, W.; Pang, C.H.; Thomas, T.; Wu, T.; Liu, S.; Yang, M. Recent Advances in Transition Metal Nitride-Based Materials for Photocatalytic Applications. *Adv. Funct. Mater.* **2021**, *31*, 2100553. [CrossRef]
20. Birkett, M.; Savory, C.N.; Fioretti, A.N.; Thompson, P.; Muryn, C.A.; Weerakkody, A.D.; Mitrovic, I.Z.; Hall, S.; Treharne, R.; Dhanak, V.R.; et al. Atypically small temperature-dependence of the direct band gap in the metastable semiconductor copper nitride Cu_3N. *Phys. Rev. B* **2017**, *95*, 115201. [CrossRef]
21. Zhu, S.; Wang, D. Photocatalysis: Basic Principles, Diverse Forms of Implementations and Emerging Scientific Opportunities. *Adv. Energy Mater.* **2017**, *7*, 1700841. [CrossRef]
22. Chen, H.; Li, X.A.; Zhao, J.; Wu, Z.; Yang, T.; Ma, Y.; Huang, W.; Yao, K. First principles study on the influence of electronic configuration of M on Cu_3NM: M = Sc, Ti, V, Cr, Mn, Fe, Co, Ni. *Comput. Theor. Chem.* **2014**, *1027*, 33–38. [CrossRef]
23. Ashraf, I.; Rizwan, S.; Iqbal, M. A Comprehensive Review on the Synthesis and Energy Applications of Nano-structured Metal Nitrides. *Front. Mater.* **2020**, *7*, 181. [CrossRef]
24. Rao, T.; Cai, W.; Zhang, H.; Liao, W. Nanostructured metal nitrides for photocatalysts. *J. Mater. Chem. C* **2021**, *9*, 5323–5342. [CrossRef]
25. Mazumder, B.; Hector, A.L. Synthesis and applications of nanocrystalline nitride materials. *J. Mater. Chem.* **2009**, *19*, 4673–4686. [CrossRef]
26. Panda, C.; Menezes, P.W.; Zheng, M.; Orthmann, S.; Driess, M. In Situ Formation of Nanostructured Core–Shell Cu_3N–CuO to Promote Alkaline Water Electrolysis. *ACS Energy Lett.* **2019**, *4*, 747–754. [CrossRef]
27. Wu, H.; Chen, W. Copper Nitride Nanocubes: Size-Controlled Synthesis and Application as Cathode Catalyst in Alkaline Fuel Cells. *J. Am. Chem. Soc.* **2011**, *133*, 15236–15239. [CrossRef]
28. Sithole, R.K.; Machogo, L.F.E.; Moloto, M.J.; Gqoba, S.S.; Mubiayi, K.P.; Van Wyk, J.; Moloto, N. One-step synthesis of Cu_3N, Cu_2S and Cu_9S_5 and photocatalytic degradation of methyl orange and methylene blue. *J. Photochem. Photobiol. A Chem.* **2020**, *397*, 112577. [CrossRef]
29. Wang, D.; Li, Y. Controllable synthesis of Cu-based nanocrystals in ODA solvent. *Chem. Commun.* **2011**, *47*, 3604–3606. [CrossRef]
30. Xi, P.; Xu, Z.; Gao, D.; Chen, F.; Xue, D.; Tao, C.-L.; Chen, Z.-N. Solvothermal synthesis of magnetic copper nitride nanocubes with highly electrocatalytic reduction properties. *RSC Adv.* **2014**, *4*, 14206–14209. [CrossRef]
31. Egeberg, A.; Warmuth, L.; Riegsinger, S.; Gerthsen, D.; Feldmann, C. Pyridine-based low-temperature synthesis of CoN, Ni_3N and Cu_3N nanoparticles. *Chem. Commun.* **2018**, *54*, 9957–9960. [CrossRef]

52. Sithole, R.K.; Kolokoto, T.; Machogo, L.F.E.; Ngubeni, G.N.; Moloto, M.J.; Van Wyk, J.; Moloto, N. Simultaneous capping and substitution of nitrogen ions of Cu$_3$N nanocrystals with sulfur ions using DDT as a co-surfactant to form chalcocite and digenite nanocrystals. *Mater. Chem. Phys.* **2020**, *251*, 123074. [CrossRef]
53. Nakamura, T.; Hayashi, H.; Hanaoka, T.-A.; Ebina, T. Preparation of Copper Nitride (Cu$_3$N) Nanoparticles in Long-Chain Alcohols at 130–200 °C and Nitridation Mechanism. *Inorg. Chem.* **2014**, *53*, 710–715. [CrossRef] [PubMed]
54. Deshmukh, R.; Zeng, G.; Tervoort, E.; Staniuk, M.; Wood, D.W.; Niederberger, M. Ultrasmall Cu$_3$N Nanoparticles: Surfactant-Free Solution-Phase Synthesis, Nitridation Mechanism, and Application for Lithium Storage. *Chem. Mater.* **2015**, *27*, 8282–8288 [CrossRef]
55. Li, X.; Hector, A.L.; Owen, J.R. Evaluation of Cu$_3$N and CuO as Negative Electrode Materials for Sodium Batteries. *J. Phys. Chem. C* **2014**, *118*, 29568–29573. [CrossRef]
56. Nakamura, T.; Hiyoshi, N.; Hayashi, H.; Ebina, T. Preparation of plate-like copper nitride nanoparticles from a fatty acid copper(II) salt and detailed observations by high resolution transmission electron microscopy and high-angle annular dark-field scanning transmission electron microscopy. *Mater. Lett.* **2015**, *139*, 271–274. [CrossRef]
57. Miura, A.; Takei, T.; Kumada, N. Synthesis of Cu$_3$N from CuO and NaNH$_2$. *J. Asian Ceram. Soc.* **2014**, *2*, 326–328. [CrossRef]
58. Szczęsny, R.; Szłyk, E.; Wiśniewski, M.A.; Hoang, T.K.A.; Gregory, D.H. Facile preparation of copper nitride powders and nanostructured films. *J. Mater. Chem. C* **2016**, *4*, 5031–5037. [CrossRef]
59. Choi, J.; Gillan, E.G. Solvothermal Synthesis of Nanocrystalline Copper Nitride from an Energetically Unstable Copper Azide Precursor. *Inorg. Chem.* **2005**, *44*, 7385–7393. [CrossRef]
60. Paniconi, G.; Stoeva, Z.; Doberstein, H.; Smith, R.I.; Gallagher, B.L.; Gregory, D.H. Structural chemistry of Cu$_3$N powders obtained by ammonolysis reactions. *Solid State Sci.* **2007**, *9*, 907–913. [CrossRef]
61. Vaughn Ii, D.D.; Araujo, J.; Meduri, P.; Callejas, J.F.; Hickner, M.A.; Schaak, R.E. Solution Synthesis of Cu$_3$PdN Nanocrystals as Ternary Metal Nitride Electrocatalysts for the Oxygen Reduction Reaction. *Chem. Mater.* **2014**, *26*, 6226–6232. [CrossRef]
62. Reichert, M.D.; White, M.A.; Thompson, M.J.; Miller, G.J.; Vela, J. Preparation and Instability of Nanocrystalline Cuprous Nitride. *Inorg. Chem.* **2015**, *54*, 6356–6362. [CrossRef] [PubMed]
63. Primc, D.; Indrizzi, L.; Tervoort, E.; Xie, F.; Niederberger, M. Synthesis of Cu$_3$N and Cu$_3$N–Cu$_2$O multicomponent mesocrystals: Non-classical crystallization and nanoscale Kirkendall effect. *Nanoscale* **2021**, *13*, 17521–17529. [CrossRef] [PubMed]
64. Mukhopadhyay, A.K.; Momin, M.A.; Roy, A.; Das, S.C.; Majumdar, A. Optical and Electronic Structural Properties of Cu$_3$N Thin Films: A First-Principles Study (LDA + U). *ACS Omega* **2020**, *5*, 31918–31924. [CrossRef]
65. Jiang, A.; Shao, H.; Zhu, L.; Ma, S.; Xiao, J. Preparation of Copper Nitride Films with Superior Photocatalytic Activity through Magnetron Sputtering. *Materials* **2020**, *13*, 4325. [CrossRef] [PubMed]
66. Sahoo, G.; Meher, S.R.; Jain, M.K. Room temperature growth of high crystalline quality Cu$_3$N thin films by modified activated reactive evaporation. *Mater. Sci. Eng. B* **2015**, *191*, 7–14. [CrossRef]
67. Ali, H.T.; Tanveer, Z.; Javed, M.R.; Mahmood, K.; Amin, N.; Ikram, S.; Ali, A.; Shah Gilani, M.R.H.; Sajjad, M.A.; Yusuf, M. A new approach for the growth of copper nitrides thin films by thermal evaporation using nitrogen as source gas. *Optik* **2021**, *245*, 167666. [CrossRef]
68. Caskey, C.M.; Richards, R.M.; Ginley, D.S.; Zakutayev, A. Thin film synthesis and properties of copper nitride, a metastable semiconductor. *Mater. Horiz.* **2014**, *1*, 424–430. [CrossRef]
69. Xiao, J.; Qi, M.; Gong, C.; Wang, Z.; Jiang, A.; Ma, J.; Cheng, Y. Crystal structure and optical properties of silver-doped copper nitride films (Cu$_3$N:Ag) prepared by magnetron sputtering. *J. Phys. D Appl. Phys.* **2018**, *51*, 055305. [CrossRef]
70. Su, C.Y.; Liu, B.H.; Lin, T.J.; Chi, Y.M.; Kei, C.C.; Wang, K.W.; Perng, T.P. Carbon nanotube-supported Cu$_3$N nanocrystals as a highly active catalyst for oxygen reduction reaction. *J. Mater. Chem. A* **2015**, *3*, 18983–18990. [CrossRef]
71. Deshmukh, R.; Schubert, U. Synthesis of CuO and Cu$_3$N Nanoparticles in and on Hollow Silica Spheres. *Eur. J. Inorg. Chem.* **2013**, *2013*, 2498–2504. [CrossRef]
72. Dongil, A.B. Recent Progress on Transition Metal Nitrides Nanoparticles as Heterogeneous Catalysts. *Nanomaterials* **2019**, *9*, 1111. [CrossRef] [PubMed]
73. Alexander, A.M.; Hargreaves, J.S.J.; Mitchell, C. The Reduction of Various Nitrides under Hydrogen: Ni$_3$N, Cu$_3$N, Zn$_3$N$_2$ and Ta$_3$N$_5$. *Top. Catal.* **2012**, *55*, 1046–1053. [CrossRef]
74. Szczęsny, R.; Hoang, T.K.A.; Dobrzańska, L.; Gregory, D.H. Solution/Ammonolysis Syntheses of Unsupported and Silica-Supported Copper(I) Nitride Nanostructures from Oxidic Precursors. *Molecules* **2021**, *26*, 4926. [CrossRef] [PubMed]
75. Rasaki, S.A.; Zhang, B.; Anbalgam, K.; Thomas, T.; Yang, M. Synthesis and application of nano-structured metal nitrides and carbides: A review. *Prog. Solid State Chem.* **2018**, *50*, 1–15. [CrossRef]
76. Tareen, A.K.; Priyanga, G.S.; Behara, S.; Thomas, T.; Yang, M. Mixed ternary transition metal nitrides: A comprehensive review of synthesis, electronic structure, and properties of engineering relevance. *Prog. Solid State Chem.* **2019**, *53*, 1–26. [CrossRef]
77. Lord, R.W.; Holder, C.F.; Fenton, J.L.; Schaak, R.E. Seeded Growth of Metal Nitrides on Noble-Metal Nanoparticles To Form Complex Nanoscale Heterostructures. *Chem. Mater.* **2019**, *31*, 4605–4613. [CrossRef]
78. Polte, J. Fundamental growth principles of colloidal metal nanoparticles—A new perspective. *CrystEngComm* **2015**, *17*, 6809–6830. [CrossRef]
79. Kieda, N.; Messing, G.L. Microfoamy particles of copper oxide and nitride by spray pyrolysis of copper–ammine complex solutions. *J. Mater. Sci. Lett.* **1998**, *17*, 299–301. [CrossRef]

1. Cremer, R.; Witthaut, M.; Neuschütz, D.; Trappe, C.; Laurenzis, M.; Winkler, O.; Kurz, H. Deposition and Characterization of Metastable Cu$_3$N Layers for Applications in Optical Data Storage. *Microchim. Acta* **2000**, *133*, 299–302. [CrossRef]
2. Kaur, N.; Choudhary, N.; Goyal, R.N.; Viladkar, S.; Matai, I.; Gopinath, P.; Chockalingam, S.; Kaur, D. Magnetron sputtered Cu$_3$N/NiTiCu shape memory thin film heterostructures for MEMS applications. *J. Nanopart. Res.* **2013**, *15*, 1468. [CrossRef]
3. Yuan, N.Y.; Wang, S.Y.; Ding, J.N.; Qiu, J.H.; Wang, X.Q.; Huang, W.H. The Componental and Morphological Characteristics of Cu$_3$N Induced by Femtosecond Laser Pulses. *Key Eng. Mater.* **2014**, *609–610*, 135–140. [CrossRef]
4. Shi, J.; Jiang, B.; Li, C.; Yan, F.; Wang, D.; Yang, C.; Wan, J. Review of Transition Metal Nitrides and Transition Metal Nitrides/Carbon nanocomposites for supercapacitor electrodes. *Mater. Chem. Phys.* **2020**, *245*, 122533. [CrossRef]
5. Zhu, L.; Cao, X.; Xiao, J.; Ma, S.; Ta, S. Structure and photocatalytic properties of TiO$_2$/Cu$_3$N composite films prepared by magnetron sputtering. *Mater. Today Commun.* **2021**, *26*, 101739. [CrossRef]
6. Kuzmin, A.; Anspoks, A.; Kalinko, A.; Timoshenko, J.; Nataf, L.; Baudelet, F.; Irifune, T. Origin of Pressure-Induced Metallization in Cu$_3$N: An X-ray Absorption Spectroscopy Study. *Phys. Status Solidi* **2018**, *255*, 1800073. [CrossRef]
7. Jiang, A.; Qi, M.; Xiao, J. Preparation, structure, properties, and application of copper nitride (Cu$_3$N) thin films: A review. *J. Mater. Sci. Technol.* **2018**, *34*, 1467–1473. [CrossRef]
8. Yee, Y.S.; Inoue, H.; Hultqvist, A.; Hanifi, D.; Salleo, A.; Magyari-Köpe, B.; Nishi, Y.; Bent, S.F.; Clemens, B.M. Copper interstitial recombination centers in Cu$_3$N. *Phys. Rev. B* **2018**, *97*, 245201. [CrossRef]
9. Ścigała, A.; Szłyk, E.; Dobrzańska, L.; Gregory, D.H.; Szczęsny, R. From binary to multinary copper based nitrides-Unlocking the potential of new applications. *Coord. Chem. Rev.* **2021**, *436*, 213791. [CrossRef]
10. Hahn, U.; Weber, W. Electronic structure and chemical-bonding mechanism of Cu$_3$N, Cu$_3$NPd, and related Cu(I) compounds. *Phys. Rev. B Condens. Matter* **1996**, *53*, 12684–12693. [CrossRef]
11. Moreno-Armenta, M.G.; Martínez-Ruiz, A.; Takeuchi, N. Ab initio total energy calculations of copper nitride: The effect of lattice parameters and Cu content in the electronic properties. *Solid State Sci.* **2004**, *6*, 9–14. [CrossRef]
12. Wang, Y.; Zhang, H.; Zhu, J.; Lü, X.; Li, S.; Zou, R.; Zhao, Y. Antiperovskites with Exceptional Functionalities. *Adv. Mater.* **2020**, *32*, 1905007. [CrossRef]
13. Lee, B.S.; Yi, M.; Chu, S.Y.; Lee, J.Y.; Kwon, H.R.; Lee, K.R.; Kang, D.; Kim, W.S.; Lim, H.B.; Lee, J.; et al. Copper nitride nanoparticles supported on a superparamagnetic mesoporous microsphere for toxic-free click chemistry. *Chem. Commun.* **2010**, *46*, 3935–3937. [CrossRef] [PubMed]
14. Zhao, J.G.; You, S.J.; Yang, L.X.; Jin, C.Q. Structural phase transition of Cu$_3$N under high pressure. *Solid State Commun.* **2010**, *150*, 1521–1524. [CrossRef]
15. Tilemachou, A.; Zervos, M.; Othonos, A.; Pavloudis, T.; Kioseoglou, J. p-Type Iodine-Doping of Cu$_3$N and Its Conversion to γCuI for the Fabrication of γCuI/Cu$_3$N p-n Heterojunctions. *Electron. Mater.* **2022**, *3*, 2. [CrossRef]
16. Zakutayev, A.; Caskey, C.M.; Fioretti, A.N.; Ginley, D.S.; Vidal, J.; Stevanovic, V.; Tea, E.; Lany, S. Defect Tolerant Semiconductors for Solar Energy Conversion. *J. Phys. Chem. Lett.* **2014**, *5*, 1117–1125. [CrossRef]
17. Kong, F.; Hu, Y.; Wang, Y.; Wang, B.; Tang, L. Structural, elastic and thermodynamic properties of anti-ReO$_3$ type Cu$_3$N under pressure from first principles. *Comput. Mater. Sci.* **2012**, *65*, 247–253. [CrossRef]
18. Gordillo, N.; Gonzalez-Arrabal, R.; Diaz-Chao, P.; Ares, J.R.; Ferrer, I.J.; Yndurain, F.; Agulló-López, F. Electronic structure of copper nitrides as a function of nitrogen content. *Thin Solid Film.* **2013**, *531*, 588–591. [CrossRef]
19. Han, N.; Liu, P.; Jiang, J.; Ai, L.; Shao, Z.; Liu, S. Recent advances in nanostructured metal nitrides for water splitting. *J. Mater. Chem. A* **2018**, *6*, 19912–19933. [CrossRef]
20. Ningthoujam, R.S.; Gajbhiye, N.S. Synthesis, electron transport properties of transition metal nitrides and applications. *Prog. Mater. Sci.* **2015**, *70*, 50–154. [CrossRef]
21. Jiang, H.; Li, X.; Zang, S.; Zhang, W. Mixed cobalt-nitrides Co$_x$N and Ta$_2$N bifunction-modified Ta$_3$N$_5$ nanosheets for enhanced photocatalytic water-splitting into hydrogen. *J. Alloys Compd.* **2021**, *854*, 155328. [CrossRef]
22. Guo, Q.; Zhao, J.; Yang, Y.; Huang, J.; Tang, Y.; Zhang, X.; Li, Z.; Yu, X.; Shen, J.; Zhao, J. Mesocrystalline Ta$_3$N$_5$ superstructures with long-lived charges for improved visible light photocatalytic hydrogen production. *J. Colloid Interface Sci.* **2020**, *560*, 359–368. [CrossRef]
23. Sun, Z.; Chen, H.; Zhang, L.; Lu, D.; Du, P. Enhanced photocatalytic H$_2$ production on cadmium sulfide photocatalysts using nickel nitride as a novel cocatalyst. *J. Mater. Chem. A* **2016**, *4*, 13289–13295. [CrossRef]
24. Chen, L.; Huang, H.; Zheng, Y.; Sun, W.; Zhao, Y.; Francis, P.S.; Wang, X. Noble-metal-free Ni$_3$N/g-C$_3$N$_4$ photocatalysts with enhanced hydrogen production under visible light irradiation. *Dalton Trans.* **2018**, *47*, 12188–12196. [CrossRef] [PubMed]
25. Wang, Y.; Wu, Y.; Sun, K.; Mi, Z. A quadruple-band metal–nitride nanowire artificial photosynthesis system for high efficiency photocatalytic overall solar water splitting. *Mater. Horiz.* **2019**, *6*, 1454–1462. [CrossRef]
26. Di Valentin, C.; Diebold, U.; Selloni, A. Doping and functionalization of photoactive semiconducting metal oxides. *Chem. Phys.* **2007**, *339*, vii–viii. [CrossRef]
27. Xu, W.; Liu, H.; Zhou, D.; Chen, X.; Ding, N.; Song, H.; Ågren, H. Localized surface plasmon resonances in self-doped copper chalcogenide binary nanocrystals and their emerging applications. *Nano Today* **2020**, *33*, 100892. [CrossRef]
28. Zhang, L.; Ding, N.; Lou, L.; Iwasaki, K.; Wu, H.; Luo, Y.; Li, D.; Nakata, K.; Fujishima, A.; Meng, Q. Localized Surface Plasmon Resonance Enhanced Photocatalytic Hydrogen Evolution via Pt@Au NRs/C$_3$N$_4$ Nanotubes under Visible-Light Irradiation. *Adv. Funct. Mater.* **2019**, *29*, 1806774. [CrossRef]

108. Brongersma, M.L.; Halas, N.J.; Nordlander, P. Plasmon-induced hot carrier science and technology. *Nat. Nanotechnol.* **2015**, *10*, 25–34. [CrossRef]
109. Lalisse, A.; Tessier, G.; Plain, J.; Baffou, G. Plasmonic efficiencies of nanoparticles made of metal nitrides (TiN, ZrN) compared with gold. *Sci. Rep.* **2016**, *6*, 38647. [CrossRef]
110. Askes, S.H.C.; Schilder, N.J.; Zoethout, E.; Polman, A.; Garnett, E.C. Tunable plasmonic HfN nanoparticles and arrays. *Nanoscale* **2019**, *11*, 20252–20260. [CrossRef]
111. Beierle, A.; Gieri, P.; Pan, H.; Heagy, M.D.; Manjavacas, A.; Chowdhury, S. Titanium nitride nanoparticles for the efficient photocatalysis of bicarbonate into formate. *Sol. Energy Mater. Sol. Cells* **2019**, *200*, 109967. [CrossRef]
112. Sajeev, A.; Paul, A.M.; Nivetha, R.; Gothandapani, K.; Gopal, T.S.; Jacob, G.; Muthuramamoorty, M.; Pandiaraj, S.; Alodhayb, A.; Kim, S.Y.; et al. Development of Cu_3N electrocatalyst for hydrogen evolution reaction in alkaline medium. *Sci. Rep.* **2022**, *12*, 2004. [CrossRef]
113. Yin, Z.; Yu, C.; Zhao, Z.; Guo, X.; Shen, M.; Li, N.; Muzzio, M.; Li, J.; Liu, H.; Lin, H.; et al. Cu_3N Nanocubes for Selective Electrochemical Reduction of CO_2 to Ethylene. *Nano Lett.* **2019**, *19*, 8658–8663. [CrossRef] [PubMed]
114. Ni, D.; Shen, H.; Li, H.; Ma, Y.; Zhai, T. Synthesis of high efficient Cu/TiO_2 photocatalysts by grinding and their size-dependent photocatalytic hydrogen production. *Appl. Surf. Sci.* **2017**, *409*, 241–249. [CrossRef]
115. Cao, S.; Jiang, J.; Zhu, B.; Yu, J. Shape-dependent photocatalytic hydrogen evolution activity over a Pt nanoparticle coupled g-C3N4 photocatalyst. *Phys. Chem. Chem. Phys.* **2016**, *18*, 19457–19463. [CrossRef] [PubMed]
116. Li, J.; Yao, C.; Kong, X.; Li, Z.; Jiang, M.; Zhang, F.; Lei, X. Boosting Hydrogen Production by Electrooxidation of Urea over 3D Hierarchical Ni_4N/Cu_3N Nanotube Arrays. *ACS Sustain. Chem. Eng.* **2019**, *7*, 13278–13285. [CrossRef]
117. Pereira, N.; Dupont, L.; Tarascon, J.M.; Klein, L.C.; Amatucci, G.G. Electrochemistry of Cu_3N with Lithium. *J. Electrochem. Soc.* **2003**, *150*, A1273. [CrossRef]
118. Chen, W.; Zhang, H.; Yang, B.; Li, B.; Li, Z. Characterization of Cu_3N/CuO thin films derived from annealed Cu_3N for electrode application in Li-ion batteries. *Thin Solid Film.* **2019**, *672*, 157–164. [CrossRef]
119. Matsuzaki, K.; Okazaki, T.; Lee, Y.-S.; Hosono, H.; Susaki, T. Controlled bipolar doping in Cu_3N (100) thin films. *Appl. Phys. Lett.* **2014**, *105*, 222102. [CrossRef]
120. Yue, G.H.; Yan, P.X.; Liu, J.Z.; Wang, M.X.; Li, M.; Yuan, X.M. Copper nitride thin film prepared by reactive radio-frequency magnetron sputtering. *J. Appl. Phys.* **2005**, *98*, 103506. [CrossRef]
121. Li, X.; Liu, Z.; Zuo, A.; Yuan, Z.; Yang, J.; Yao, K. Properties of Al-doped copper nitride films prepared by reactive magnetron sputtering. *J. Wuhan Univ. Technol. Mater. Sci. Ed.* **2007**, *22*, 446–449. [CrossRef]
122. Du, Y.; Yin, Y.; Wang, J.; Wang, Z.; Li, C.; Baunack, S.; Ma, L.; Schmidt, O.G. Nanoporous Copper Pattern Fabricated by Electron Beam Irradiation on Cu_3N Film for SERS Application. *Phys. Status Solidi* **2019**, *256*, 1800378. [CrossRef]
123. Song, R.; Zhang, L.; Zhu, F.; Li, W.; Fu, Z.; Chen, B.; Chen, M.; Zeng, H.; Pan, D. Hierarchical Nanoporous Copper Fabricated by One-Step Dealloying Toward Ultrasensitive Surface-Enhanced Raman Sensing. *Adv. Mater. Interfaces* **2018**, *5*, 1800332. [CrossRef]
124. Jaque, D.; Richard, C.; Viana, B.; Soga, K.; Liu, X.; García Solé, J. Inorganic nanoparticles for optical bioimaging. *Adv. Opt. Photon.* **2016**, *8*, 1–103. [CrossRef]
125. Wang, Y.; Song, W.; Ruan, W.; Yang, J.; Zhao, B.; Lombardi, J.R. SERS Spectroscopy Used To Study an Adsorbate on a Nanoscale Thin Film of CuO Coated with Ag. *J. Phys. Chem. C* **2009**, *113*, 8065–8069. [CrossRef]
126. Wu, L.; Tan, J.; Chen, S.; Liu, X. Catalyst-enhanced micro-galvanic effect of Cu_3N/Cu-bearing NiTi alloy surface for selective bacteria killing. *Chem. Eng. J.* **2022**, *447*, 137484. [CrossRef]
127. Blaznik, U.; Krušič, S.; Hribar, M.; Kušar, A.; Žmitek, K.; Pravst, I. Use of Food Additive Titanium Dioxide (E171) before the Introduction of Regulatory Restrictions Due to Concern for Genotoxicity. *Foods* **2021**, *10*, 1910. [CrossRef]
128. Shanker, U.; Rani, M.; Jassal, V. Degradation of hazardous organic dyes in water by nanomaterials. *Environ. Chem. Lett.* **2017**, *15*, 623–642. [CrossRef]

Article

Sunlight-Driven Photocatalytic Degradation of Methylene Blue with Facile One-Step Synthesized Cu-Cu$_2$O-Cu$_3$N Nanoparticle Mixtures

Patricio Paredes [1], Erwan Rauwel [1], David S. Wragg [2], Laetitia Rapenne [3], Elias Estephan [4], Olga Volobujeva [5] and Protima Rauwel [1,*]

[1] Institute of Forestry and Engineering Sciences, Estonian University of Life Sciences, Kreutzwaldi 56/1, 51014 Tartu, Estonia; patricio.paredes@emu.ee (P.P.); erwan.rauwel@emu.ee (E.R.)
[2] Department of Chemistry and SMN, University of Oslo, 0315 Oslo, Norway; d.s.wragg@smn.uio.no
[3] Grenoble Institute of Engineering, LMGP, University Grenoble Alpes, CNRS, F-38000 Grenoble, France; laetitia.rapenne@grenoble-inp.fr
[4] Laboratory of Bioengineering and Biosciences, LBN, Univ Montpellier, 34193 Montpellier, France
[5] Institute of Materials and Environmental Technology, Tallinn University of Technology, 19086 Tallinn, Estonia; olga.volobujeva@taltech.ee
* Correspondence: protima.rauwel@emu.ee

Abstract: Sunlight-driven photocatalytic degradation is an effective and eco-friendly technology for the removal of organic pollutants from contaminated water. Herein, we describe the one-step synthesis of Cu-Cu$_2$O-Cu$_3$N nanoparticle mixtures using a novel non-aqueous, sol-gel route and their application in the solar-driven photocatalytic degradation of methylene blue. The crystalline structure and morphology were investigated with XRD, SEM and TEM. The optical properties of the as-prepared photocatalysts were investigated with Raman, FTIR, UV-Vis and photoluminescence spectroscopies. The influence of the phase proportions of Cu, Cu$_2$O and Cu$_3$N in the nanoparticle mixtures on the photocatalytic activity was also investigated. Overall, the sample containing the highest quantity of Cu$_3$N exhibits the highest photocatalytic degradation efficiency (95%). This enhancement is attributed to factors such as absorption range broadening, increased specific surface of the photocatalysts and the downward band bending in the p-type semiconductors, i.e., Cu$_3$N and Cu$_2$O. Two different catalytic dosages were studied, i.e., 5 mg and 10 mg. The higher catalytic dosage exhibited lower photocatalytic degradation efficiency owing to the increase in the turbidity of the solution.

Keywords: photocatalysis; nanoparticles; sunlight; Cu$_3$N; dye degradation; semiconductor band bending

1. Introduction

Presently, wastewater treatment is one of the most critical issues, owing to the release of large amounts of industrial effluents into water bodies [1]. In that regard, the use of organic dyes in food, papermaking, cosmetics, pharmaceuticals and textiles poses a threat to the environment [2]. Presently, different aqueous remediation methods against organic pollutants are being used, including advanced oxidation processes [3]. Today, heterogeneous photocatalysis is considered an efficient and environmentally friendly method for the degradation or removal of water-soluble organic pollutants [4,5]. To that end, several nanomaterials based on oxides and nitrides have been synthesized and studied in order to enhance the degradation of aqueous contaminants using photocatalytic processes. In particular, Cu-based nanomaterials have attracted interest due to the earth abundance of Cu, which allows for tackling issues related to the sustainability and cost-effectiveness of photocatalytic processes [6,7]. Therefore, cuprous oxide (Cu$_2$O) and copper nitride (Cu$_3$N) semiconductors, with narrow band gaps of 1.2–2.5 eV for Cu$_2$O [8–10] and 0.2–2 eV for Cu$_3$N [11], are seen as promising visible-light-activated photocatalysts.

In recent years, scientists have provided some insights into the photocatalytic degradation mechanism of organic dyes using Cu-based nanomaterials. Norzaee et al. reported a 90% degradation efficiency of aniline after 90 min under a UV-C lamp with CuO nanoparticles [12]. Yu et al. synthesized Cu_2O nanoparticles using the liquid-phase reduction method, which degraded 83.2% of fluroxypyr under a 500 W metal halide lamp [13]. Sithole et al. reported a photocatalytic degradation of 61% for methylene blue and 89% for methylene orange by Cu_3N nanocubes after 180 min and 240 min, respectively, using a solar simulator [14]. Furthermore, metallic Cu nanoparticles produce free-electron resonance under visible light excitation and are therefore potential photocatalysts. For instance, Zhang et al. demonstrated that the surface plasmon resonance (SPR) of Cu nanoparticles engenders a spontaneous photocatalytic water-splitting [15]. In fact, plasmonic metal nanoparticles combined with other luminescent nanoparticles tend to augment the overall visible absorption range and also act as a sink for excited electrons that, in turn, reduces excitonic recombination [16,17]. Even though there are some reports on the semiconductor-metal-based photocatalyst-heterostructures [18], to the best of our knowledge, the synthesis of Cu-Cu_2O-Cu_3N nanoparticle mixtures applied to photocatalytic dye degradation has not yet been reported.

In the present work, a novel synthesis route that allows varying the proportions of the crystalline phases of Cu, Cu_2O and Cu_3N using a non-aqueous sol-gel method was devised. The samples were characterized with powder X-ray diffraction (PXRD), scanning electron microscopy (SEM), transmission electron microscopy (TEM), Fourier transform infrared (FTIR), Raman, UV-Vis and photoluminescence spectroscopies. The objective is to understand the influence of the nanoparticle proportions and the catalyst dosage on the photocatalytic properties. Herein, the photocatalytic performances of the samples in degrading methylene blue (MB) dye under solar radiation was evaluated using a commercial hand-held Lovibond photometer, which also allows on-site analysis of dye-contaminated aqueous media in real-life situations.

2. Materials and Methods

2.1. Materials

Copper (II) nitrate trihydrate 99% ($Cu(NO_3)_2 \cdot 3H_2O$) was purchased from Acros Organics; 1-octadecene (90%), oleylamine (90%) and MB were purchased from Thermo Scientific, and 2-propanol was purchased from Honeywell. All chemicals were of analytical grade and used as received without further purification.

2.2. Synthesis of Cu-Based Nanocomposite

The Cu-Cu_2O-Cu_3N nanoparticle mixtures were synthesized using a non-aqueous sol-gel route inside a glovebox (controlled N_2 atmosphere), and the reagents were sealed in an autoclave for the synthesis [19,20]. In a typical experiment, 0.3 g of $Cu(NO_3)_2 \cdot 3H_2O$ was dissolved in 10 mL of Octadecene. Then, 10 mL of oleylamine was added and magnetically stirred at 110 °C for 20 min until a homogenous solution was obtained. Subsequently, the solution was placed inside an autoclave and heated in an oven at different reaction intervals of 3, 6, 12 and 24 h at a temperature of 280 °C, and the samples were named Cu-3, Cu-6, Cu-12 and Cu-24, respectively. The autoclave was then removed from the oven, and the reaction mixture was allowed to cool down to room temperature. Finally, the precipitated nanomaterials were isolated using centrifugation, washed twice with isopropanol and dried.

2.3. Characterization

The crystalline phase, proportions and particle sizes were examined with PXRD using a Bruker D8 Discover diffractometer (Bruker AXS, Karlsruhe, Germany) with CuKα1 radiation (λ = 0.15, 406 nm) selected with a Ge (111) monochromator and LynxEye detector. Phases were identified using the crystallography open database [21] in Bruker EVA version 6.0. Diffraction patterns were fitted using the Rietveld method in TOPAS version

6 [22] to obtain phase weight percentages and crystallite sizes (Scherrer method with full profile fit, k = 0.89). The peak shape was modeled using a fundamental parameters approach and the cubic lattice parameter, scale and Lorentzian size broadening were refined for three phases: copper (I) nitride (COD 1010167), cuprite (COD 1000063) and metallic copper (COD 9013014). The microstructure of the samples was characterized with the use of transmission electron microscope (TEM) JEOL 2010 LaB$_6$ TEM (JEOL, Japan) operating at 200 kV in TEM mode and providing a point-to-point resolution of 1.9 Å. The surface morphology was investigated with a high-resolution scanning electron microscope HR-SEM Zeiss Merlin (Carl Zeiss Microscopy, Munich, Germany), and the chemical composition was determined using an energy dispersive X-ray analysis (EDS) system Bruker EDX-XFlash6/30 detector (Bruker, Oxford, UK) with an acceleration voltage of 4 kV for SEM and 10 kV for EDX analysis. The elemental composition quantification was performed using PB/ZAF standard less mode. Another SEM, EVO LS15 Zeiss (Carl Zeiss Microscopy, Germany), working in backscattered and secondary electron mode at a voltage of 15 kV was used to study the morphology and topographic characteristics of the Cu-based nanocomposites. The vibrational properties were studied using a WITec Confocal Raman Microscope System alpha 300R (WITec Inc., Ulm, Germany). Excitation in confocal Raman microscopy is generated with a frequency-doubled Nd:YAG laser (New-port, Irvine, CA, USA) at a wavelength of 532 nm, with 50 mW maximum laser output power in a single longitudinal mode. The incident laser beam is focused onto the sample with a 60× NIKON having a numerical aperture of 1.0 with a working distance of 2.8 mm (Nikon, Tokyo, Japan). The acquisition time of a single spectrum was set to 0.5 s. The chemical bonds were studied using a Fourier transform infrared (FTIR) spectrometer (Nicolet is10 Thermo Scientific, Driesch, Germany) in the range of 360–1100 cm^{-1}. The optical absorbance of the nanocomposites was determined using a VWR UV-VIS spectrometer (UV-1600PC, USA) in the range of 350–1100 nm. Finally, photoluminescence spectroscopy was carried out at room temperature on the nanopowders with excitation wavelengths of 365 nm and 532 nm using diode lasers LSM-365A and LSM-533A LED (Ocean insight, Orlando, FL, USA) with specified output powers of 10 mW and 1.96 mW, respectively. The emission was collected using a FLAME ES UV-Vis spectrometer (Ocean optics, USA) with a spectral resolution of 1.34 nm.

2.4. Adsorption Experiments

The calibration curve of MB using the Lambert–Beers law is available in Figure S1, which was obtained with a Lovibond MD 610 photometer at the MB characteristic absorption peak of 660 nm. The MB adsorption study on the photocatalysts was carried out at room temperature under dark conditions. For this, 5 mg of the absorbent was added to 15 mL of a MB (5 mg/L) stock solution. Aliquots were carefully obtained using a pipet for analysis. The amount of MB adsorbed was calculated using Equation (1):

$$Q_t = \frac{V(C_o - C_e)}{m} \left(\frac{mg}{g}\right), \tag{1}$$

where Q_t (mg/g) is the amount of dye (MB) adsorbed/unit weight of the sample; C_0—initial concentration of MB (mg/L); C_e—concentration of MB at equilibrium time (mg/L); V—volume of solution (L) and m—the weight of the sample (mg). The percentage of MB adsorbed is given by R (%) as is given in Equation (2):

$$R = \frac{(C_0 - C_t)}{m} 100\%, \tag{2}$$

where C_0 and C_t are the initial and final concentrations of MB in the solution, respectively.

The MB adsorption behavior of the nanoparticle mixtures was also verified for the pseudo-second-order kinetic model, which assumes that the rate-limiting step is the interaction between two reagent particles. The equation of this model is illustrated as Equation (3):

$$\frac{t}{Q_t} = \frac{1}{k_2 Q_e^2} + \frac{t}{Q_e}, \tag{3}$$

where k_2 is the equilibrium rate constant of the pseudo-second-order (g/mg min). The linear fit between the t/Q_t and contact time (t) can be approximated as pseudo-second order kinetics.

2.5. Photocatalytic Experiments

The photocatalytic activity of the Cu-Cu_2O-Cu_3N nanoparticle mixtures was evaluated by monitoring the photodegradation of MB under solar radiation in the open air. In a typical experiment, 5 mg and 10 mg of the photocatalyst were dispersed in 20 mL of the MB aqueous solution with a concentration of 5 mg/L. For evaluating the photocatalytic activity after adsorption, the suspension was placed in the dark for 2 h in order to attain adsorption–desorption equilibrium before being exposed to solar radiation in beakers with and without a glass cover. The glass cover served as a filter to block UV radiation from the sun and allowed us to solely evaluate the visible-light, solar-driven photocatalysis.

3. Results and Discussion

3.1. Structure and Morphology

The variation in the crystalline phase proportions of Cu_2O, Cu_3N and Cu in the samples as a function of synthesis time was studied using PXRD analysis (see Figure 1). The PXRD patterns of samples Cu-3, Cu-6 and Cu-12 exhibit the characteristic PXRD peaks of Cu_2O, Cu_3N and metallic Cu structures in variable proportions (see Table 1). This result clearly suggests that the Cu^I species were generated with the reduction of Cu^{II} by oleylamine upon heating in reductive environments. In the PXRD pattern, peaks at 29.68°, 36.47°, 42.3°, 61.4° and 73.6°, correspond to (110), (111), (200), (220) and (311) respectively reflections of Cu_2O belonging to the cubic $Pn\bar{3}m$ phase (COD 1000063). The diffraction peaks visible at 41.12° and 47.84° correspond to the crystal planes (111) and (200) respectively, of Cu_3N nanocrystals belonging to the $Pm\bar{3}m$ space group (COD 1010167). The proportion of Cu metal nanoparticles increases with the synthesis time, and sample Cu-24 consists of pure metallic Cu nanoparticles, indicating a complete reduction of Cu^I to Cu^0. The PXRD peaks visible at 43.26°, 50.42° and 74.12° correspond to the (111), (200) and (220), respectively, planes of Cu-metal face-centered cubic crystal structure $Fm\bar{3}m$ (COD 9013014). The differences in phase composition are due to changes in the synthesis reactions, influenced by the synthesis time. In fact, as the synthesis time increases, a large decomposition or/and desorption of the oleylamine capping agent is likely, leading to the further reduction of Cu_2O and Cu_3N into Cu metal [23]. The PXRD data indicate that the thermal decomposition of $Cu(NO_3)_2 \cdot 3H_2O$ in octadecene and olyelamine at 280 °C results in the precipitation of Cu-Cu_2O-Cu_3N after 3 h. Then, as the synthesis time increases, the Cu_3N and Cu_2O phases are reduced to Cu metal.

Table 1. Phase ratios of Cu_2O, Cu_3N and Cu in the samples and their crystallite size distribution calculated from the PXRD patterns.

	Cu-3		Cu-6		Cu-12		Cu-24	
Crystalline Phase	Crystallite Size (nm)	Crystalline Phase (%)	Crystallite Size (nm)	Crystalline Phase (%)	Crystallite Size (nm)	Crystalline Phase (%)	Crystallite Size (nm)	Crystalline Phase (%)
Cu	50	32.1	55.7	44.6	54	90.3	71	100
Cu_2O	14.8	49.5	26.7	54.3	30	5.5	-	-
Cu_3N	2.8	18	30	1.1	29	4.2	-	-

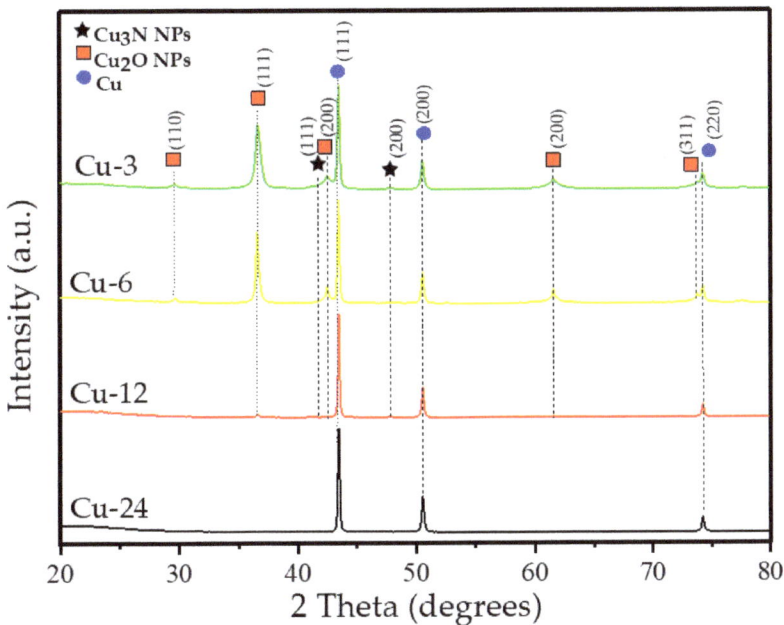

Figure 1. PXRD patterns of the samples.

The crystallite size and weight percentages of Cu_2O, Cu_3N and Cu nanoparticles in the samples are calculated from Rietveld fits of the PXRD data. Table 1 reports the average crystallite diameter of each phase and the weight % composition of the samples from Rietveld fits to their PXRD patterns. The calculations indicate that for Cu-3, the particle sizes of Cu_2O and Cu_3N are 15 nm and 3 nm, respectively. As the reaction time increases for Cu-6 and Cu-12, the nanoparticles adopt an average size of 30 nm for both the Cu_2O and Cu_3N phases. After 24 h (Cu-24), only the diffraction peaks characteristic of Cu metal are observed. The Cu nanoparticles have an average size of 50 nm in samples Cu-3, Cu-6 and Cu-12, but for Cu-24, in which it is the only phase present, the average particle size increases slightly to 70 nm.

Figures 2 and 3 are the typical SEM and TEM images of the samples. In Figure 2a,b, corresponding to the Cu-24 sample, the pure Cu nanoparticles exhibit a homogeneous powder morphology. On the other hand, Figure 2c, showing the Cu-3 sample, contains a blend of all three phases. However, two main morphologies are visible, including the powder morphology of the Cu nanoparticles in Figure 2a,b. The existence of sphere-like nanoparticles obtained after 3 h of synthesis can be attributed to Cu_2O nanoparticles, considering their phase proportion of ~50%. These spheres are agglomerates of smaller nanoparticles, given their granular surfaces. For this sample, the Cu_3N nanoparticles are not discernable because of their small average particle size of 7 nm, according to the PXRD results. With an increase in synthesis time, i.e., for the Cu-6 and Cu-12 samples (Figure 2d–f), additional morphologies are visible. Some reports suggest that the thermal decomposition of oleylamine depends on the synthesis time, which plays an important role in controlling the size and shape of the nanocrystals [24,25]. Thus, in addition to the spherical and powder morphologies corresponding mostly to Cu_2O and Cu nanoparticles, tetrahedral and cubic morphologies are also observed. In fact, both Cu_2O and Cu_3N can present cube and pyramidal morphologies. In the case of Cu_2O, several morphologies, i.e., cube, truncated octahedron and tetrapods are possible with varying the quantity of ammonia, hydroxyl groups and water in the solution [26]. In our synthesis, the quantity of ammonia in the reaction mixture under solvothermal conditions, along with the optimum reaction time, determines not only the quantity of Cu_3N and Cu_2O nanoparticles precipitated but also their morphology [24].

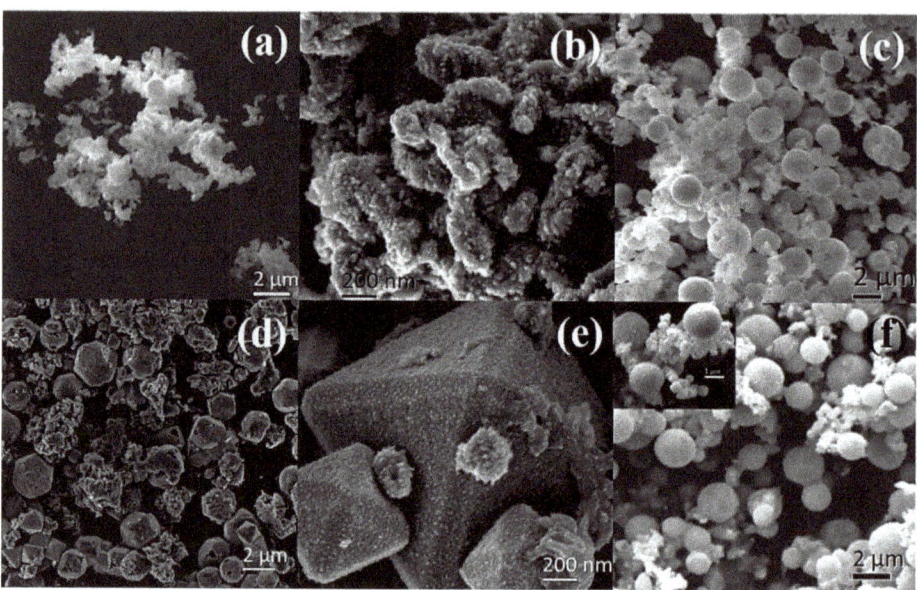

Figure 2. SEM images of (**a,b**) Cu-24, (**c**) Cu-3, (**d**) and (**e**) Cu-6, (**f**) Cu-12 (the inset is a higher magnification image of all the 3 morphologies, i.e., powder, sphere and cube).

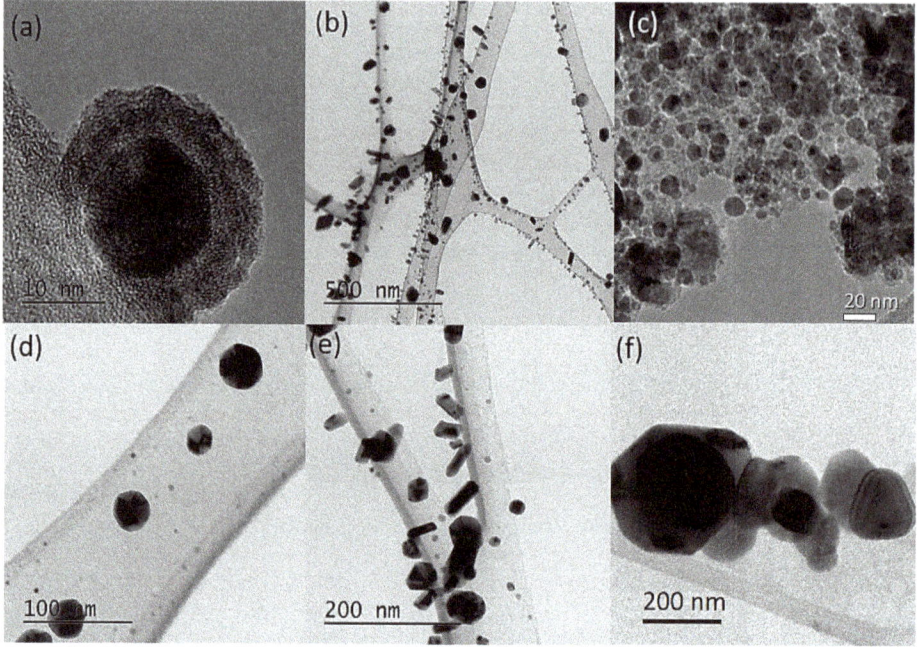

Figure 3. TEM images of the Cu_2O, Cu_3N and Cu nanoparticle mixture samples synthesized at different reaction times (**a,d**) Cu-3, (**b,e**) Cu-6, (**c**) Cu-12 and (**f**) Cu-24.

Furthermore, EDX was performed on several particles for the Cu-6 sample, in order to distinguish Cu_3N from Cu_2O. It should be noted that EDX is not suitable to quantify lighter elements, such as oxygen and nitrogen. However, variations in the EDX peak intensities of N and O from various particles could provide an indication of N/O-rich or poor compositions. As seen in the EDX spectrum in Figure S2, the truncated octahedral particles manifest a relatively intense oxygen peak compared to the nitrogen peak and are, therefore, most likely Cu_2O.

TEM was carried out in order to study the particle sizes and morphologies of the Cu-based nanoparticles. It is noteworthy that the TEM images on their own are not sufficient to determine the nanoparticle phase. However, techniques such as XRD coupled with Rietveld refinement provide the average crystallite size for each phase. The size of nanoparticles from TEM was compared with PXRD data from Table 1 in order to estimate the phase of the nanoparticles. Figure 3a,d shows the micrographs of the Cu-3h sample. In Figure 3a, a nanoparticle of size 15 nm is most likely Cu_2O, in agreement with the XRD results. However, a shell is visible on its surface (Figure 3a), probably due to the incomplete reaction of oleylamine. Several studies have shown that the presence of oleylamine surfactant modifies the optical properties of the materials. For instance, ZnO/ZnCdSe alloy synthesized with oleylamine manifested a blueshift in the absorption spectrum and a red shift in the emission spectrum [27]. A similar trend was also observed for CuO nanoparticles synthesized with oleylamine [28]. In Figure 3d, the contrast of the TEM images reveals the presence of twin boundaries, which are typical of metallic nanoparticles such as Cu with a particle size of ~50 nm. In addition, there are smaller nanoparticles in the background of ~5 nm in size, as shown in Figure 3d, which are most likely Cu_3N (Table 1).

As the reaction time increases, changes in particle sizes and morphologies are visible. Additionally, the capping agent that was present in the Cu-3 sample is absent at longer synthesis intervals. In Figure 3b,e, the TEM images of the Cu-6 sample reveal different morphologies for Cu_2O, Cu_3N and Cu nanoparticles. The cuboidal nanoparticles would most probably correspond to Cu_2O, as the crystalline proportion of Cu_3N is very low (~1.1%). For the Cu-12 sample (Figure 3c), the TEM micrograph reveals spherical nanoparticles of ~50 nm that correspond to the Cu phase, whereas smaller nanoparticles of ~20 nm could be both Cu_3N and Cu_2O. Finally, the TEM image of the Cu-24 sample in Figure 3f displays pure Cu nanoparticles with sizes ranging from 50 nm to 300 nm.

3.2. Elemental Characterization

The relationship between the physical and chemical properties of the nanocomposites was probed using Raman spectroscopy, as shown in Figure 4a. In principle, the presence of Cu nanoparticles cannot be ascertained with Raman spectroscopy as metals possess negative-real and positive-imaginary dielectric constants and also exhibit surface plasmon resonance [29]. However, in the present case, Cu metal nanoparticles from the Cu-24 sample oxidized under the green laser beam into cupric oxide (CuO) [29]. The peak at 280 cm^{-1} is assigned in the literature to the stretching vibrational mode of CuO. This peak is present in all the samples due to the presence of Cu nanoparticles that systematically undergo oxidation under the laser. It should be noted that CuO is not present in the as-synthesized nanocomposites (Figure 1). However, the small redshift in this peak and the broadening in the bandwidth at around 290 cm^{-1} for the Cu-3 sample could be attributed to the small nanoparticles in the Cu_3N phase (~3 nm). The Raman peaks at around 152 cm^{-1}, 216 cm^{-1} and 515 cm^{-1} are fully consistent with peaks of Cu_2O [30]. According to Wei et al., the peak at about 216 cm^{-1} could be attributed to the second-order overtones $2\Gamma_{12}$, and the other weaker peaks at 152 cm^{-1}, 515 cm^{-1} and 624 cm^{-1} correspond to Γ_{15} of oxygen vacancies; the fourth-order overtone $4\Gamma_{12}$ and the red-allowed mode Γ_{15} are phonon vibrations of Cu_2O [31]. The relative intensity of the peak at 515 cm^{-1} decreases with the synthesis time due to the decreasing quantity of Cu_2O. According to Sajeev et al., the peak located at 624 cm^{-1} corresponds to the stretching and bending in the Cu-N bond from the Cu_3N

phase [32]. The relative intensities of this peak for Cu-6 and Cu-12 are almost the same due to the low amounts of Cu_3N present in these samples. However, for the Cu-3 sample there is an increase in the relative intensity of this peak owing to a higher amount of the Cu_3N phase (18%). In addition, the band between 151 and 171 cm^{-1} corresponds also to the Cu-Cu dimer, which is prominent in the Cu-24 sample [33].

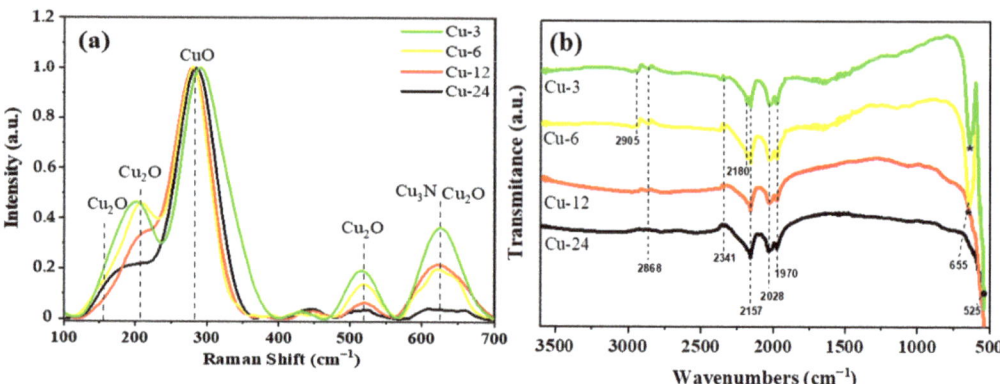

Figure 4. (a) Normalized Raman spectra and (b) FTIR spectra of the mixtures of the nanocomposites.

The FT-IR spectra, as shown in Figure 4b, highlight the vibrational bands of the organic and inorganic moieties present in the nanocomposites. FT-IR indicates changes in the framework configurations of the nanocomposites via shifts in the bands as a function of synthesis time. In the FT-IR spectra of the Cu-3 and Cu-6 samples, the peak located at 2868 cm^{-1} corresponds to the stretching vibrational mode of the Cu-O bonds from the Cu_2O crystalline phase [34,35]. Moreover, a decrease in the peak intensity signifies a decrease in Cu_2O content in the samples. This peak is no longer present in samples synthesized at longer synthesis intervals. Furthermore, the characteristic peak located at 655 cm^{-1} in the Cu-3 and Cu-6 samples can also be ascribed to the intrinsic lattice mode vibration in Cu-N bonds from the Cu_3N nanoparticles present in the nanocomposites [32]. However, for the Cu-12 sample, the peak at 655 cm^{-1} is not visible due to a very low fraction in the Cu_3N phase (1.1%), as well as for the Cu-24 sample that corresponds to pure Cu-metal nanoparticles. The main stretching vibration in pure Cu nanoparticles was found at 520 cm^{-1} (represented by •) [36]. The FTIR spectra of the Cu-3 and Cu-6 samples also exhibit a small peak at ~2905 cm^{-1} that can be attributed to the stretching vibration in the C-H bonds from the oleylamine [37]. Free N_2 has a N≡N stretching at around 2331 cm^{-1} [38]; in contrast, because N_2 coordinated with a metal atom is a weak Lewis base, this coordination shifts the stretching peak to 1970−2180 cm^{-1} [38].

3.3. Optical Properties

UV-Vis absorption spectra of the nanocomposites show a broad absorption peak located at ~600 nm that can be assigned to the SPR band of Cu nanoparticles (Figure 5a) [39]. This peak is prominent in the Cu-24 sample, which corresponds to a pure phase of Cu metal. For samples synthesized at shorter intervals, the SPR of the Cu-metal phase shows a redshift. It is reported that the SPR band of Cu-metal nanoparticles shifts as a function of the Cu-nanoparticle size, shape, dielectric properties and the presence of other phases in the sample [40–42]. For instance, Mott et al. reported a redshift in the SPR of Cu as the particle size increases [39]. In our study, the red shift in the SPR of Cu is attributed to the presence of the Cu_2O and Cu_3N phases, with additional absorption wavelengths in the red part of the visible spectrum and the near-infrared, respectively. For the samples that contain a higher amount of Cu_3N nanoparticles, i.e., Cu-3 and Cu-6, a characteristic absorption peak is located at around 1020 nm, but for the Cu-12 sample, this peak is around 980 nm. These corroborate with the reported band gap of ~1.2 eV for Cu_3N [43,44]. The absorption peaks visible at around 870 nm (1.4 eV) in the Cu-3, Cu-6 and Cu-12 samples correspond to the absorption edges of Cu_2O nanoparticles, corroborating with the range of the reported bandgap, i.e., 1.2 eV–2 eV [10,45].

The emission properties of the nanocomposites were examined using room temperature photoluminescence spectroscopy. Under UV-excitation of 365 nm (Figure 5c), the emission peak maximum of all samples is between 580 nm and 620 nm, corroborating with the plasmonic emission of Cu-metal nanoparticles. The fluorescence is ascribed to the radiative recombination of electrons in the s-p conduction band below the Fermi level with the holes in the d-band of Cu. In addition, the asymmetrical emission spectra have a low energy tail related to intraband transitions [46]. Changes in the intensity of the intraband emission are a result of changes in the coupling of Cu with various nanocomposites, owing to electron transfer mechanisms.

Since we performed solar-driven photocatalysis, and the maximum intensity of the solar radiation is around 550 nm, we therefore chose a 533 nm excitation source to study the emission properties of the samples (Figure 5d). The photoluminescence emission spectra of Cu and Cu_2O have been adequately studied in the literature [47]. However, very few reports exist on the photoluminescence emission of Cu_3N using a visible light excitation source. In fact, in most of the studies, a UV excitation source with very high emission intensity was used. Yeshchenko et al. studied photoluminescence emission localized at 660 nm in Cu metal nanoparticles with an excitation wavelength of 355 nm [46]. Basavalingaiah et al. reported the photoluminescence emission of Cu_2O with an excitation wavelength of 250 nm, showing two-emission peaks at 712 nm and 564 nm [48]. Sithole et al. studied the photoluminescence spectra of Cu_3N nanoparticles with an emission around 486 nm using an excitation wavelength of 200 nm [14].

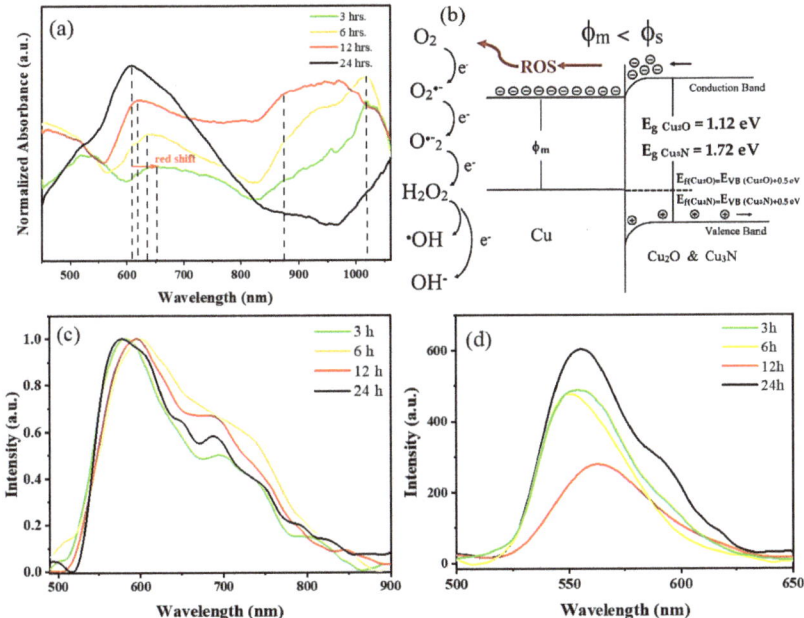

Figure 5. (a) Normalized UV-Vis absorption spectra of the nanoparticle mixtures, (b) downward band bending in p-type semiconductors, along with electron accumulation and transfer to the oxygen molecules generating ROS. The bandgaps of Cu_3N and Cu_2O were determined using the Tauc plots in Figure S4. Their band edges were then determined using the energy levels in references [49,50]. Photoluminescence emission spectra of the samples with an excitation wavelength of (c) 365 nm (normalized emission) and (d) 533 nm.

In the photoluminescence spectra shown in Figure 5c,d, the emission is dominated by the Cu plasmonic emission for all the samples irrespective of the presence of other phases. PL emissions corresponding to Cu_3N and Cu_2O were once again not observed. For the Cu-24 sample, the pure Cu metal nanoparticles exhibit a strong emission peak at 554 nm, very close to the photoluminescence band of bulk cooper (560 nm) [46]. Nevertheless, a decrease in the emission intensity along with energy shifts are also notable in the spectra, influenced by the presence of secondary phases and differences in nanoparticle sizes. According to the literature, the blueshift in the photoluminescence spectrum of the Cu-based nanoparticles is indicative of the presence of surface traps, while a redshift towards the band edge emission indicates defect-free nanoparticles [14]. The changes in the emission peak positions and intensities can be attributed to electron transfer mechanisms between various phases and changes in radiative recombination influenced by the ambient as well as semiconductor band bending. In fact, band bending is a phenomenon that occurs at the interface of a semiconductor and a metal. The nature of the band bending depends on the work functions of the two materials. In our case, the work function of Cu (4 eV) is lower than the work functions of Cu_3N (5.06 eV) and Cu_2O (5 eV) [51,52]. Therefore, a downward band bending in the p-type Cu_2O and Cu_3N occurs due to the formation of a Schottky junction with Cu nanoparticles [53–56]. As a consequence, the downward band bending reduces the radiative recombination in Cu_3N and Cu_2O in the band bending region (Figure 5b). Upon photoexcitation, the band bending leads to an accumulation of electrons on the surface of Cu_3N and Cu_2O that are subsequently transferred to Cu nanoparticles. At the same time, holes are directed toward the volume of the nanoparticle from the band-bending region. Consequently, the excitonic recombination is suppressed in Cu_3N and Cu_2O. On the other hand, owing to the electron transfer to Cu nanoparticles, the emission of Cu metal nanoparticles is enhanced. Simultaneously, electrons on the surfaces of Cu_2O and Cu_3N would also undergo non-radiative recombination, e.g., via transfer to chemisorbed oxygen molecules, when photoexcited in ambient conditions. The highest emission is observed in the Cu-24 sample containing pure Cu metal nanoparticles, while the lowest is attributed to the Cu-12 sample with Cu and Cu_2O phases. For photocatalysis, a low emission or longer excitonic recombination lifetime is important in order to ensure electron transfer from the nanoparticle to the dye solution [54,56].

4. Adsorption Kinetics

Adsorption kinetics were studied at different time intervals in order to quantify the adsorption rate and elucidate the mechanism of adsorption, i.e., physical or chemical adsorption. Adsorption kinetics describe the rate of release of a sorbate from an aqueous solution to a solid-phase interface. The isotherms of MB adsorbed on the surface of the different nanoparticle mixture samples are show in Figure 6a. In dark conditions, the MB adsorbed on the surface of the nanoparticles exhibits an abrupt increase up to the first 100 min after which it reaches a plateau. This initial uptake is due to the abundance of active sites on the surface of the nanoparticles, which decreases over time, i.e., when solute concentration gradient becomes very high. When C_e/C_0 reaches a plateau, it suggests that all active sites have been filled and no additional adsorption is possible. Furthermore, the adsorption–desorption equilibrium is achieved after ~2 h. This trend indicates the layer-by-layer (monolayer to multi-layer) adsorption of MB on the surface of the Cu-based nanoparticles [57]. The MB removal for the Cu-3, Cu-6, Cu-12 and Cu-24 samples is 8.04%, 7.58%, 7.16% and 5.77%, respectively, after 250 min. Additionally, Q_t is higher for samples containing mixed phases of Cu, Cu_2O and Cu_3N and lower for the pure Cu phase nanoparticles.

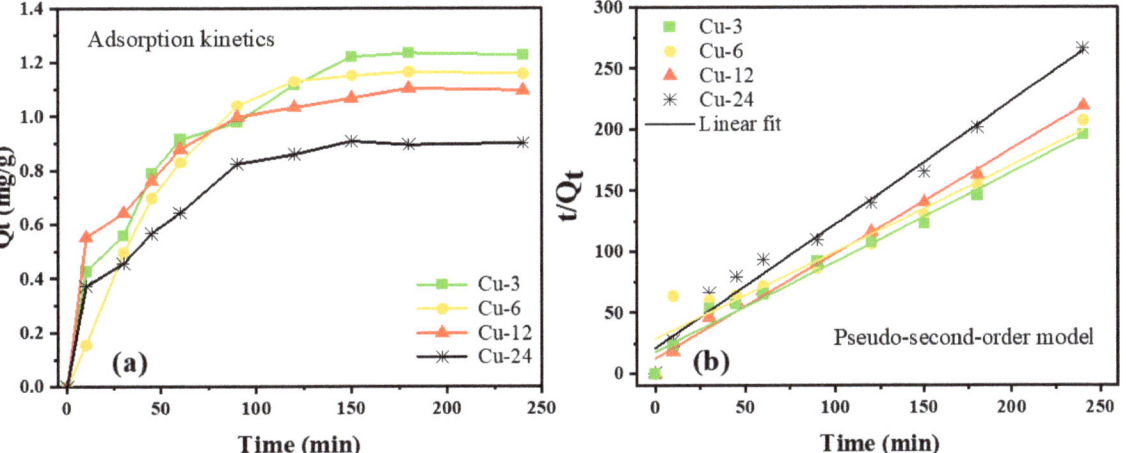

Figure 6. Adsorption kinetics (**a**) and the pseudo-second-order kinetic (**b**) plots for 5 mg of the nanocomposite in 15 mL aqueous solution containing a 5 mg/L concentration of MB.

For further investigation, the pseudo-second-order model was also applied in order to interpret the adsorption kinetics (Figure 6b). In the pseudo-second-order model, we assume that (i) the adsorbate concentration is constant in time and (ii) the total number of binding sites depends on the amount adsorbed at equilibrium. Furthermore, this model is most adapted to chemical sorption involving valence forces through sharing or exchanging of electrons between adsorbent and sorbate, as in the case of organic pollutants. A linear relationship with high correlation coefficients (R^2) is observed between t/Q_t and t, indicating the applicability of the pseudo-second-order model to describe the adsorption process of MB with the nanoparticle mixtures. In Table 2, the values of the correlation coefficient R^2 suggest that the adsorption tendency corroborates with the pseudo-second-order kinetic model with a slight improvement for samples containing larger amounts of Cu nanoparticles (Cu-12 and Cu-24). The amount of MB adsorbed per gram of the photocatalyst (Q_e) is the highest for the Cu-3 and Cu-6 samples. The rate constant K_2 for pseudo-second-order was the lowest for the Cu-6 sample followed by the Cu-3 sample, indicating quicker reaction kinetics for these samples containing higher amounts of Cu_3N and Cu_2O phases. In fact, the co-precipitation of the Cu phase during synthesis indicates that both Cu_2O and Cu_3N were synthesized in Cu-rich conditions, likely resulting in oxygen and nitrogen vacancies on their surface. These positively charged defects are active areas for chemical adsorption and improved absorption kinetics.

Table 2. Adsorption kinetic parameters of MB by the nanocomposites obtained using the pseudo-second-order kinetic Equation (3).

Sample	Pseudo-Second-Order		
	$Q_{e,fitted}$ (mg/g)	K_2 (g/mg min)	R^2
Cu-3	1.3089	0.0296	0.9783
Cu-6	1.4080	0.0174	0.9361
Cu-12	1.1629	0.0588	0.9821
Cu-24	0.9862	0.0431	0.9920

5. Sunlight-Driven Photocatalysis

Visible light photocatalytic activity of the nanoparticle mixture samples was subsequently evaluated by measuring the removal or degradation of MB from an aqueous solution under sunlight radiation with a UV index of 4–5. The photodegradation of MB was monitored with the normalized change in its concentration (C_t/C_0) as a function of the irradiation time (t), as shown in Figure 7. All the solutions tested consisted of 5 mg/L of MB in distilled water except for the control sample, which did not contain any nanoparticles. Under solar radiation, the photobleaching in the control sample is observed. However, the tests with nanoparticles showed that the photocatalytic degradation of MB is higher than the photobleaching in the control sample and reaches 55% degradation after 4 h of sunlight exposure. Furthermore, an equilibrium is achieved after 6 h, and the photocatalytic degradation rate decreases. MB degradation of 96%, 95%, 93% and 92% using the Cu-3, Cu-6, Cu-12 and Cu-24 samples, respectively, is obtained using 5 mg of nanocomposites for each experiment. All experiments were carried out in beakers without a lid. The higher degradation efficiency observed for the Cu-3 sample can be ascribed to the augmented visible light absorption range, owing to a higher amount of Cu_3N nanoparticles in the sample. The presence of these phases broadens the absorption spectra up to the infrared. In addition, the high specific surface of the Cu_3N and Cu_2O nanoparticles is directly linked to their average particle sizes of 2.9 nm and 15 nm, respectively. As the synthesis time increases, the particles enlarge, and as a consequence, the specific surface is lowered. In addition, the SPR of Cu nanoparticles would create charge polarization that can influence the absorption and scattering of visible light, in turn, enhancing the photocatalytic degradation efficiency of Cu_3N and Cu_2O [58].

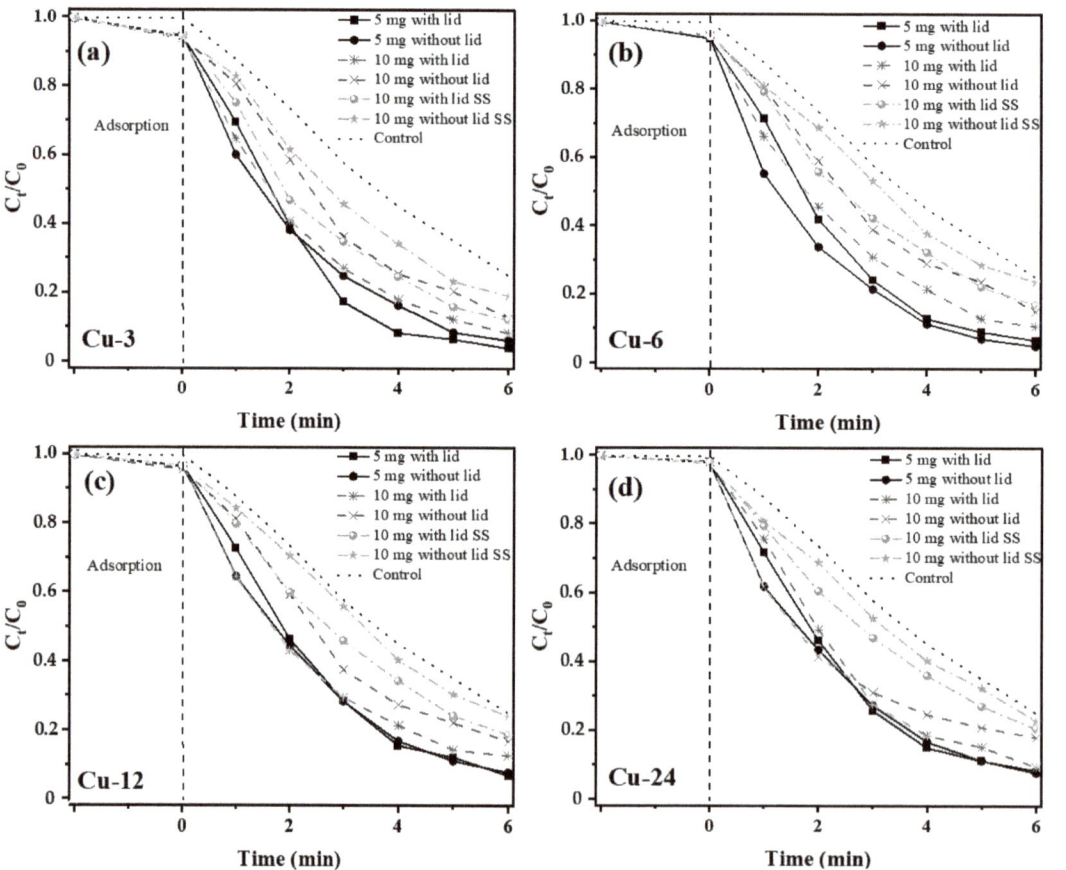

Figure 7. Normalized MB degradation curves under sunlight radiation with (**a**) Cu-3, (**b**) Cu-6, (**c**) Cu-12 and (**d**) Cu-24 in 20 mL aqueous solution with an MB concentration of 5 mg/L.

The photocatalytic degradation efficiency depends on the active surface area of the photocatalyst. For higher amounts of catalysts, i.e., 10 mg, the dye-degradation efficiency decreases to 87%, 85%, 83% and 82% for the Cu-3, Cu-6, Cu-12 and Cu-24 samples, respectively, at the end of 6 h. It has been reported that the photocatalytic degradation efficiency increases as the amount of photocatalyst increases up to an optimum quantity of catalyst, after which a decrease in degradation is observed [59]. In this particular study, the nanocomposites are hydrophobic and remain on the surface of the aqueous solution. For 5 mg of catalyst, a thin layer of nanoparticles is formed on the surface of the liquid, which thickens with an increase in the catalyst amount, making the surface opaque. Additionally, only the topmost layers in the catalyst surface are exposed to sunlight, which may not be in contact with the aqueous solution. This implies that the charge transfer from the nanoparticles to the organic pollutants in the aqueous solution may be less efficient. This shielding effect due to the increase in the quantity of the photocatalyst could explain the lower photocatalytic activity. In addition, at high concentrations, the photocatalyst nanoparticles have a tendency to aggregate, which, in turn, reduces the number of active sites. Khataee et al. reported similar results, in which degradation efficiency increases from 14% to 57% with a catalyst dosage from 0.25 to 1 g/L, followed by a decrease in the degradation efficiency by 50% with a further increase in the catalyst dosage up to 2 g/L [60]. In fact, for very high concentrations of catalysts, the increase in turbidity of the suspension decreases the light penetration and the photodegradation efficiency [61].

The photocatalytic dye degradation in several Cu-based nanomaterials is presented in Table 3, for comparison. In general, their dye degradation efficiency is very high, depending upon the dye, excitation source and catalyst load. However, most of the photocatalytic degradation activity is carried out under high-power UV- light, which showed degradation efficiencies of up to 90% for various organic dyes. Most of the studies used higher catalyst amounts owing to higher C_0. Under sunlight or solar simulator, the degradation efficiencies were lower ranging between 65 and 80%. In comparison, the nanoparticle mixtures in this study manifest very high degradation efficiencies under sunlight, reaching 96% for Cu-3, consisting of the highest quantity of Cu_3N nanoparticles.

Table 3. Comparison of photocatalytic performance of recent works using Cu-based nanomaterials and the present study.

Catalyst	Dye	Light Source	Initial Concentration	Catalyst Dosage	Degradation Rate (%)	Ref.
CuO-Cu_2O	MB Methylene Orange	150 W metal halide lamp	5 mg/L	0.2 g	80 50	[8]
CuO	Aniline	UV-C lamp	50 mg/L	0.01–0.1 g	90	[12]
Cu_2O	Fluroxypyr	500 W metal halide lamp	11.17 mg/L	0.1 g	83	[13]
Cu_3N	MB Methylene Orange	Solar simulator	20 mg/L	0.1 g	61 89	[14]
Au-Cu_3N	MB	250 W UV lamp	25 mg/L	15 mg	84	[25]
Cu_2O/CuO/Cu	MB Rhodamine B	Sun light	20 mg/L	40 mg	65 60	[62]
Cu-ZnO	Methylene Orange Indigo Carmine Rhodamine B	30 W UV-light lamp	10 mg/L	25 mg	91 92 90	[63]
Cu-3 Cu-6 Cu-12 Cu-24	MB	Sun light	5 mg/L	5 mg	96 95 93 92	Present study

In this section, the effect of blocking UV-radiation, in order to exclusively study visible solar-driven photocatalytic degradation of MB, is analyzed. The absorption spectra of the nanoparticle mixtures provide an additional peak at ~300 nm (Figure S3), which corresponds to the absorption peak of oleylamine, corroborating our FTIR studies [64]. The presence of this organic moiety shields the nanoparticle surface and could have a negative effect on their photocatalytic efficiencies. For this, a glass lid is placed on the beakers during sunlight exposure. Figure S5 provides the transmittance characteristics of distilled water (DI), MB solution and the glass cover/lid. As observed, DI transmits from the UV to infrared with an absorption peak at around 980 nm. The transmittance of DI with MB is very similar to the transmittance of DI alone, except for a strong absorption peak at 660 nm corresponding to MB absorption. However, the glass lid transmits uniformly from 400 nm to 1100 nm, i.e., from the visible to the infrared regions. The purpose of the lid was therefore to limit the photoexcitation wavelengths from 400 nm to 1100 nm, allowing us to study MB degradation under visible light. For the 5 mg catalyst loading, the lid did not change the photocatalytic degradation efficiency significantly. On the other hand, for the 10 mg loading, an overall reduction in the photocatalytic efficiency is observed compared to the 5 mg loading, irrespective of the lid. Considering the 10 mg loading with and without the lid, the presence of the lid improves the degradation efficiency by 10% with MB degradations of 92%, 89%, 88% and 90% for the Cu-3, Cu-6, Cu-12 and Cu-24 samples, respectively. In order to reduce the shielding effect mentioned before, the 10 mg catalyst-loaded MB solution was sonicated (represented as SS on the graphs in Figure 7) in order to homogeneously disperse the nanoparticles in the solution. After sonication, the turbidity of the solution increased, which consequently decreased the photocatalytic degradation by 10% compared to the non-sonicated samples. In fact, the MB degradation with the lid was 83%, 88%, 81% and 79%, and without the lid, it was 76%, 81%, 76% and 77%, for the Cu-3, Cu-6, Cu-12 and Cu-24 samples, respectively. Since, both Cu_2O and Cu_3N are p-type semiconductors, when in contact with metallic Cu, air or water, they undergo downward band bending, creating an energy barrier for the photogenerated holes [65]. Band bending is an important mechanism for tuning the surface properties of a semiconductor. For p-type semiconductors in contact with O-rich environments, such as in aqueous media, the conduction band and valence band undergo downward band bending. This phenomenon tends to reduce the excitonic radiative recombination in the band-bending region [66]. In nanomaterials with a high surface-to-volume ratio, it results in an overall decrease in radiative recombination. In such situations, excitonic relaxation occurs via non-radiative pathways. Furthermore, downward band bending results in electron accumulation near the surface of the nanoparticles on photoexcitation. Subsequently, the transfer of electrons to oxygen molecules and hydroxyl radicals in contact with the nanoparticle surface occurs in aqueous media via a type II mechanism of dye degradation. Thereupon, oxidizing species, such as singlet oxygen, superoxide, hydroxyl radical and ion, are generated that degrade organic pollutants, such as MB, as shown in Figure 5b. Even though, dyes undergo photobleaching under the appropriate radiation via type I and II mechanisms, the presence of a semiconductor photocatalyst produces a high number of electrons that are transferred to the oxygen molecules that subsequently create a higher amount of ROS, which, in turn, enhances the dye degradation. After several oxidations of the oxygen molecule to either the hydroxyl radical or ions that degrade the dye, they finally condense into water, which is the final product of the photocatalytic process. Even though Cu nanoparticles themselves demonstrate good photocatalytic efficiency via efficient production of ROS, the samples with Cu_3N tend to possess a higher catalytic efficiency of ~4%. However, the catalytic efficiency of Cu nanoparticles decreases with higher catalyst loading. Additionally, samples containing Cu_3N tend to present higher adsorption kinetics than the Cu-24 sample containing only Cu nanoparticles. Therefore, the combined effect of adsorption and photocatalysis is required in order to obtain a higher dye-degradation efficiency.

6. Conclusions

In summary, we have successfully synthesized Cu-based nanocomposites with different ratios of Cu, Cu_2O and Cu_3N using non-aqueous sol-gel routes by varying the synthesis time. These were successfully applied to the visible-light-driven photocatalytic degradation of MB. The presence of the plasmonic Cu metal nanoparticles in these samples induces downward band bending in the p-type Cu_2O and Cu_3N, creating electron accumulation on their surfaces. Therefore, the mechanism of dye degradation in addition to photobleaching is the generation of reactive oxygen species on the transfer of electrons from the nanoparticle surface to the oxygen and hydroxyl radicals present in the aqueous medium. In narrow band gap semiconductors, the rapid excitonic recombination rate that tends to decrease the photocatalytic efficiency is circumvented via the band bending effect in the p-type semiconductor at the metal interface as well as at the aqueous interface. Other parameters that enhance the photocatalytic activity include the increase in the absorption range of the photocatalyst, their morphology, higher specific surface, optimum catalyst dosage and the homogeneity of the catalyst dispersion in the solution. This study showed that the presence of Cu_3N extends the light absorption range to the near-infrared region, with an effective absorption range from 600 nm to 1000 nm for the nanoparticle mixtures adapted to visible light photocatalysis. The photocatalytic degradation is also the highest for the sample containing nanoparticles with the highest specific surface, i.e., Cu_2O and Cu_3N. However, the stability of these nanoparticles still needs to be explored as Cu_3N and Cu can undergo oxidation and Cu_2O can further oxidize to CuO. Another challenge remains in applying these nanomaterials to larger-scale setups and reclaiming the catalysts after exhaustion. Therefore, our future work will consist of understanding the stability of these nanocomposites for various cycles of dye degradation by immobilizing them on appropriate supports to facilitate their recovery.

Supplementary Materials: The following supporting information can be downloaded at: https://www.mdpi.com/article/10.3390/nano13081311/s1, Figure S1: Calibration curve of methylene blue with the Lovibond photometer; Figure S2: EDX of the truncated octahedron in sample 6 h; Figure S3: UV-Vis absorption spectrum of the 6 h sample with an absorption peak at ~300 nm; Figure S4: Tauc plot of Cu-3. Figure S5: UV-Vis spectra of distilled water (DI), methylene blue (MB) and the glass cover.

Author Contributions: Conceptualization: P.P. and P.R.; methodology: P.P., P.R. and E.R.; validation, P.R. and E.R.; formal analysis, P.P., D.S.W., L.R. and E.E.; investigation, P.P., P.R., E.R., D.S.W., L.R., E.E. and O.V.; resources, P.R. and E.R.; data curation, P.P.; writing—original draft preparation, P.P.; writing—review and editing, P.P., P.R. and E.R.; supervision, P.R. and E.R.; project administration, P.R.; funding acquisition, P.R. and E.R. All authors have read and agreed to the published version of the manuscript.

Funding: This research has been supported by the European Regional Development Fund project grant number TK134 "EQUiTANT", Eesti Maaülikool (EMÜ Bridge Funding (P200030TIBT)), and T210013TIBT "PARROT mobility program" from Campus France, EMÜ Astra project EMBio "Value-chain based bio-economy".

Institutional Review Board Statement: Not applicable.

Informed Consent Statement: Not applicable.

Data Availability Statement: Not applicable.

Conflicts of Interest: The authors declare no conflict of interest.

References

Shindhal, T.; Rakholiya, P.; Varjani, S.; Pandey, A.; Ngo, H.H.; Guo, W.; Ng, H.Y.; Taherzadeh, M.J. A critical review on advances in the practices and perspectives for the treatment of dye industry wastewater. *Bioengineered* **2021**, *12*, 70–87. [CrossRef] [PubMed]

Sultana, K.A.; Islam, M.T.; Silva, J.A.; Turley, R.S.; Hernandez-Viezcas, J.A.; Gardea-Torresdey, J.L.; Noveron, J.C. Sustainable synthesis of zinc oxide nanoparticles for photocatalytic degradation of organic pollutant and generation of hydroxyl radical. *J. Mol. Liq.* **2020**, *307*, 112931. [CrossRef]

3. Gutierrez-Mata, A.G.; Velazquez-Martínez, S.; Álvarez-Gallegos, A.; Ahmadi, M.; Hernández-Pérez, J.A.; Ghanbari, F.; Silva-Martínez, S. Recent Overview of Solar Photocatalysis and Solar Photo-Fenton Processes for Wastewater Treatment. *Int. J. Photoenergy* **2017**, *2017*, 8528063. [CrossRef]
4. Iervolino, G.; Zammit, I.; Vaiano, V.; Rizzo, L. *Limitations and Prospects for Wastewater Treatment by UV and Visible-Light-Active Heterogeneous Photocatalysis: A Critical Review*; Springer International Publishing: Berlin/Heidelberg, Germany, 2020; pp. 225–264. [CrossRef]
5. Ren, G.; Han, H.; Wang, Y.; Liu, S.; Zhao, J.; Meng, X.; Li, Z. Recent Advances of Photocatalytic Application in Water Treatment: A Review. *Nanomaterials* **2021**, *11*, 1804. [CrossRef] [PubMed]
6. Chen, Y.; Wu, Y.; Zhang, Y.; Huang, S.; Lv, H.; Chen, J.; Zhai, Y.; Cheng, J.; Pan, J. Cu-based heterostructure photocatalysts derived from Cu sludge and municipal sewage sludge for efficient degradation of 2,4-dichlorophenol. *Chem. Eng. J.* **2022**, *429*, 132140. [CrossRef]
7. Dasineh Khiavi, N.; Katal, R.; Kholghi Eshkalak, S.; Masudy-Panah, S.; Ramakrishna, S.; Jiangyong, H. Visible Light Driven Heterojunction Photocatalyst of CuO–Cu$_2$O Thin Films for Photocatalytic Degradation of Organic Pollutants. *Nanomaterials* **2019**, *9*, 1011. [CrossRef]
8. Bayat, F.; Sheibani, S. Enhancement of photocatalytic activity of CuO-Cu$_2$O heterostructures through the controlled content of Cu$_2$O. *Mater. Res. Bull.* **2022**, *145*, 111561. [CrossRef]
9. Mkhalid, I.A.; Shawky, A. Cu-supported Cu$_2$O nanoparticles: Optimized photodeposition enhances the visible light photodestruction of atrazine. *J. Alloys Compd.* **2021**, *853*, 157040. [CrossRef]
10. Luo, Z.; Jiang, H.; Li, D.; Hu, L.; Geng, W.; Wei, P.; Ouyang, P. Improved photocatalytic activity and mechanism of Cu$_2$O/N–TiO$_2$ prepared by a two-step method. *RSC Adv.* **2014**, *4*, 17797–17804. [CrossRef]
11. Paredes, P.; Rauwel, E.; Rauwel, P. Surveying the Synthesis, Optical Properties and Photocatalytic Activity of Cu$_3$N Nanomaterials. *Nanomaterials* **2022**, *12*, 2218. [CrossRef]
12. Norzaee, S.; Djahed, B.; Khaksefidi, R.; Mostafapour, F.K. Photocatalytic degradation of aniline in water using CuO nanoparticles. *J. Water Supply Res. Technol.-Aqua* **2017**, *66*, 178–185. [CrossRef]
13. Yu, X.; Kou, S.; Zhang, J.; Tang, X.; Yang, Q.; Yao, B. Preparation and characterization of Cu$_2$O nano-particles and their photocatalytic degradation of fluroxypyr. *Environ. Technol.* **2018**, *39*, 2967–2976. [CrossRef]
14. Sithole, R.K.; Machogo, L.F.E.; Moloto, M.J.; Gqoba, S.S.; Mubiayi, K.P.; Van Wyk, J.; Moloto, N. One-step synthesis of Cu$_3$N, Cu$_2$S and Cu$_9$S$_5$ and photocatalytic degradation of methyl orange and methylene blue. *J. Photochem. Photobiol. A Chem.* **2020**, *397*, 112577. [CrossRef]
15. Zhang, P.; Liu, H.; Li, X. Photo-reduction synthesis of Cu nanoparticles as plasmon-driven non-semiconductor photocatalyst for overall water splitting. *Appl. Surf. Sci.* **2021**, *535*, 147720. [CrossRef]
16. Wang, Q.; Liu, Z.; Zhang, S.; Cui, Y.; Gao, S.; Wang, Y. Hydrothermal deposition of Cu$_2$O-Ag nanoparticles co-sensitized TiO$_2$ nanotube arrays and their enhanced photoelectrochemical performance. *Sep. Purif. Technol.* **2019**, *211*, 866–872. [CrossRef]
17. Huang, H.; Zhao, J.; Weng, B.; Lai, F.; Zhang, M.; Hofkens, J.; Roeffaers, M.B.J.; Steele, J.A.; Long, J. Site-Sensitive Selective CO$_2$ Photoreduction to CO over Gold Nanoparticles. *Angew. Chem. Int. Ed.* **2022**, *61*, e202204563. [CrossRef]
18. Ben Saber, N.; Mezni, A.; Alrooqi, A.; Altalhi, T. A review of ternary nanostructures based noble metal/semiconductor for environmental and renewable energy applications. *J. Mater. Res. Technol.* **2020**, *9*, 15233–15262. [CrossRef]
19. Rauwel, E.; Galeckas, A.; Soares, M.R.; Rauwel, P. Influence of the Interface on the Photoluminescence Properties in ZnO Carbon-Based Nanohybrids. *J. Phys. Chem. C* **2017**, *121*, 14879–14887. [CrossRef]
20. Rauwel, E.; Galeckas, A.; Rauwel, P.; Fjellvåg, H. Unusual Photoluminescence of CaHfO$_3$ and SrHfO$_3$ Nanoparticles. *Adv. Funct. Mater.* **2012**, *22*, 1174–1179. [CrossRef]
21. Grazulis, S.; Chateigner, D.; Downs, R.T.; Yokochi, A.F.T.; Quiros, M.; Lutterotti, L.; Manakova, E.; Butkus, J.; Moeck, P.; Le Bail, A. Crystallography Open Database—An open-access collection of crystal structures. *J. Appl. Crystallogr.* **2009**, *42*, 726–729. [CrossRef]
22. Coelho, A. TOPAS and TOPAS-Academic: An optimization program integrating computer algebra and crystallographic objects written in C++. *J. Appl. Crystallogr.* **2018**, *51*, 210–218. [CrossRef]
23. Primc, D.; Indrizzi, L.; Tervoort, E.; Xie, F.; Niederberger, M. Synthesis of Cu$_3$N and Cu$_3$N–Cu$_2$O multicomponent mesocrystals: Non-classical crystallization and nanoscale Kirkendall effect. *Nanoscale* **2021**, *13*, 17521–17529. [CrossRef]
24. Mourdikoudis, S.; Liz-Marzán, L.M. Oleylamine in Nanoparticle Synthesis. *Chem. Mater.* **2013**, *25*, 1465–1476. [CrossRef]
25. Barman, D.; Paul, S.; Ghosh, S.; De, S.K. Cu$_3$N Nanocrystals Decorated with Au Nanoparticles for Photocatalytic Degradation of Organic Dyes. *ACS Appl. Nano Mater.* **2019**, *2*, 5009–5019. [CrossRef]
26. Won, Y.-H.; Stanciu, L.A. Cu$_2$O and Au/Cu$_2$O Particles: Surface Properties and Applications in Glucose Sensing. *Sensors* **2012**, *12*, 13019–13033. [CrossRef]
27. Chen, Y.-A.; Chou, K.-H.; Kuo, Y.-Y.; Wu, C.-Y.; Hsiao, P.-W.; Chen, P.-W.; Yuan, S.-H.; Wuu, D.-S. Formation of ZnO/Zn0.5Cd0.5Se Alloy Quantum Dots in the Presence of High Oleylamine Contents. *Nanomaterials* **2019**, *9*, 999. [CrossRef]
28. Mbewana-Ntshanka, N.G.; Moloto, M.J.; Mubiayi, P.K. Role of the amine and phosphine groups in oleylamine and trioctylphosphine in the synthesis of copper chalcogenide nanoparticles. *Heliyon* **2020**, *6*, e05130. [CrossRef]
29. Sahai, A.; Goswami, N.; Kaushik, S.D.; Tripathi, S. Cu/Cu$_2$O/CuO nanoparticles: Novel synthesis by exploding wire technique and extensive characterization. *Appl. Surf. Sci.* **2016**, *390*, 974–983. [CrossRef]
30. Liu, Y.; Ren, F.; Shen, S.; Fu, Y.; Chen, C.; Liu, C.; Xing, Z.; Liu, D.; Xiao, X.; Wu, W.; et al. Efficient enhancement of hydrogen production by Ag/Cu$_2$O/ZnO tandem triple-junction photoelectrochemical cell. *Appl. Phys. Lett.* **2015**, *106*, 123901. [CrossRef]

31. Wei, W.; Xu, B.; Huang, Q. Controllable Synthesis and Catalytic Property of Novel Copper Oxides (CuO and Cu$_2$O) Nanostructures. *Int. J. Mater. Sci. Appl.* **2016**, *5*, 18–22. [CrossRef]
32. Sajeev, A.; Paul, A.M.; Nivetha, R.; Gothandapani, K.; Gopal, T.S.; Jacob, G.; Muthuramamoorty, M.; Pandiaraj, S.; Alodhayb, A.; Kim, S.Y.; et al. Development of Cu$_3$N electrocatalyst for hydrogen evolution reaction in alkaline medium. *Sci. Rep.* **2022**, *12*, 2004. [CrossRef] [PubMed]
33. Todaro, M.; Alessi, A.; Sciortino, L.; Agnello, S.; Cannas, M.; Gelardi, F.M.; Buscarino, G. Investigation by Raman Spectroscopy of the Decomposition Process of HKUST-1 upon Exposure to Air. *J. Spectrosc.* **2016**, *2016*, 8074297. [CrossRef]
34. Li, Y.; Wang, Y.; Wang, M.; Zhang, J.; Wang, Q.; Li, H. A molecularly imprinted nanoprobe incorporating Cu$_2$O@Ag nanoparticles with different morphologies for selective SERS based detection of chlorophenols. *Microchim. Acta* **2019**, *187*, 59. [CrossRef] [PubMed]
35. Zayyoun, N.; Bahmad, L.; Laânab, L.; Jaber, B. The effect of pH on the synthesis of stable Cu$_2$O/CuO nanoparticles by sol–gel method in a glycolic medium. *Appl. Phys. A* **2016**, *122*, 488. [CrossRef]
36. Ullah, N.; Ullah, A.; Rasheed, S. Green synthesis of copper nanoparticles using extract of Dicliptera Roxburghiana, their characterization and photocatalytic activity against methylene blue degradation. *Lett. Appl. NanoBioSci.* **2020**, *9*, 897–901. [CrossRef]
37. Mikami, K.; Kido, Y.; Akaishi, Y.; Quitain, A.; Kida, T. Synthesis of Cu$_2$O/CuO Nanocrystals and Their Application to H$_2$S Sensing. *Sensors* **2019**, *19*, 211. [CrossRef]
38. Nakamura, T.; Hayashi, H.; Hanaoka, T.A.; Ebina, T. Preparation of copper nitride (Cu$_3$N) nanoparticles in long-chain alcohols at 130–200 °C and nitridation mechanism. *Inorg. Chem.* **2014**, *53*, 710–715. [CrossRef]
39. Mott, D.; Galkowski, J.; Wang, L.; Luo, J.; Zhong, C.-J. Synthesis of Size-Controlled and Shaped Copper Nanoparticles. *Langmuir* **2007**, *23*, 5740–5745. [CrossRef]
40. Atarod, M.; Nasrollahzadeh, M.; Sajadi, S.M. Green synthesis of a Cu/reduced graphene oxide/Fe$_3$O$_4$ nanocomposite using Euphorbia wallichii leaf extract and its application as a recyclable and heterogeneous catalyst for the reduction of 4-nitrophenol and rhodamine B. *RSC Adv.* **2015**, *5*, 91532–91543. [CrossRef]
41. Valodkar, M.; Modi, S.; Pal, A.; Thakore, S. Synthesis and anti-bacterial activity of Cu, Ag and Cu–Ag alloy nanoparticles: A green approach. *Mater. Res. Bull.* **2011**, *46*, 384–389. [CrossRef]
42. Mallick, S.; Sanpui, P.; Ghosh, S.S.; Chattopadhyay, A.; Paul, A. Synthesis, characterization and enhanced bactericidal action of a chitosan supported core–shell copper–silver nanoparticle composite. *RSC Adv.* **2015**, *5*, 12268–12276. [CrossRef]
43. Dorranian, D.; Dejam, L.; Sari, A.H.; Hojabri, A. Effect of nitrogen content on optical constants of copper nitride thin films prepared by DC magnetron reactive sputtering. *J. Theor. Appl. Phys.* **2009**, *3*, 37–41.
44. Sahoo, G.; Meher, S.R.; Jain, M.K. Band gap variation in copper nitride thin films. In Proceedings of the International Conference on Advanced Nanomaterials & Emerging Engineering Technologies, Chennai, India, 24–26 July 2013; pp. 540–542.
45. Chinnaiah, K.; Maik, V.; Kannan, K.; Potemkin, V.; Grishina, M.; Gohulkumar, M.; Tiwari, R.; Gurushankar, K. Experimental and Theoretical Studies of Green Synthesized Cu$_2$O Nanoparticles Using Datura Metel L. *J. Fluoresc.* **2022**, *32*, 559–568. [CrossRef]
46. Yeshchenko, O.A.; Bondarchuk, I.S.; Losytskyy, M.Y. Surface plasmon enhanced photoluminescence from copper nanoparticles: Influence of temperature. *J. Appl. Phys.* **2014**, *116*, 054309. [CrossRef]
47. Li, J.; Mei, Z.; Liu, L.; Liang, H.; Azarov, A.; Kuznetsov, A.; Liu, Y.; Ji, A.; Meng, Q.; Du, X. Probing Defects in Nitrogen-Doped Cu$_2$O. *Sci. Rep.* **2014**, *4*, 7240. [CrossRef]
48. Basavalingaiah, K.R.; Udayabhanu; Harishkumar, S.; Nagaraju, G. Synthesis of Cu$_2$O/Ag Composite Nanocubes with Promising Photoluminescence and Photodegradation Activity over Methylene Blue Dye. *Adv. Mater. Lett.* **2019**, *10*, 832–838. [CrossRef]
49. Cheng, Z.; Qi, W.; Pang, C.H.; Thomas, T.; Wu, T.; Liu, S.; Yang, M. Recent Advances in Transition Metal Nitride-Based Materials for Photocatalytic Applications. *Adv. Funct. Mater.* **2021**, *31*, 2100553. [CrossRef]
50. Li, H.-B.; Xie, X.; Wang, W.; Cheng, Y.; Wang, W.-H.; Li, L.; Liu, H.; Wen, G.; Zheng, R. Room-temperature ferromagnetism in nanocrystalline Cu/Cu$_2$O core-shell structures prepared by magnetron sputtering. *APL Mater.* **2013**, *1*, 042106. [CrossRef]
51. Wang, L.-C.; Liu, B.-H.; Su, C.-Y.; Liu, W.-S.; Kei, C.-C.; Wang, K.-W.; Perng, T.-P. Electronic Band Structure and Electrocatalytic Performance of Cu$_3$N Nanocrystals. *ACS Appl. Nano Mater.* **2018**, *1*, 3673–3681. [CrossRef]
52. Deuermeier, J.; Liu, H.; Rapenne, L.; Calmeiro, T.; Renou, G.; Martins, R.; Muñoz-Rojas, D.; Fortunato, E. Visualization of nanocrystalline CuO in the grain boundaries of Cu$_2$O thin films and effect on band bending and film resistivity. *APL Mater.* **2018**, *6*, 096103. [CrossRef]
53. Kumar, V.; O'Donnell, S.C.; Sang, D.L.; Maggard, P.A.; Wang, G. Harnessing Plasmon-Induced Hot Carriers at the Interfaces With Ferroelectrics. *Front Chem* **2019**, *7*, 299. [CrossRef] [PubMed]
54. Zhang, Y.; Huang, Y.; Zhu, S.-S.; Liu, Y.-Y.; Zhang, X.; Wang, J.-J.; Braun, A. Covalent S-O Bonding Enables Enhanced Photoelectrochemical Performance of Cu$_2$S/Fe$_2$O$_3$ Heterojunction for Water Splitting. *Small* **2021**, *17*, 2100320. [CrossRef] [PubMed]
55. Zhang, Y.; Yuan, S.-Y.; Zou, Y.; Li, T.-T.; Liu, H.; Wang, J.-J. Enhanced charge separation and conductivity of hematite enabled by versatile NiSe$_2$ nanoparticles for improved photoelectrochemical water oxidation. *Appl. Mater. Today* **2022**, *28*, 101552. [CrossRef]
56. Yuan, S.-Y.; Jiang, L.-W.; Hu, J.-S.; Liu, H.; Wang, J.-J. Fully Dispersed IrOx Atomic Clusters Enable Record Photoelectrochemical Water Oxidation of Hematite in Acidic Media. *Nano Lett.* **2023**, *23*, 2354–2361. [CrossRef]
57. Mrunal, V.K.; Vishnu, A.K.; Momin, N.; Manjanna, J. Cu$_2$O nanoparticles for adsorption and photocatalytic degradation of methylene blue dye from aqueous medium. *Environ. Nanotechnol. Monit. Manag.* **2019**, *12*, 100265. [CrossRef]

58. Tonda, S.; Kumar, S.; Shanker, V. Surface plasmon resonance-induced photocatalysis by Au nanoparticles decorated mesoporous g-C_3N_4 nanosheets under direct sunlight irradiation. *Mater. Res. Bull.* **2016**, *75*, 51–58. [CrossRef]
59. Yunus, N.N.; Hamzah, F.; So'aib, M.S.; Krishnan, J. Effect of Catalyst Loading on Photocatalytic Degradation of Phenol by Using N, S Co-doped TiO_2. *IOP Conf. Ser. Mater. Sci. Eng.* **2017**, *206*, 012092. [CrossRef]
60. Khataee, A.; Darvishi Cheshmeh Soltani, R.; Hanifehpour, Y.; Safarpour, M.; Gholipour Ranjbar, H.; Joo, S.W. Synthesis and Characterization of Dysprosium-Doped ZnO Nanoparticles for Photocatalysis of a Textile Dye under Visible Light Irradiation. *Ind. Eng. Chem. Res.* **2014**, *53*, 1924–1932. [CrossRef]
61. Soltani, N.; Saion, E.; Hussein, M.Z.; Erfani, M.; Abedini, A.; Bahmanrokh, G.; Navasery, M.; Vaziri, P. Visible Light-Induced Degradation of Methylene Blue in the Presence of Photocatalytic ZnS and CdS Nanoparticles. *Int. J. Mol. Sci.* **2012**, *13*, 12242–12258. [CrossRef]
62. Uma, B.; Anantharaju, K.S.; Malini, S.S.; More, S.; Vidya, Y.S.; Meena, S.; Surendra, B.S. Synthesis of novel heterostructured Fe-doped Cu_2O/CuO/Cu nanocomposite: Enhanced sunlight driven photocatalytic activity, antibacterial and supercapacitor properties. *Ceram. Int.* **2022**, *48*, 35834–35847. [CrossRef]
63. Karthik, K.V.; Raghu, A.V.; Reddy, K.R.; Ravishankar, R.; Sangeeta, M.; Shetti, N.P.; Reddy, C.V. Green synthesis of Cu-doped ZnO nanoparticles and its application for the photocatalytic degradation of hazardous organic pollutants. *Chemosphere* **2022**, *287*, 132081. [CrossRef]
64. Schöttle, C.; Bockstaller, P.; Gerthsen, D.; Feldmann, C. Tungsten nanoparticles from liquid-ammonia-based synthesis. *Chem. Commun.* **2014**, *50*, 4547–4550. [CrossRef]
65. Kibria, M.G.; Chowdhury, F.A.; Zhao, S.; AlOtaibi, B.; Trudeau, M.L.; Guo, H.; Mi, Z. Visible light-driven efficient overall water splitting using p-type metal-nitride nanowire arrays. *Nat. Commun.* **2015**, *6*, 6797. [CrossRef]
66. Stevanovic, A.; Büttner, M.; Zhang, Z.; Yates, J.T., Jr. Photoluminescence of TiO_2: Effect of UV Light and Adsorbed Molecules on Surface Band Structure. *J. Am. Chem. Soc.* **2012**, *134*, 324–332. [CrossRef]

Disclaimer/Publisher's Note: The statements, opinions and data contained in all publications are solely those of the individual author(s) and contributor(s) and not of MDPI and/or the editor(s). MDPI and/or the editor(s) disclaim responsibility for any injury to people or property resulting from any ideas, methods, instructions or products referred to in the content.

Article

Revealing the Dependency of Dye Adsorption and Photocatalytic Activity of ZnO Nanoparticles on Their Morphology and Defect States

Yuri Hendrix [1], Erwan Rauwel [2], Keshav Nagpal [1], Ryma Haddad [3], Elias Estephan [4], Cédric Boissière [3] and Protima Rauwel [1,*]

[1] Institute of Forestry and Engineering Sciences, Estonian University of Life Sciences, 51014 Tartu, Estonia; yurihendrix@gmail.com (Y.H.); keshav.nagpal@student.emu.ee (K.N.)
[2] Institute of Veterinary Medicine and Animal Sciences, Estonian University of Life Sciences, 51006 Tartu, Estonia; erwan.rauwel@emu.ee
[3] Laboratoire de Chimie de la Matière Condensée de Paris (LCMCP), Collège de France, CNRS, Sorbonne Université, 75005 Paris, France; ryma.haddad@sorbonne-universite.fr (R.H.); cedric.boissiere@upmc.fr (C.B.)
[4] Laboratoire Bioinginirie et Nanoscience (LBN), University of Montpellier, 34193 Montpellier, France; elias.estephan@umontpellier.fr
* Correspondence: protima.rauwel@emu.ee; Tel.: +372-731-3322

Abstract: ZnO is an effective photocatalyst applied to the degradation of organic dyes in aqueous media. In this study, the UV-light and sunlight-driven photocatalytic activities of ZnO nanoparticles are evaluated. A handheld Lovibond photometer was purposefully calibrated in order to monitor the dye removal in outdoor conditions. The effect of ZnO defect states, i.e., the presence of zinc and oxygen defects on the photocatalytic activity was probed for two types of dyes: fuchsin and methylene blue. Three morphologies of ZnO nanoparticles were deliberately selected, i.e., spherical, facetted and a mix of spherical and facetted, ascertained via transmission electron microscopy. Aqueous and non-aqueous sol-gel routes were applied to their synthesis in order to tailor their size, morphology and defect states. Raman spectroscopy demonstrated that the spherical nanoparticles contained a high amount of oxygen vacancies and zinc interstitials. Photoluminescence spectroscopy revealed that the facetted nanoparticles harbored zinc vacancies in addition to oxygen vacancies. A mechanism for dye degradation based on the possible surface defects in facetted nanoparticles is proposed in this work. The reusability of these nanoparticles for five cycles of dye degradation was also analyzed. More specifically, facetted ZnO nanoparticles tend to exhibit higher efficiencies and reusability than spherical nanoparticles.

Keywords: nanoparticles; ZnO; green photocatalysis; surface defects; ROS

1. Introduction

Removal of toxic chemicals, such as dyes from industrial effluents requires innovative treatment methods. Dyes can be toxic and carcinogenic in high accumulated doses in the body [1–4]. Furthermore, other than the dye itself, molecular side-groups, e.g., toluene generated from the dyes during natural degradation processes are also harmful compounds [5]. In that regard, dye degradation in the presence of a photocatalyst appears to be an efficient and eco-friendly method that produces simple compounds, such as water and carbon dioxide. Photocatalytic nanomaterials such as TiO_2 and ZnO are capable of fully degrading a large variety of organic and some inorganic molecules [6–13]. The dye-degradation mechanism involves electron transfer from the nanoparticle to the oxygen species, biomolecules or other organic compounds in its vicinity. While some concerns about the toxicity of semiconductor nanoparticles due to the creation of reactive oxygen species (ROS) have emerged [14–17], ZnO nanoparticles tend to be relatively safe, presenting cytotoxicity only in large quantities on inhalation and skin exposure [14–16]. In places with an abundance of natural sunlight, photocatalysis has shown to be an excellent option in terms of cost-effectiveness and sustainability. However, photocatalytic treatment of wastewater still needs further research in areas such as the degradation rates of different molecules, the influence of solar irradiation and their reusability. More particularly, the effect of dye adsorption by the nanoparticles on degradation efficiency is still ambiguous.

ZnO is a wide-bandgap semiconductor with a direct bandgap of 3.37 eV and is considered a highly efficient photocatalyst for wastewater treatment [7,8]. Absorption of light in semiconductors leads to the creation of excitons. The excited electron–hole pairs react with hydroxyl and oxygen molecules in their vicinity to form ROS that degrade organic molecules, such as dyes. In wide-bandgap semiconductors, the longer recombination lifetimes of excitons, augments the probability of ROS production through electron transfer. However, a wide bandgap also implies photo-absorption in the UV region, which in turn renders them unsuitable for visible-light photocatalysis.

Previous reports have demonstrated that the ZnO-nanoparticle synthesis route has an enormous impact on its photocatalytic activity [7,8,18–25]. Synthesis routes tailor the morphology, size and specific surface of the nanoparticle. Besides the shape, the photo catalytic activity also depends on the crystalline quality of the material or the presence of defect states [18–21,25]. Since the lifetime of excitons is short (~322 ps) for bulk ZnO, only those able to reach the surface contribute to the photocatalytic activity through the creation of ROS [26]. Furthermore, defects in the material, such as oxygen (V_O) and zinc vacancies (V_{Zn}) act as traps and prolong the lifetime of the excitons [27]. Nevertheless, only defects on the surface of the nanoparticle participate in photocatalysis. On the other hand, volume-related defects in ZnO act as traps and recombination centers and prevent the photogenerated electron from reaching the surface. Therefore, the reduction of defects in general has been of interest for several photocatalytic water-splitting applications using oxynitride materials [28,29]. This was achieved through the increase in surface area of the photocatalysts. Other strategies for more efficient photocatalytic water splitting include adding a metal co-catalyst that acts as an electron trap and discourages radiative recombinations [30]. Nevertheless, in small nanoparticles, surface defects are dominant due to the high surface-to-volume ratio, while in larger nanoparticles, both surface and volume defects are present. Surface defects can also adsorb oxide radicals and other molecules [18,19]. In particular, V_O on the surface of the nanoparticles tends to capture excitons that are easily transferred to the organic molecules adsorbed by them; therefore, these defect sites act as active centers for dye degradation.

In this work, we investigate the influence of the synthesis routes on the photocatalytic activity of ZnO nanoparticles through their size and morphology. Photocatalytic activity is known to be dependent on the morphology, nanoparticle size and defect states of ZnO. The adsorption and degradation of basic fuchsin ($C_{19}H_{17}N_3 \cdot HCl$) and methylene blue ($C_{16}H_{18}ClN_3S$) by the ZnO nanoparticles are evaluated. For this purpose, the Lovibond handheld photometer, especially suited to on-site analysis (Lovibond photometer MD 600) was calibrated for both dyes. The aim was to devise an on-site technique for facile and rapid dye-concentration analyses. Besides experiments under UV-light at the lab-scale, the samples were also exposed to direct sunlight in order to assess their photocatalytic activity in outdoor conditions. In principle, a photocatalyst should be reusable for multiple cycles of dye degradation. Therefore, the reusability of the nanoparticles for several cycles of dye degradation is an important parameter to monitor in order to ensure sustainability, which is also addressed in this work. All in all, this study evaluates the ability of ZnO nanoparticles synthesized via three different routes to remove dyes with optimum turnover rates in a sustainable and practical way.

2. Methods

2.1. Materials and Reagents

Zinc acetate $Zn(CH_3CO_2)_2$ (99.99%, Aldrich, St. Louis, MO, USA), benzyl amine (≥99.0%, Aldrich), zinc acetate anhydrous (99.9%, Aldrich), absolute ethanol (≥99.5%, Honeywell, Charlotte, NC, USA), NaOH (99.9%, Aldrich).

2.2. Synthesis of ZnO Nanoparticles

2.2.1. Synthesis of ZnO A

Inside a glovebox (>1 ppm H_2O), 3.41 mmol (626 mg) zinc acetate $Zn(CH_3CO_2)_2$ (99.99%, Aldrich) was added to 20 mL benzyl amine (≥99.0%, Aldrich). The reaction mixture was transferred to a stainless-steel autoclave and carefully sealed. Thereafter, the autoclave was taken out of the glovebox and heated in a furnace at 300 °C for 2 days. The resulting milky suspension was centrifuged; the precipitate was thoroughly washed with ethanol and dichloromethane and subsequently dried in air at 60 °C.

2.2.2. Synthesis of ZnO B

For the synthesis of ZnO B nanoparticles, 183.48 mg zinc acetate anhydrous (99.9%, Aldrich) was dissolved in 20 mL of absolute ethanol (≥99.5%, Honeywell). The solution was maintained under continuous magnetic stirring at 65 °C until the precursor was fully dissolved. Later, 0.125 M of NaOH (99.9%, Aldrich) in 20 mL absolute ethanol was added dropwise to obtain a 1:2.5 molar ratio. Thereafter, the mixture was maintained at 65 °C under continuous magnetic stirring for 2 h. The resulting dispersion was centrifuged at 4500 rpm for 6 min and dried for 24 h at 60 °C in air.

2.2.3. Synthesis of ZnO C

Firstly, 219.5 mg of zinc acetate dihydrate was dissolved in 20 mL of aqueous ethanol (70%). Then, the solution was maintained at 60 °C with a water bath under continuous magnetic stirring until a transparent solution was obtained. An amount of 0.125 M of NaOH (99.9%, Aldrich) in 20 mL aqueous ethanol (70%) was added drop-wise to obtain a 1:2.5 molar ratio. The aqueous ethanol was made with absolute ethanol (≥99.5%, Honeywell) and distilled water. The solution was maintained at 60 °C for another 2 h. The resulting dispersion was centrifuged and dried in air at 60 °C.

2.3. Characterization

X-ray diffraction patterns were collected in Bragg–Brentano geometry using a Bruker D8 Discover diffractometer (Bruker AXS, Karlsruhe, Germany) with CuKα1 radiation (λ = 0.15406 nm) selected with a Ge (111) monochromator and LynxEye detector. Transmission electron microscopy (TEM) was carried out on a Tecnai G2 F20 equipped with a 200 kV field emission gun (FEG) and a Jeol 200Cx equipped with a LaB_6 emitter. High-resolution TEM provides a point-to-point resolution of ~2.4 Å. Characteristics such as specific surface and porosity were measured during N_2 adsorption–desorption experiments using a surface area and pore size analyzer (BELSORP mini2, Osaka, Japan). Samples were degassed overnight at a temperature of 110 °C. The specific surface and pore size distribution were determined from the linear part of the Brunauer–Emmet–Teller (BET) equation limited by Rouquerol's rules in a range of relative pressures p/p_0 from 0.05 to 0.35. The Barrett–Joyner–Halenda (BHJ) method was used on the desorption branch of the isotherm for calculating the pore size distribution. The optical absorbance of the ZnO samples was determined using an UV-Vis UV-1600PC spectrophotometer in the 300–700 nm region. The bandgap of the ZnO samples was subsequently calculated with Tauc plots. Photoluminescence (PL) spectroscopy was carried out on ZnO powders at room temperature with an excitation wavelength of 365 nm of a LSM-365A LED (Ocean insight, Orlando, FL, USA) with a specified output power of 10 mW. The emission was collected using a FLAME UV-Vis spectrometer (Ocean optics, Orlando, FL, USA) with spectral resolution of 1.34 nm. Raman spectra were collected using a WITec Confocal Raman Microscope System alpha 300R

(WITec Inc., Ulm, Germany). Excitation in confocal Raman microscopy was generated using a frequency-doubled Nd:YAG laser (Newport, Irvine, CA, USA) at wavelengths of 532 nm and 633 nm with 50 mW maximum laser output power in a single longitudinal mode. The incident laser beam was focused onto the sample through a 20× NIKON (Nikon, Tokyo, Japan). The acquisition time and accumulation time for a single spectrum were set to 10 s.

2.4. Photocatalytic Degradation

The photocatalytic degradation of two different dyes, i.e., basic fuchsin (general-purpose grade, Fisher Chemical, Hampton, NH, USA) and methylene blue (high purity, biological stain, Thermos Scientific, Waltham, MA, USA) were investigated. Three different configurations were tested. The measurements were performed with a handheld Lovibond photometer (MD 600) which was calibrated against the UV-Vis spectrophotometer (UV-1600PC) as shown in the supporting data Figure S1. The dye removal was tracked by measuring the absorbance at or close to the main absorption peaks (i.e., 560 nm for basic fuchsin and 660 nm for methylene blue) at regular intervals for 3 h. The amount of residual dye in the solution is directly proportional to the absorbance. The removed amount can be calculated from the difference in the initial and residual dye concentrations taking into account the mass of the catalyst and the volume of the solution taken. This removal can be due to adsorption of the dye molecules onto the nanoparticles, as well as degradation of the molecules. The remaining dye concentration can be extrapolated from the calibration curve of Figure S1 as the absorbance is directly proportional to the concentration of the dye following Beers–Lambert law.

2.4.1. Photocatalytic Activity under UV-Light

For experiments in laboratory, 1 mg of each ZnO sample and 10 mL of 5 ppm dye solution were added to a Petri dish and placed under a UV-lamp of 365 nm excitation, consisting of four 9 W bulbs. The emission spectrum of the UV-lamp is provided in Figure S2 of the supporting information. To evaluate the reusability of the samples, the same experiment was repeated four times for each sample with a new solution.

2.4.2. Photocatalytic Activity under Sunlight

The second configuration was used to test the sample under solar irradiation in open air, on the roof of the Tehnikainstituut, Kreutzwaldi 56/1, Tartu, Estonia. For this, 1 mg of ZnO nanoparticles was mixed with a 10 mL aqueous solution containing 5 ppm of dye in a Petri dish. The irradiance and UV-index of the corresponding days and location are shown in Figure S3. All the experiments were performed on days with UV-index between 4–5. In order to counteract the evaporation of the solvent, a constant volume of 10 mL was maintained by adding distilled water before each absorbance measurement.

2.4.3. Adsorption of Dye and Its Influence on the Photocatalytic Activity under UV-Light

The third configuration enabled distinguishing the type of dye-removal process, i.e., photocatalytic degradation or adsorption. It also enabled evaluation of the influence of the adsorbed dye on the photocatalytic activity under UV light. The decrease in dye concentration after mixing 1 mg of ZnO nanoparticles with a 10 mL aqueous solution containing 5 ppm of dye in a Petri dish was monitored at several intervals. Photocatalysis was performed on these samples, first, immediately after attaining adsorption equilibrium and second, after 24 h of adsorption.

In order to elucidate the mechanisms involved in photocatalysis after dye adsorption, further experiments were carried out. After reaching an adsorption equilibrium, the nanoparticles were isolated from the dye solution via centrifugation. They were then exposed for three hours to the UV lamp, after which they were returned to the dye solution to check their reusability. This procedure was repeated 4 times. The procedure adopted to evaluate the adsorption and photocatalytic activity of the ZnO nanoparticles is provided in Figure 1.

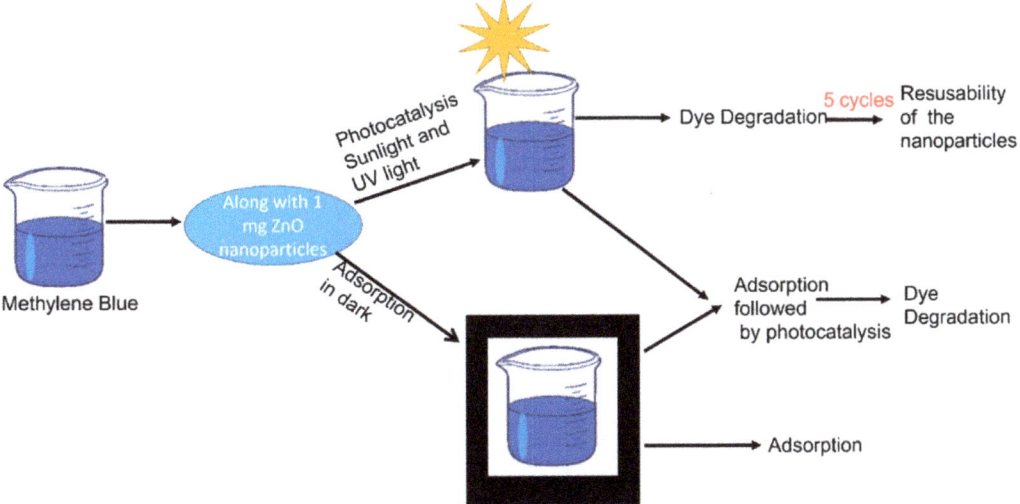

Figure 1. Procedure followed for evaluating the adsorption, photocatalytic efficiency and adsorption plus photocatalytic efficiency of ZnO A, ZnO B and ZnO C.

3. Results
3.1. Crystal Structure and Morphology

Diffractograms of the ZnO samples are shown in Figure 2a. The peaks (100), (002), (101), (102), and (110) correspond to the hexagonal Wurtzite structure (a = 3.25 Å and c = 5.20 Å) of ZnO (JCPDS, Card Number 36-1451). No secondary phases are visible in the XRD patterns, indicating that only single-phase ZnO nanoparticles were precipitated. The peak for ZnO B at 26° belongs to unreacted zinc acetate precursor [31].

The TEM images in Figure 2b–g provide an overview of the particle sizes and morphologies of the three ZnO samples. All three samples present different morphologies and particle sizes, due to differences in synthesis conditions (i.e., solvent, precursor, synthesis temperature and time). Among the three samples, ZnO A in Figure 2b displays the largest particle sizes with sharp edges and facets in hexagonal and triangular nanoparticles. The higher magnification image of ZnO A in Figure 2e consists of hexagonal and triangular-facetted nanoparticles. On the other hand, ZnO B, shown in Figure 2c, consists of much smaller agglomerates of nanoparticles that are spherical, exhibiting a porous structure. The HRTEM image of ZnO B in Figure 2f, illustrates spherical nanoparticles with an average size of 5 nm. The size of ZnO C nanoparticles is intermediate to the other two samples and nanoparticles with both sharp and smooth edges are visible in Figure 2d. Figure 2g is a higher magnification image of ZnO C, where several large hexagonally facetted nanoparticles are visible, along with less defined and smaller nanoparticles. The specific surface of the nanoparticles was calculated using the average size of the nanoparticles provided in the size distribution histograms in Figure S4. ZnO A, ZnO B and ZnO C have average particles sizes of 112 nm, 5 nm and 23 nm, respectively [20,21,31]. The hexagonally shaped nanoparticles were approximated to a sphere for specific surface calculations. Subsequently, the number of nanoparticles in 1 mg of the catalyst loading was calculated after determining the volume of the sphere, which was multiplied by its surface area. The theoretical specific surfaces of ZnO A, ZnO B and ZnO C are therefore 9.56 m^2/g, 214 m^2/g and 22.7 m^2/g, respectively. Subsequently, the surface areas of dry powders were deduced from N$_2$ physisorption using the BET method shown in Figure 2h–j [32]. The insets are the N$_2$ physisorption isotherms. In particular, the isotherm of ZnO A is poorly resolved due to its very low surface area, even though we used more than one gram of powder, suggesting that the detection limit of the apparatus was reached. Sample ZnO A presents a type II adsorption isotherm characteristic of macroporous materials. Sample ZnO B presents an isotherm with a hysteresis loop in between type IV-H2 and IV-H3 characteristics of the presence of mesopores, probably due to the packing of the small ZnO particles observed in the TEM image of Figure 2. Finally, sample ZnO C presents an isotherm with a hysteresis loop above 0.9 in p/p$_0$ of type IV-H3 which can be attributed to mesopores with slit-shaped porosities. Parameters such as the specific surface (S_{BET}), BET constant (C_{BET}), BET particle size (d_{BET}), monolayer adsorption volume (V_m) and pore volume (V_p) are tabulated in Table 1. ZnO B presented the largest specific surface of 72 m^2/g and the specific surface of ZnO C of 24 m^2/g. In addition, the volume of the monolayer adsorption decreased with increasing particle size, owing to the decrease in the specific surface. The pore volume (V_p) of samples ZnO B and C are approximately equal. Furthermore, ZnO B displayed a clear mesoporosity compared to ZnO C with a pore size distribution around 10 nm (inset of Figure 2i). For ZnO C, the average pore size distribution is higher at ~46 nm with minimum and maximum pore sizes of 10 nm and 60 nm (inset of Figure 2j), respectively. On the other hand, ZnO A displayed near-zero mesoporosity, suggesting that ZnO A consisted of large particles and that their packing creates a porosity larger than 50 nm and could not be evaluated by the present method. We noticed that for ZnO A, the presence of facetted and sharp-edged nanoparticles, could also promote their tight packing in dry-powder form. Our study however consisted of employing a colloidal dispersion of the nanoparticles in an aqueous dye solution that promotes the homogeneous distribution of nanoparticles without agglomeration. Therefore, the specific surfaces of the dried powders obtained from BET may not be entirely relevant for this present study, considering the fact that we are evaluating the photocatalytic activities of ZnO nanoparticles as colloidal suspensions in aqueous dyes.

Table 1. Parameters obtained from BET for the three samples.

Sample	S_{BET} (m^2/g)	C_{BET}	$d_{BET} = \frac{6000}{S_{BET} \times \rho_{ZnO}}$ (nm)	d_{TEM} (nm)	V_m (cm^3/g)	V_p (cm^3/g)
ZnO A	0.5	74	>50	114	1.2×10^{-7}	0
ZnO B	72	20	14.8	5	1.66×10^{-5}	0.23
ZnO C	24	22	44.2	47	5.5×10^{-6}	0.22

ZnO B was prepared with absolute ethanol and an anhydrous zinc acetate precursor. On the other hand, ZnO C was prepared with aqueous ethanol and a dihydrate acetate precursor. Thus, the difference between these two synthesis methods is the presence of water molecules in the solvent and precursor for ZnO C. These water molecules increase the reaction rate of the hydrothermal synthesis route and in turn, engender the precipitation of larger nanoparticles [21]. On the other hand, ZnO A, which was synthesized via non-aqueous sol-gel routes using benzyl amine as a solvent was devoid of water. The characteristics of these nanoparticles are tabulated in Table 2. Therefore, the type of solvent plays a crucial role in the morphology of the synthesized nanoparticles and influences other properties, as given below.

Table 2. Description of ZnO nanoparticles.

Sample	Diameter (nm)	Shape	Synthesis Route (Solvent)	Temp (°C)	Reaction Time (h)
ZnO A	80–140	Hexagon, tetrahedron, octahedron with sharp edges and facets.	Non-aqueous (Benzylamine)	300	48
ZnO B	5	Spheres	Non-aqueous (Ethanol)	65	2
ZnO C	27	Spheres, facetted nanoparticles	Aqueous (Ethanol (70%) and water (30%))	60	2

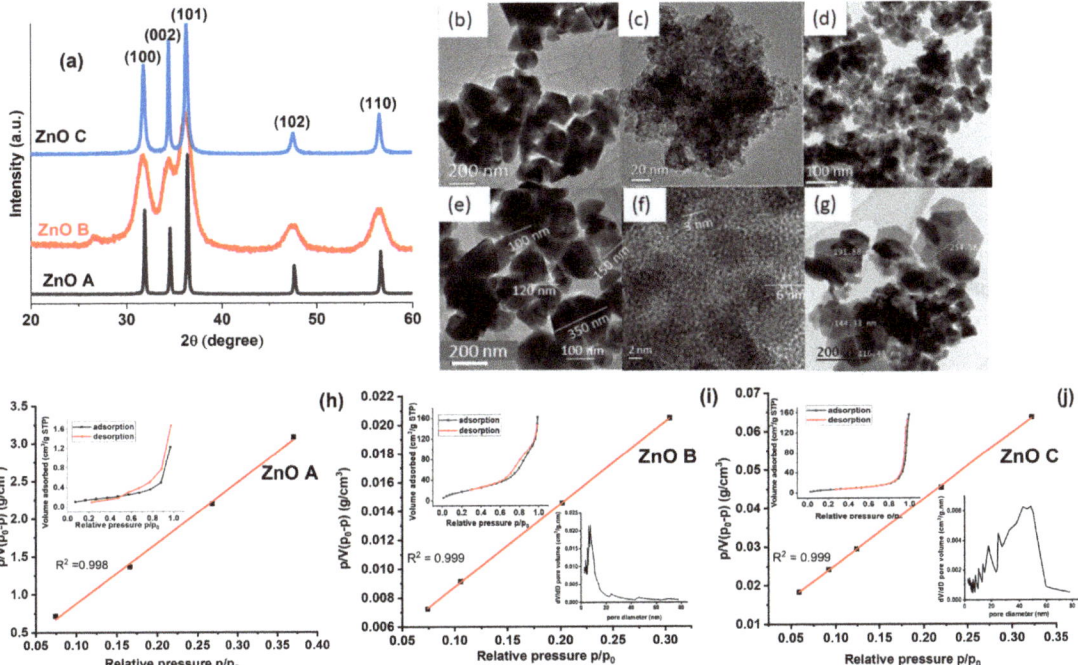

Figure 2. (**a**) X-ray diffraction patterns of the ZnO nanoparticles (JCPDS, card number 36-1451). TEM images of (**b**,**e**) ZnO A, (**c**,**f**) ZnO B, (**d**,**g**) ZnO C. (**h**–**j**) BET plots for N_2 adsorption on ZnO A, ZnO B and ZnO C. Insets are the full adsorption isotherms for the three samples, along with their pore size distributions.

3.2. Optical Characteristics

The Raman spectra for the samples were acquired within the range of 200 cm^{-1} and 800 cm^{-1} in Figure 3. For the Samples ZnO B and ZnO C, the laser excitation was 532 nm. On the other hand, due to the high fluorescence exhibited by ZnO A under the green laser excitation, a red laser with an excitation wavelength of 633 nm was used. In fact, 532 nm corresponds to the defect level absorption in ZnO, which produced a fluorescence signal that in turn, dissimulated the Raman signal. Starting from lower frequencies, Raman signatures under 300 cm^{-1} are assigned to the vibrations of Zn_i, and those above 300 cm^{-1} are assigned to the vibrations of oxygen atoms [33]. The peak at 275 cm^{-1} has been attributed to Zn_i or Zn_i clustering and is only visible in samples ZnO B and ZnO C [34,35]. The peak visible at ~320 cm^{-1} is a multiphonon scattering mode of E_{2H}–E_{2L} and is related to the crystalline quality of the sample. In fact, E_{2L} involves the vibration of the heavy Zn sublattice [36], and the E_{2H} at 440 cm^{-1} corresponds to lattice–oxygen vibrations of Wurtzite ZnO. The relative intensity of this peak is the highest for ZnO A, owing to the large particle size. The other phonon modes obtained at ~585 cm^{-1} and ~667 cm^{-1} correspond to $E_1(LO)$ and $E_1(TO)$ modes, respectively [37]. In general, the E_{2H}, E_{2H}–E_{2L}, $E_1(LO)$ modes involve the oxygen component of ZnO; especially, $E_1(LO)$ corresponds to oxygen-related defects [38]. For ZnO C, the E_{2H} mode has shifted to higher wave numbers, indicating a more stable lattice–oxygen configuration, or a lower amount of V_O. On the other hand, the E_{2H} peak for the other two samples has shifted to slightly lower wavenumbers, indicative of a higher number of V_O [31]. The $A_1(TO)$ with $E_1(TO)$ modes reveal variations in lattice bonds affected by local changes induced by intrinsic defects. The $E_1(LO)$ mode is usually dominant in doped ZnO samples, where Zn is replaced by a metallic dopant that modifies the local charge distribution owing to the presence of V_{Zn} or Zn_i [39].

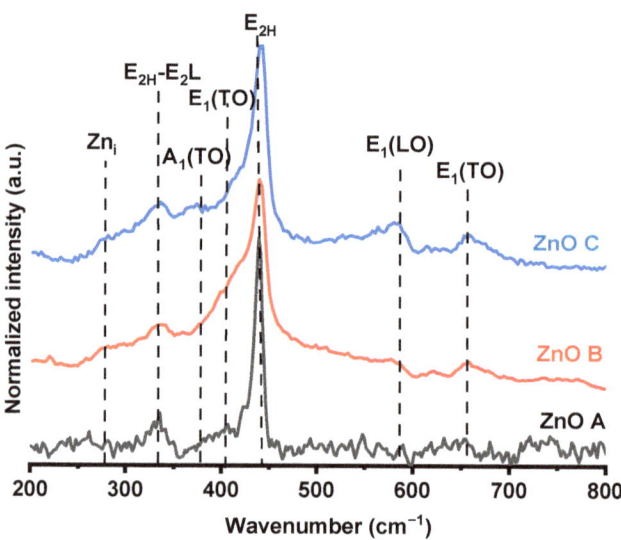

Figure 3. Normalized Raman spectra of the ZnO samples in the range of 100–800 cm^{-1}.

UV-Vis absorption spectroscopy was carried out followed by Tauc plots (Shown in Figure 4) in order to determine the absorbance range and band gaps of ZnO nanoparticles. For all the samples, the band gap lies between 3.07 eV and 3.26 eV, which is typical for ZnO. For ZnO B and ZnO C (Figure 4b,c), the band tail tapers off at ~1.75 eV and 2 eV in the visible region, respectively. Variations in the band gaps for ZnO nanoparticles are likely due to the differences in the synthesis conditions [20,40]. In addition, errors in the band gap estimation from Tauc plots may be due to the presence of organic moieties as a result of the synthesis routes [41]. In general, Tauc plot allows accurate estimation of band gaps for bare semiconductors.

Photoluminescence emission spectroscopy is necessary to understand the various radiative and non-radiative recombination mechanisms that could hinder or promote the photocatalytic activity of ZnO. The emission properties of the three ZnO samples are presented in Figure 4d. These emissions correspond to typical emission bands of ZnO with variations in intensities of certain emission signatures. In general, the ZnO emission spectra consist of two distinct bands called the near-band emission (NBE) and the defect level emission (DLE). NBE is in the UV region corresponds to band-to-band transitions [42]; DLE in the visible region arises from transitions between the conduction band or donor states to defect or band gap states. These defect states correspond to V_O, Zn_i, and V_{Zn} in these samples. ZnO emissions at ~2.5 eV and 2.2 eV correspond to volume (V_O^+) and surface (V_O^{++}) oxygen vacancy components, respectively, while the peak at ~3 eV corresponds to either Zn_i or V_{Zn} [20]. In the Raman spectrum of ZnO A in Figure 3, the vibrations corresponding to Zn_i are absent. In addition, in a previous study it was demonstrated that this sample was Zn-deficient and therefore, the emission at 3 eV is mostly likely due to V_{Zn} [43]. For ZnO B and C, redshifts of the NBE are due to shallow donor states, confirmed by the presence of Zn_i in their Raman spectra.

The ratio of the NBE to DLE is an indicator of the crystalline quality of ZnO. Therefore, ZnO C, with the highest ratio, exhibits the best crystalline quality among the three samples. Similarly, ZnO B displays a negligible DLE, which consists of mostly surface-related V_O^{++}. Despite the fact that the solvent and precursors used in the synthesis of ZnO B were non-aqueous and anhydrous, the presence of NaOH in the reaction serves as an oxygen supplying source via a hydrolytic reaction, which can reduce the quantity of V_O^+/V_O^{++}. For ZnO A, the volume-related V_O^+ emission at 2.5 eV dominates the PL spectrum followed by the V_O^{++} emission. Therefore, the presence of both types of oxygen vacancies can be correlated to the non-aqueous and non-hydrolytic synthesis route creating an oxygen-deficient synthesis environment.

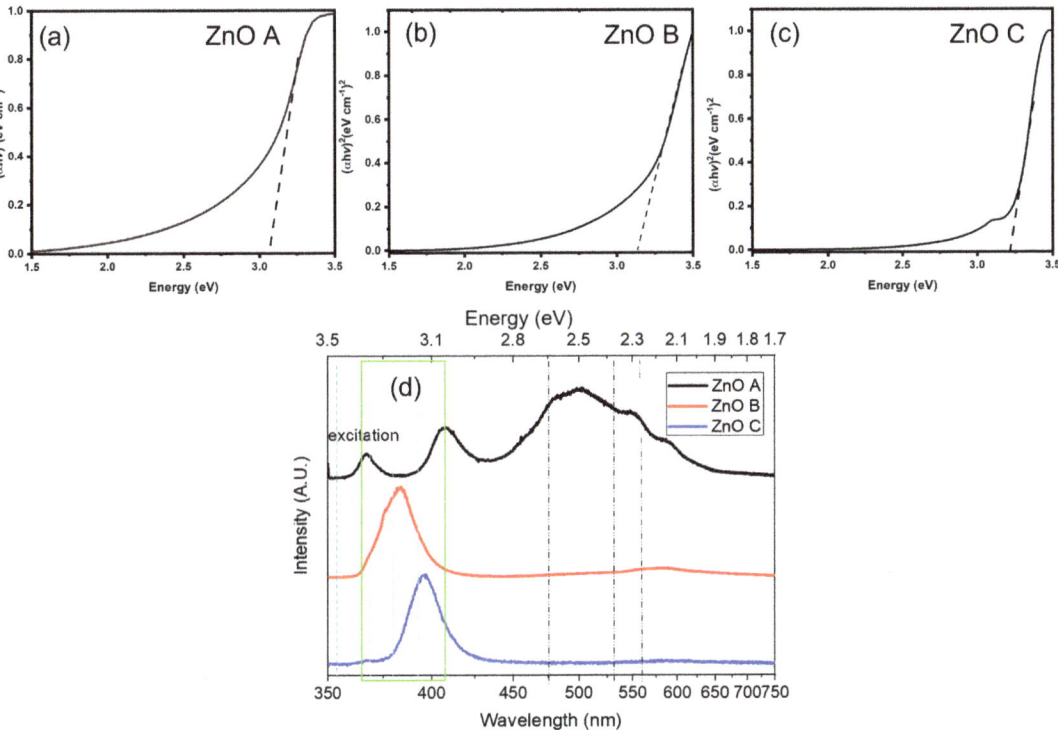

Figure 4. Tauc plots of (**a**) ZnO A, (**b**) ZnO B and (**c**) ZnO C and (**d**) photoluminescence emission spectra of ZnO A, B and C under 365 nm excitation.

3.3. Photocatalytic Dyedegradation

The ZnO nanoparticles were applied to the photocatalytic degradation of two dyes: methylene blue and basic fuchsin. Dye degradation or removal is noted as C/C_0, where C is the concentration of the dye after its removal and C_0 is the initial concentration. Depending on the experiment, the reduction of C/C_0 is considered an outcome of both, adsorption of dye molecules, as well as their degradation. In Figure 4a,d, the photocatalytic activity of the three ZnO samples is provided under UV light. For fuchsin removal, the photocatalysts are most efficient in the initial hours with the highest removal rate, as in Figure 5a. This tendency is more prominent for ZnO B-containing nanoparticles with the largest specific surface, displaying the lowest C/C_0 (i.e., 57.3% at 60 min) or the highest removal rate. However, the removal rate of fuchsin by ZnO B slows down significantly after the first hour. On the other hand, ZnO A has a more constant removal rate, while also having the lowest C/C_0 after three hours. The different tendencies in the removal rates of fuchsin with ZnO B and ZnO C compared to ZnO A could be a combination of both adsorption and photocatalytic degradation processes. Furthermore, the lower removal rate for ZnO B after the first hour suggests a shielding of the surface of the photocatalysts owing to the adsorption of dye molecules, as explained further in the manuscript.

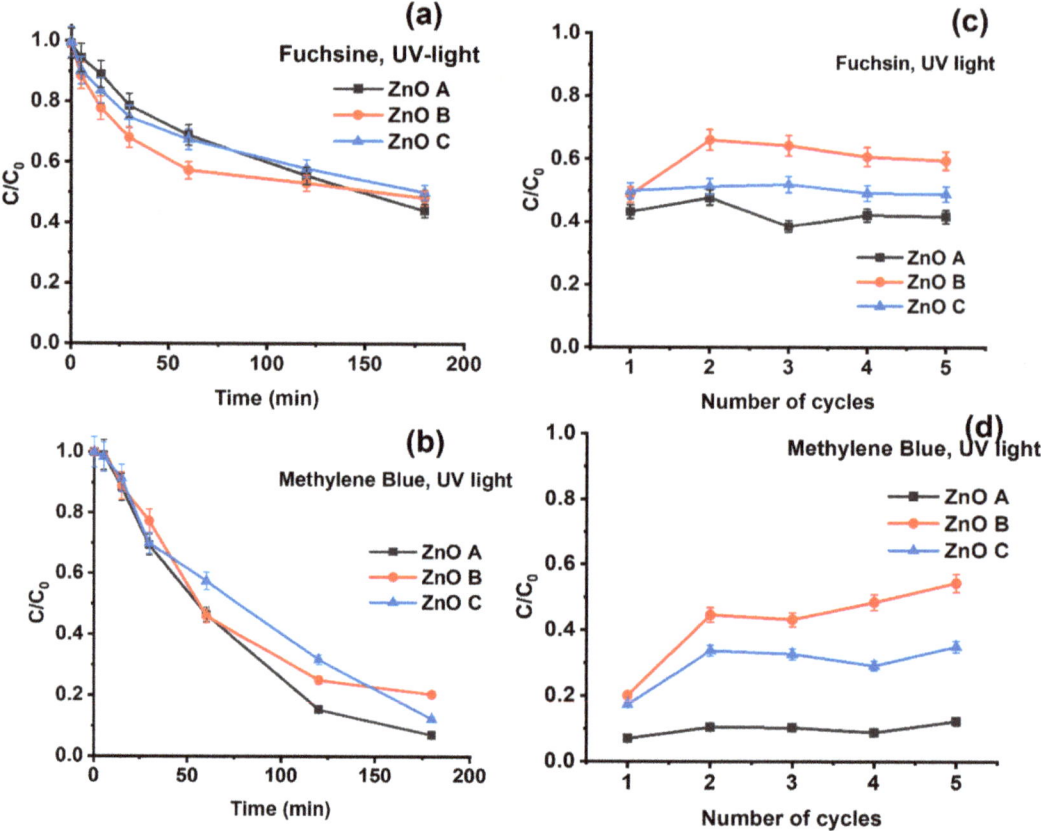

Figure 5. Photocatalytic degradation of a 10 mL solution containing 5 ppm of (**a**) basic fuchsin and of (**b**) methylene blue using 1 mg of ZnO nanoparticles under 365 nm UV light. Photocatalytic degradation of (**c**) basic fuchsin and of (**d**) methylene blue for five cycles of photocatalytic degradation. During each cycle of photocatalysis, the nanoparticles were exposed for 3 h to UV light. The nanoparticles were centrifuged and dried before the next cycle.

For methylene blue in Figure 5b, on the other hand, the initial high removal rate for the smallest ZnO nanoparticles (i.e., ZnO B and ZnO C) is absent. During the first 30 min, the dye degradation tendency is similar, i.e., linear for all the three samples. After the initial 30 min, the removal rate decreases slowly but remains higher than for fuchsin. In general, methylene blue is more actively degraded by the ZnO nanoparticles than fuchsin. Once again, ZnO A shows the most efficient degradation reaching removal rates of 90% after 3 h followed by ZnO C with 88% and ZnO B with 80%.

The reusability of the samples and the sustainability of the photocatalytic process were evaluated for five cycles of dye degradation under 365 nm UV light (Figure 5c,d). For ZnO A, i.e., the sample with the largest crystal size, no significant reduction in photocatalytic activity was observable after five cycles. In addition, ZnO C shows similar removal efficiencies for fuchsin in each cycle of Figure 5c. However, for methylene blue, the removal efficiency decreases after the first utilization, and then shows the same efficiency for four consecutive cycles, as seen in Figure 5d. ZnO C shows a lower efficiency than ZnO A, but a slightly higher efficiency than ZnO B. On the other hand, ZnO B shows a large decline in dye-removal efficiency after the first cycle for both fuchsin and methylene blue. One possible explanation for this decline could be the agglomeration of the nanoparticles due to drying between cycles or the adsorption of dye that degrades the surface of the nanoparticles and, in turn, decreases the photocatalytic degradation efficiency. In addition, the attachment of the dye on to the surface of the nanoparticles can alter the electrostatic charges on their surfaces and promote their agglomeration, which eventually results in lowering the effective specific surface.

Even though the morphology of the nanoparticles plays an important role, other reasons for the decrease in the removal capacity could be the passivation of surface defects via dye adsorption. This surface shielding could explain the lower photocatalytic activities of ZnO B and ZnO C. On the other hand, the presence of sharp edges in ZnO A of Figure 2b, limits shielding because edges cannot be covered, exposing highly active sites on the edges for catalytic reactions. Ni et al. [20], reported that edges are an important part of nanoparticles in the field of catalysis. In their study, they showed that atoms on edge sites are more active due to different chemical environments. They compared their activity with atoms from the facets and showed that atoms on edges promote chemical reactions with higher efficiency. Hejral et al. [37] also investigated the influence of nanoparticle shape in heterogeneous catalysis. They observed that the efficiency of catalytic reactions increases with the number of available edges on the surfaces.

On the other hand, facets in ZnO nanoparticles can be polar or non-polar. For example, hexagonal nanoparticles contain six polar and two non-polar facets [44]. Polar facets are either Zn- or O-terminated, and harbor the corresponding surface defect, i.e., V_{Zn} or V_O [45]. Since the dyes used in this study are cationic, they would most likely attach to the negatively charged V_{Zn}. Nevertheless, in aqueous media, hydroxyl groups or oxygen radicals would also be adsorbed on the surface of these nanoparticles at V_O^{++} that are positively charged, turning their surfaces electronegative and serving as functional groups to attach cationic dye molecules through covalent bonds. These edges are present to some extent in ZnO C, but in high amounts in ZnO A (Figure 2b,d). Therefore, for ZnO A, both V_{Zn} and V_O serve as anchoring sites for oxygen radicals and dye molecules. Consequently, despite having the lowest specific surface, ZnO A shows the highest efficiency after five cycles of photocatalytic degradation, without loss in efficiency. In addition, ZnO A has the highest number of volume-related defects, owing to the synthesis conditions that cannot be passivated or shielded, unlike surface defects. However, it is very unlikely that volume-related defects participate in the dye-degradation process. ZnO B harbours surface-related V_O^{++} and therefore, attracts hydroxyl groups or oxygen radicals. On the other hand, ZnO C that possesses the lowest amount of surface defects due to improved oxidation of the ZnO lattice during its synthesis, consists of shallow donors of Zn_i and a negligible amount of surface V_O^{++}. Some of the nanoparticles in ZnO C present sharp edges in Figure 2d, while others present no defined shape. The presence of these sharp edges is probably the explanation of its slightly higher photocatalytic activity compared to ZnO B.

The same experimental conditions, as in the previous section were applied to the study of the photocatalytic degradation of the two dyes under sunlight exposure. It is noteworthy that both fuchsin and methylene blue undergo degradation under sunlight exposure even in the absence of a photocatalyst (control sample), via the process of photobleaching Nevertheless, the presence of ZnO nanoparticles increases the dye-removal rate. For fuchsin, the presence of ZnO nanoparticles increased dye removal by 16–33% depending on the sample, and for methylene blue it increased by 15%.

In all cases, the degradation follows similar tendencies for the control sample, as well as the samples with the photocatalysts. In the initial stages, a fast decrease in dye concentration is observed during the first 40 min in the presence of the photocatalysts in Figure 6a,b. However, after this initial high removal rate, all curves then follow the same tendency as the control sample, implying that photobleaching becomes the dominant mechanism for dye degradation. Nevertheless, the dye-degradation efficiency is higher by 45% (ZnO A), 33% (ZnO B) and 25% (ZnO C) for fuchsin at the end of 3 h compared to the control sample; the maximum dye-removal efficiency was ~75% for ZnO A. In Figure 5a, ZnO B and ZnO C in fuchsin show lower removal rates under sunlight compared to ZnO A, similar to UV light exposure in Figure 5a. It should be noted that photobleaching in the absence of a photocatalyst is not significant under the UV light of 365 nm, implying that photocatalysis is the only mechanism for the decrease in C/C_0 under UV light and in the presence of ZnO. This suggests that photons of other wavelengths are responsible for photobleaching under sunlight, which, naturally, are the wavelengths corresponding to the main absorbance peaks of the dye (i.e., 660 nm for methylene blue and 550 nm for basic fuchsin).

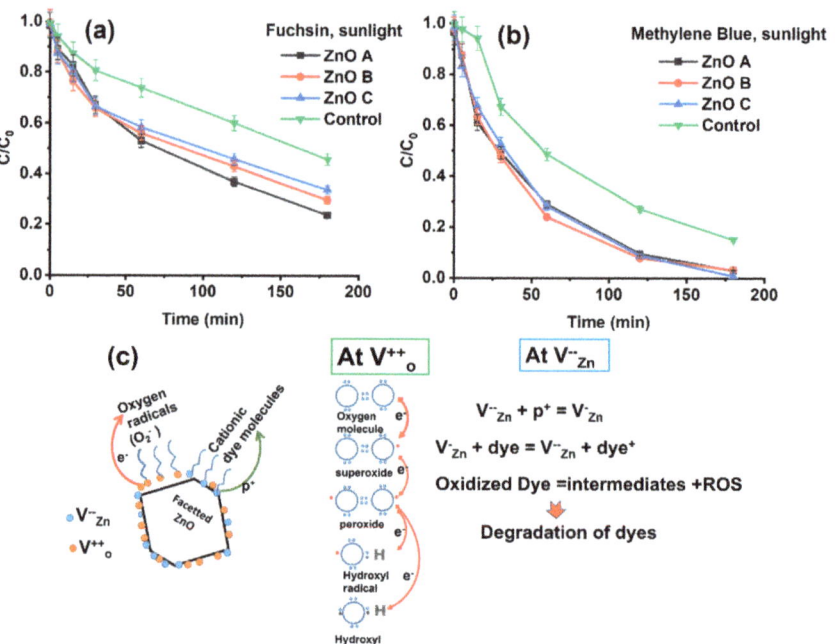

Figure 6. Photocatalytic degradation of a 10 mL solution containing 5 ppm (**a**) basic fuchsin and of (**b**) methylene blue with the three different ZnO samples (1 mg) and photobleaching of the control sample during 3 h of sunlight exposure and (**c**) mechanism of electron transfer from ZnO surface defects on sharp edges and the formation of ROS. The ZnO hexagonal nanoparticle considers only non-polar ZnO surfaces harboring surface defects for the sake of clarity.

For both, fuchsin and methylene blue, the photobleaching depends on the formation of singlet oxygen under oxygen-rich conditions, as photobleaching in de-oxygenated environments is less prominent or even suppressed [46]. In fact, the enhanced photobleaching of methylene blue compared to fuchsin is probably due to the longer lifetimes of the singlet state of the former [22]. This allows a higher number of electrons in the excited singlet-state to undergo intersystem crossing into the triplet state. At this point, they can either follow a type I or a type II photochemical route or de-excite via phosphorescence, along with a change in their spin multiplicity. Type I reactions involve the interaction or electron transfer from the triplet state to biomolecules or organic species. However, like several dyes, triplet states can react with the O_2 molecule in the ground state or sensitize the production of singlet oxygen in a type II reaction [47]. These oxidizing species, viz., singlet oxygen, superoxides, peroxides, hydroxyl ions and radicals are the main components involved in the dye degradation process as shown in Figure 6c [48,49]. In the presence of electron donors, such as from the conduction band of ZnO on photoexcitation, the dye degradation is enhanced, due to the enhanced production of ROS. In particular, the high number of electrons transferred to oxygen radicals create hydroxyl radicals that finally condense into water.

Photocatalytic studies using ZnO nanorods have shown that the aspect ratio of the nanorods affects the surface defects, which in turn influences the photocatalytic activity of the ZnO nanorods [50]. It has to be considered that nanomaterials with higher aspect ratios tend to have lower quantities of surface defects, unless deliberately induced during the growth process. In contract, nanoparticles of very small sizes harbor surface defects, irrespective of the synthesis condition due to very large specific surfaces. Here, ZnO A with distinct facets harboring V_{Zn} or V_O, appears to have enhanced photocatalytic degradation properties. Under photoexcitation, electrons are trapped at V_O^+/V_O^{++}, and holes are trapped at V_{Zn}^{--} on the surface of the nanoparticle, as shown in Figure 6c. Then, the trapped electron is transferred to the oxygen molecule or radical, creating ROS via a type II mechanism. Simultaneously, holes are trapped at V_{Zn}; since the dyes studied are cationic, they have an affinity to the negatively charged V_{Zn} that accepts an electron from the dye molecule, leading to the oxidation of the dye. However, the transfer of the electron from the dye to ZnO nanoparticles can only occur in the singlet state of the dye [51]. The dye oxidation or degradation process is then finalized by the produced ROS. The presence of both types of defects, along with sharp edges appears to be the main reason for a more efficient dye-degradation compared to ZnO B and ZnO C that harbor mainly V_O^{++}.

Since ZnO is a n-type semiconductor, in an aqueous environment, hydroxyl groups and O_2 radicals attach to the surface of these nanoparticles and produce an upward band-bending [52]. In consequence, a more efficient separation of the photogenerated charges is possible. Another outcome is the suppression of the band-to-band transition due to the increase in the depletion region size around the nanoparticle, whereupon the probability of electron capture at defect sites increases [31]. The upward-band-bending phenomenon is therefore beneficial to the photocatalytic activity.

The photocatalytic dye degradation of methylene blue in Figure 6b is ~20% higher for all the ZnO samples after 3 h of sunlight exposure, and reaches almost 100% degradation for ZnO C. In the present case, the shielding effect due to adsorption of dye molecules is not observed, most probably due to the lack of affinity of the dye to the ZnO nanoparticle surface. Another important effect that increases the rate and efficiency of dye degradation is the wider excitation spectrum of sunlight ranging from the infrared to the UV. While ZnO normally does not absorb visible light due to its large bandgap, the defect states present in nanoparticles enable absorption of photons of visible light around 550 nm in ZnO A (Figure 4a). In fact, a wavelength of 550 nm also corresponds to the most intense emission wavelength of the solar spectrum.

3.4. Adsorption Dependent Photocatalytic Degradation

The decrease in fuchsin concentration in the presence of the ZnO nanoparticles under dark conditions highlights a clear adsorption of the dye by ZnO nanoparticles (Figure 6a). Interestingly, ZnO B and ZnO C both appear to have similar high adsorption capacities compared to ZnO A, for which it is significantly lower. This lower adsorption is likely due to the lower surface-to-volume ratio and its smooth crystalline surfaces or facets with certain low-adsorption crystal faces [23,24]. In addition, surface defects can play an important role in the adsorption of dye molecules [18]. Due to its high specific surface, ZnO B will adsorb the highest amount of dye molecules. Regardless of the ZnO morphology and defect states, it appears that methylene blue is not adsorbed by the ZnO nanoparticles under dark conditions for up to 3 h in Figure 7b.

The effect of the preliminary adsorption of dye molecules on the photocatalytic activity of ZnO nanoparticles was investigated. Figure 7c,d show the photocatalytic degradation curves of the three different ZnO samples after 24 h of adsorption in dark conditions. Unlike the photocatalytic degradation curves in Figure 5, C/C_0 begins at 0.90 for ZnO A, 0.65 for ZnO B and 0.625 for ZnO C, due to the preliminary adsorption of fuchsin in Figure 7c. However, in the case of methylene blue, only 12% of the dye was adsorbed for all the three samples after 24 h, (Figure 7d), with C/C_0 starting at 0.88.

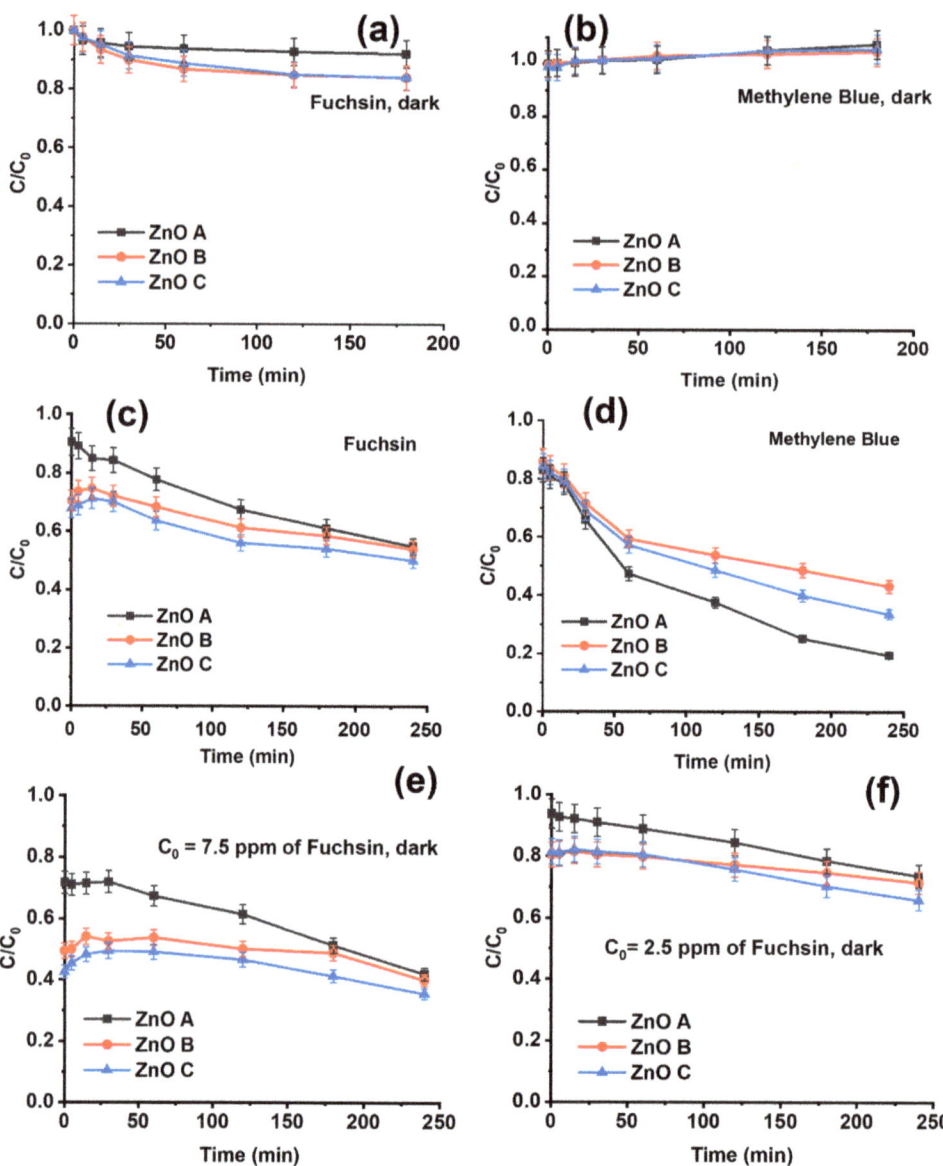

Figure 7. Adsorption study under dark conditions with 10 mL aqueous solution of 5 ppm (**a**) basic fuchsin, and (**b**) methylene blue using 1 mg of the three different ZnO samples for 3 h. Photocatalytic degradation under UV light of a 10 mL aqueous solution containing 5 ppm of dye with 1 mg of ZnO nanoparticles, after reaching adsorption equilibrium (24 h) for (**c**) basic fuchsin, (**d**) methylene blue and for other basic-fuchsin dye concentrations of (**e**) 7.5 ppm and (**f**) 2.5 ppm.

The higher adsorption of fuchsin molecules on the surface of ZnO B and ZnO C influences their photocatalytic degradation properties significantly. In Figure 7c, an increase in the fuchsin dye concentration is visible immediately under exposure to UV-light, i.e., an increase in C/C_0, except for ZnO A that showed very low adsorption kinetics in Figure 7a. The reason for the increase in dye concentration under UV exposure is the probable desorption of fuchsin from the surface of ZnO B and ZnO C samples. As basic fuchsin is a cationic molecule, extra surface holes and radicalization of hydroxyl groups under UV exposure can weaken the ionic bonds on the surface, releasing loosely bound molecules. The presence of dye molecule on the surface of the samples acts as a shield against photocatalytic activity, which explains the low removal in the initial stages for ZnO B and ZnO C samples. For methylene blue on the other hand, the desorption was not observable due to low adsorption in Figure 7d. Nevertheless, the overall photocatalytic activity appears less efficient, when photocatalysis follows adsorption.

In Figure 7e,f, we observe that the adsorption capacity does not depend on the dye concentration in the solution. In fact, similar quantities of fuchsin were adsorbed by the different ZnO samples in all three experiments (including the 5 ppm fuchsin concentration in Figure 7c). The adsorption capacity of ZnO A is 0.5 µg/mg and 1.5 µg/mg for both ZnO B and ZnO C samples. Thus, the adsorption capacity is solely determined by the number of active sites available on the ZnO nanoparticle surface, regulated by the synthesis conditions that influence the specific surface and surface defects.

In order to study the possible desorption of the dye molecules from the ZnO nanoparticle surface, as well as the delay in the photocatalytic removal of the dye after adsorption, the ZnO nanoparticles were removed from the fuchsin solution. Then, the ZnO nanoparticles were exposed to UV light for three hours to degrade the fuchsin-dye molecules adsorbed on their surfaces. After this treatment, the fuchsin solution was returned to the Petri dish containing the UV-treated ZnO nanoparticles. Figure 8a shows the photocatalytic activity under UV light of the remaining fuchsin using the UV-treated ZnO nanoparticles. In fact, the UV treatment degraded the adsorbed dye molecules, which consequently suppressed both the initial delay in photocatalytic activity and fuchsin desorption. This clearly suggests that the desorption of dye is responsible for the initial increase in dye concentration in the early stages of photocatalysis. However, both the initial adsorption and the photocatalytic degradation of fuchsin by ZnO B and ZnO C in Figure 8a are significantly lower than in Figure 7, suggesting that dye molecules may have modified the surface of the nanoparticles.

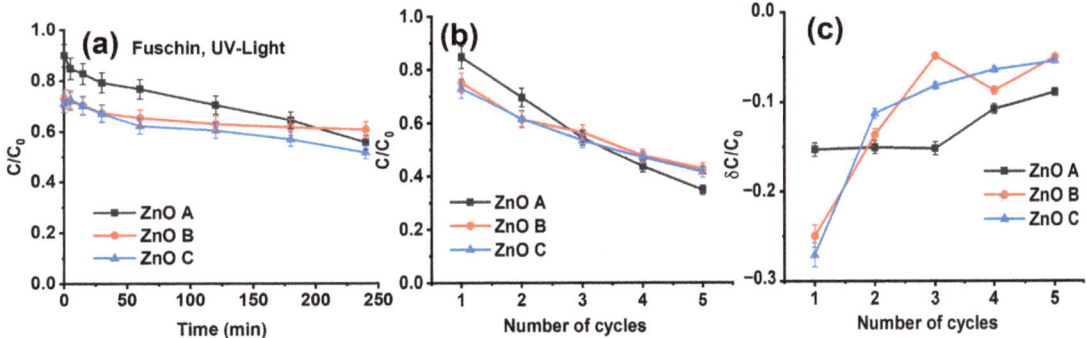

Figure 8. Photocatalytic activity of 1 mg of ZnO nanoparticles in 10 mL aqueous solution of 5 ppm fuchsin under UV light. Prior to photocatalysis, adsorption was carried out for 24 h followed by separating the nanoparticles from the solution, drying them and exposing them for 3 h to UV light. (**a**) The nanoparticles were then returned to the original solution to assess their photocatalytic activity. To check their reusability, the nanoparticles were then isolated and exposed to UV light for 3 h to clean their surfaces, and then returned to the original solution containing fuchsin. The procedure was repeated for four cycles of adsorption and UV-light exposure. (**b**) The cumulative dye removal and (**c**) the C/C_0 difference for each cycle.

In Figure 8b,c, the experiments consist of several cycles of fuchsin-dye adsorption followed by removal of the adsorbed dye molecules by exposing the ZnO nanoparticles to UV light after isolating them from the dye solution. After a UV-light treatment of 3 h, the treated ZnO nanoparticles were returned to the dye solution for further adsorption. For each cycle, the concentration of fuchsin in the solution decreased as shown in Figure 8b.

Therefore, adsorbed fuchsin is indeed degraded during UV exposure. However, as shown in Figure 8c, the adsorption of fuchsin after the first cycle for ZnO B and ZnO C decreases significantly. For ZnO A, the adsorption is the same for the first three cycles and reduces slightly for cycles 4 and 5. Nevertheless, ZnO A has the highest adsorption capacity after five cycles, among the three samples. In fact, ZnO B and ZnO C exhibited the best adsorption capacity during the first test, but their adsorption capacity decreased after UV treatment, showing a lower reusability. Between these two samples, ZnO C presents the lowest adsorption per cycle that is reduced with every subsequent cycle. This result suggests that the ZnO A, owing to its sharp edges and facets, is more robust and probably exhibits a different surface chemistry than samples ZnO B and ZnO C, which needs to be further investigated. These results suggest that the adsorption of dye molecules on ZnO before photocatalysis tends to give only an initial boost in dye removal. However, these nanoparticles tend to perform less efficiently when adsorption precedes photocatalysis, which, in turn, adversely affects their reusability.

Table 3 summarizes the photocatalytic degradation efficiencies of various nanoparticles, including commercial ZnO nanoparticles against MB. There are no reports of fuchsin degradation with pure ZnO nanoparticles, to the best of our knowledge. In general, all the presented ZnO nanoparticles tend to efficiently degrade MB. Nevertheless, in other studies, the quantity of nanoparticles is higher, i.e., ranging from 15 mg to 4 g compared to the present study, where only 1 mg of nanoparticle was used. The C_0 of MB in all the solutions was between 5–10 mg/L, similar to the present study. Our nanoparticles appear to efficiently degrade under both UV light and sunlight, with a higher degradation efficiency under sunlight. In particular, ZnO A demonstrates a very high dye degradation efficiency for 1 mg of the photocatalyst after 3 h.

Table 3. Comparison of the photocatalytic efficiency of various ZnO nanoparticles against MB.

Photocatalyst	Illumination	C/C_0	Crystal Size (nm)	Time (min)	C_0	Mass	Ref
ZnO (commercial Nanograph, Chicago, IL, USA)	UV lamp	0.3	30–50	-	5	4 g	[53]
ZnO (green synthesized)	sunlight	0	200	90	5	2 g	[54]
ZnO (commercial Aldrich)	sunlight	0.03	50	140	10	25 mg	[55]
ZnO (commercial Sigma Aldrich, St. Louis, MO, USA)	Xenon lamp	0.03	50	150	10	25 mg	[56]
ZnO (green synthesized)	UV lamp	0.03	50	100	5	15 mg	[57]
ZnO A (this work)	Sunlight/UV light	0/0.08	100–300	180	5	1 mg	
ZnO B (this work)	Sunlight/UV light	0/0.15	5	180	5	1 mg	
ZnO C (this work)	Sunlight/UV light	0/0.22	50–200	180	5	1 mg	

4. Conclusions

In this work, we have investigated the photocatalytic activity of ZnO nanoparticles synthesized via three different sol-gel routes, which determined their shape, size and defect states. Solution-based synthesis routes are favorable, owing to the facile handling and controllability of the reaction mechanisms. These nanoparticles were successfully applied to UV-light- and solar-driven photocatalysis of fuchsin and methylene blue. The differences in the morphology due to the synthesis conditions resulted in bandgap states that were the determining factors in the photocatalytic efficiency of these nanoparticles. In fact, the defect states in ZnO create bandgap states, acting as traps that improve charge separation, instrumental in creating ROS. Even though, smaller ZnO nanoparticles with high specific surfaces are extremely efficient photocatalysts; nevertheless, their reusability for several cycles of dye degradation remains limited. On the other hand, ZnO nanoparticles with sharp edges and facets demonstrate a much higher dye-degradation efficiency, reaching nearly 100% in some cases. Clearly, the presence of V_{Zn} along with V_O enhances the photocatalytic degradation. In addition, their reusability emphasizes the sustainability of the photocatalytic process. Adsorption preceding photocatalysis appears to be an effective method to initially boost dye removal. However, the dye molecules adhering to the surface of the spherical nanoparticles tend to modify their surface chemistry and prove detrimental to their photocatalytic efficiency and reusability. In that regard, facetted ZnO nanoparticles appear to be good candidates for the removal of more recalcitrant organic pollutants, such as nitrates and pharmaceuticals. Future works will consider a methodology to immobilize these nanoparticles on a support in order to facilitate their recovery after exhaustion for several cycles of dye degradation.

Supplementary Materials: The following supporting information can be downloaded at: https://www.mdpi.com/article/10.3390/nano13131998/s1, Figure S1: Calibration curves with UV-Vis spectrometer and Lovibond photometer for (a) basic fuchsin at 560 nm and (b) methylene blue at 660 nm; Figure S2: Emission spectrum of the UV-lamp with maximum intensity at 365 nm and an output power; Figure S3: (a) Solar irradiance and (b) UV index during the experiments under direct sunlight. These parameters were almost identical in order to ensure that the photocatalytic degradation was carried out in similar conditions in order to compare the photobleaching, as well as the photocatalytic activity. Methylene blue and fuchsin were manipulated on different days and the red and black curves indicate the solar irradiance and UV index on those days.; Figure S4: Size distribution histograms obtained from TEM images of Figure 2 for ZnO A, ZnO B and ZnO C.

Author Contributions: Conceptualization, Y.H.; methodology, Y.H.; validation, Y.H., P.R. and E.R.; formal analysis, Y.H., E.E., K.N., R.H. and C.B.; investigation, Y.H., K.N., E.R. and P.R.; resources, P.R. and E.R.; data curation, Y.H.; writing—original draft preparation, Y.H.; writing—review and editing, Y.H., P.R. and E.R.; supervision, P.R.; project administration, P.R.; funding acquisition, P.R. and E.R. All authors have read and agreed to the published version of the manuscript.

Funding: This research has been supported by the European Regional Development Fund project, Archimedes foundation, grant number TK134 "EQUiTANT", Eesti Maaülikool (EMÜ Bridge Funding (P200030TIBT)), and T210013TIBT "PARROT mobility program". We thank the EU-H2020 research and innovation program under grant agreement no. 1029 supporting the Transnational Access Activity within the framework NFFA-Europe.

Data Availability Statement: Not applicable.

Acknowledgments: Laetitia Rapenne is acknowledged for some TEM analyses.

Conflicts of Interest: The authors declare no conflict of interest.

References

1. Gupta, V.K.; Mittal, A.; Gajbe, V.; Mittal, J. Adsorption of basic fuchsin using waste materials—Bottom ash and deoiled soya—As adsorbents. *J. Colloid Interface Sci.* **2008**, *319*, 30–39. [CrossRef]
2. Rafatullah, M.; Sulaiman, O.; Hashim, R.; Ahmad, A. Adsorption of methylene blue on low-cost adsorbents: A review. *J. Hazard. Mater.* **2010**, *177*, 70–80. [CrossRef]
3. Jjagwe, J.; Olupot, P.W.; Menya, E.; Kalibbala, H.M. Synthesis and Application of Granular Activated Carbon from Biomass Waste Materials for Water Treatment: A Review. *J. Bioresour. Bioprod.* **2021**, *6*, 292–322. [CrossRef]
4. Obey, G.; Adelaide, M.; Ramaraj, R. Biochar derived from non-customized matamba fruit shell as an adsorbent for wastewater treatment. *J. Bioresour. Bioprod.* **2022**, *7*, 109–115. [CrossRef]
5. Houas, A.; Lachheb, H.; Ksibi, M.; Elaloui, E.; Guillard, C.; Herrmann, J.-M. Photocatalytic degradation pathway of methylene blue in water. *Appl. Catal. B Environ.* **2001**, *31*, 145–157. [CrossRef]
6. Black, D.M. *Bibliography of Work on the Photocatalytic Removal of Hazardous Compounds from Water and Air*; National Renewable Energy Lab.: Golden, CO, USA, 1994.
7. Lee, K.M.; Lai, C.W.; Ngai, K.S.; Juan, J.C. Recent developments of zinc oxide based photocatalyst in water treatment technology: A review. *Water Res.* **2016**, *88*, 428–448. [CrossRef] [PubMed]
8. Becker, J.; Raghupathi, K.R.; St. Pierre, J.; Zhao, D.; Koodali, R.T. Tuning of the Crystallite and Particle Sizes of ZnO Nanocrystalline Materials in Solvothermal Synthesis and Their Photocatalytic Activity for Dye Degradation. *J. Phys. Chem. C* **2011**, *115*, 13844–13850. [CrossRef]
9. Yang, Z.; Zhang, J.; Zhang, W.; Zhou, Q.; Shen, J.; Huang, Y. Stearic acid modified Co2+-doped ZnO: The construction of micro-nano structure for its superhydrophobic performance and visible-light photocatalytic degradation of methylene blue. *J. Clean. Prod.* **2023**, *382*, 135391. [CrossRef]

10. Bhapkar, A.; Prasad, R.; Jaspal, D.; Shirolkar, M.; Gheisari, K.; Bhame, S. Visible light driven photocatalytic degradation of methylene blue by ZnO nanostructures synthesized by glycine nitrate auto combustion route. *Inorg. Chem. Commun.* **2023**, *148*, 110311. [CrossRef]
11. Perez-Cuapio, R.; Alberto Alvarado, J.; Juarez, H.; Sue, H.J. Sun irradiated high efficient photocatalyst ZnO nanoparticles obtained by assisted microwave irradiation. *Mater. Sci. Eng. B* **2023**, *289*, 116263. [CrossRef]
12. Das, T.K.; Ghosh, S.K.; Das, N.C. Green synthesis of a reduced graphene oxide/silver nanoparticles-based catalyst for degradation of a wide range of organic pollutants. *Nano-Struct. Nano-Objects* **2023**, *34*, 100960. [CrossRef]
13. Das, T.K.; Das, N.C. Advances on catalytic reduction of 4-nitrophenol by nanostructured materials as benchmark reaction. *Int. Nano Lett.* **2022**, *12*, 223–242. [CrossRef]
14. Chang, Y.-N.; Zhang, M.; Xia, L.; Zhang, J.; Xing, G. The Toxic Effects and Mechanisms of CuO and ZnO Nanoparticles. *Materials* **2012**, *5*, 2850–2871. [CrossRef]
15. Vandebriel, R.; de Jong, W. A review of mammalian toxicity of ZnO nanoparticles. *Nanotechnol. Sci. Appl.* **2012**, *5*, 61–71. [CrossRef] [PubMed]
16. Mohammed, Y.H.; Holmes, A.; Haridass, I.N.; Sanchez, W.Y.; Studier, H.; Grice, J.E.; Benson, H.A.E.; Roberts, M.S. Support for the Safe Use of Zinc Oxide Nanoparticle Sunscreens: Lack of Skin Penetration or Cellular Toxicity after Repeated Application in Volunteers. *J. Investig. Dermatol.* **2019**, *139*, 308–315. [CrossRef]
17. Skocaj, M.; Filipic, M.; Petkovic, J.; Novak, S. Titanium dioxide in our everyday life; is it safe? *Radiol. Oncol.* **2011**, *45*, 227–247. [CrossRef]
18. Liu, F.; Leung, Y.H.; Djurišić, A.B.; Ng, A.M.C.; Chan, W.K. Native Defects in ZnO: Effect on Dye Adsorption and Photocatalytic Degradation. *J. Phys. Chem. C* **2013**, *117*, 12218–12228. [CrossRef]
19. Gurylev, V.; Perng, T.P. Defect engineering of ZnO: Review on oxygen and zinc vacancies. *J. Eur. Ceram. Soc.* **2021**, *41*, 4977–4996. [CrossRef]
20. Rauwel, E.; Galeckas, A.; Soares, M.R.; Rauwel, P. Influence of the Interface on the Photoluminescence Properties in ZnO Carbon-Based Nanohybrids. *J. Phys. Chem. C* **2017**, *121*, 14879–14887. [CrossRef]
21. Nagpal, K.; Rapenne, L.; Wragg, D.S.; Rauwel, E.; Rauwel, P. The role of CNT in surface defect passivation and UV emission intensification of ZnO nanoparticles. *Nanomater. Nanotechnol.* **2022**, *12*, 18479804221079419. [CrossRef]
22. Nassar, S.J.M.; Wills, C.; Harriman, A. Inhibition of the Photobleaching of Methylene Blue by Association with Urea. *ChemPhotoChem* **2019**, *3*, 1042–1049. [CrossRef]
23. Kaneti, Y.V.; Zhang, Z.; Yue, J.; Zakaria, Q.M.D.; Chen, C.; Jiang, X.; Yu, A. Crystal plane-dependent gas-sensing properties of zinc oxide nanostructures: Experimental and theoretical studies. *Phys. Chem. Chem. Phys.* **2014**, *16*, 11471–11480. [CrossRef]
24. Nicholas, N.J.; Franks, G.V.; Ducker, W.A. Selective Adsorption to Particular Crystal Faces of ZnO. *Langmuir* **2012**, *28*, 7189–7196. [CrossRef] [PubMed]
25. Ni, B.; Wang, X. Face the Edges: Catalytic Active Sites of Nanomaterials. *Adv. Sci.* **2015**, *2*, 1500085. [CrossRef] [PubMed]
26. Zhong, Y.; Djurišić, A.B.; Hsu, Y.F.; Wong, K.S.; Brauer, G.; Ling, C.C.; Chan, W.K. Exceptionally Long Exciton Photoluminescence Lifetime in ZnO Tetrapods. *J. Phys. Chem. C* **2008**, *112*, 16286–16295. [CrossRef]
27. Rao, F.; Zhu, G.; Zhang, W.; Xu, Y.; Cao, B.; Shi, X.; Gao, J.; Huang, Y.; Huang, Y.; Hojamberdiev, M. Maximizing the Formation of Reactive Oxygen Species for Deep Oxidation of NO via Manipulating the Oxygen-Vacancy Defect Position on $(BiO)_2CO_3$. *ACS Catal.* **2021**, *11*, 7735–7749. [CrossRef]
28. Dong, B.; Qi, Y.; Cui, J.; Liu, B.; Xiong, F.; Jiang, X.; Li, Z.; Xiao, Y.; Zhang, F.; Li, C. Synthesis of $BaTaO_2N$ oxynitride from Ba-rich oxide precursor for construction of visible-light-driven Z-scheme overall water splitting. *Dalton Trans.* **2017**, *46*, 10707–10713. [CrossRef]
29. Maeda, K.; Lu, D.; Domen, K. Solar-Driven Z-scheme Water Splitting Using Modified $BaZrO_3$–$BaTaO_2N$ Solid Solutions as Photocatalysts. *ACS Catal.* **2013**, *3*, 1026–1033. [CrossRef]
30. Chen, K.; Xiao, J.; Vequizo, J.J.M.; Hisatomi, T.; Ma, Y.; Nakabayashi, M.; Takata, T.; Yamakata, A.; Shibata, N.; Domen, K. Overall Water Splitting by a $SrTaO_2N$-Based Photocatalyst Decorated with an Ir-Promoted Ru-Based Cocatalyst. *J. Am. Chem. Soc.* **2023**, *145*, 3839–3843. [CrossRef]
31. Nagpal, K.; Rauwel, E.; Estephan, E.; Soares, M.R.; Rauwel, P. Significance of Hydroxyl Groups on the Optical Properties of ZnO Nanoparticles Combined with CNT and PEDOT:PSS. *Nanomaterials* **2022**, *12*, 3546. [CrossRef]
32. Rauwel, E.; Nilsen, O.; Rauwel, P.; Walmsley, J.C.; Frogner, H.B.; Rytter, E.; Fjellvåg, H. Oxide Coating of Alumina Nanoporous Structure Using ALD to Produce Highly Porous Spinel. *Chem. Vap. Depos.* **2012**, *18*, 315–325. [CrossRef]
33. Khachadorian, S.; Gillen, R.; Choi, S.; Ton-That, C.; Kliem, A.; Maultzsch, J.; Phillips, M.R.; Hoffmann, A. Effects of annealing on optical and structural properties of zinc oxide nanocrystals. *Phys. Status Solidi B* **2015**, *252*, 2620–2625. [CrossRef]
34. Wang, J.B.; Huang, G.J.; Zhong, X.L.; Sun, L.Z.; Zhou, Y.C.; Liu, E.H. Raman scattering and high temperature ferromagnetism of Mn-doped ZnO nanoparticles. *Appl. Phys. Lett.* **2006**, *88*, 252502. [CrossRef]
35. Gluba, M.A.; Nickel, N.H.; Karpensky, N. Interstitial zinc clusters in zinc oxide. *Phys. Rev. B* **2013**, *88*, 245201. [CrossRef]
36. Zeferino, R.S.; Flores, M.B.; Pal, U. Photoluminescence and Raman Scattering in Ag-doped ZnO Nanoparticles. *J. Appl. Phys.* **2011**, *109*, 014308. [CrossRef]
37. Hejral, U.; Franz, D.; Volkov, S.; Francoual, S.; Strempfer, J.; Stierle, A. Identification of a Catalytically Highly Active Surface Phase for CO Oxidation over PtRh Nanoparticles under Operando Reaction Conditions. *Phys. Rev. Lett.* **2018**, *120*, 126101. [CrossRef]

38. Gao, Q.; Dai, Y.; Li, C.; Yang, L.; Li, X.; Cui, C. Correlation between oxygen vacancies and dopant concentration in Mn-doped ZnO nanoparticles synthesized by co-precipitation technique. *J. Alloys Compd.* **2016**, *684*, 669–676. [CrossRef]
39. Souissi, A.; Amlouk, M.; Khemakhem, H.; Guermazi, S. Deep analysis of Raman spectra of ZnO:Mo and ZnO:In sprayed thin films along with LO and TA+LO bands investigation. *Superlattices Microstruct.* **2016**, *92*, 294–302. [CrossRef]
40. Davis, K.; Yarbrough, R.; Froeschle, M.; White, J.; Rathnayake, H. Band gap engineered zinc oxide nanostructures via a sol–gel synthesis of solvent driven shape-controlled crystal growth. *RSC Adv.* **2019**, *9*, 14638–14648. [CrossRef]
41. Makuła, P.; Pacia, M.; Macyk, W. How To Correctly Determine the Band Gap Energy of Modified Semiconductor Photocatalysts Based on UV–Vis Spectra. *J. Phys. Chem. Lett.* **2018**, *9*, 6814–6817. [CrossRef]
42. Gong, Y.; Andelman, T.; Neumark, G.F.; O'Brien, S.; Kuskovsky, I.L. Origin of defect-related green emission from ZnO nanoparticles: Effect of surface modification. *Nanoscale Res. Lett.* **2007**, *2*, 297. [CrossRef]
43. Rauwel, E.; Galeckas, A.; Rauwel, P.; Sunding, M.F.; Fjellvåg, H. Precursor-Dependent Blue-Green Photoluminescence Emission of ZnO Nanoparticles. *J. Phys. Chem. C* **2011**, *115*, 25227–25233. [CrossRef]
44. Wang, X.; Yin, L.; Liu, G.; Wang, L.; Saito, R.; Lu, G.Q.; Cheng, H.-M. Polar interface-induced improvement in high photocatalytic hydrogen evolution over ZnO–CdS heterostructures. *Energy Environ. Sci.* **2011**, *4*, 3976–3979. [CrossRef]
45. Wöll, C. The chemistry and physics of zinc oxide surfaces. *Prog. Surf. Sci.* **2007**, *82*, 55–120. [CrossRef]
46. Diaspro, A.; Chirico, G.; Usai, C.; Ramoino, P.; Dobrucki, J. Photobleaching. In *Handbook of Biological Confocal Microscopy*; Pawley, J.B., Ed.; Springer: Boston, MA, USA, 2006; pp. 690–702. [CrossRef]
47. Pena Luengas, S.L.; Marin, G.H.; Aviles, K.; Cruz Acuña, R.; Roque, G.; Rodríguez Nieto, F.; Sanchez, F.; Tarditi, A.; Rivera, L.; Mansilla, E. Enhanced singlet oxygen production by photodynamic therapy and a novel method for its intracellular measurement. *Cancer Biother. Radiopharm.* **2014**, *29*, 435–443. [CrossRef] [PubMed]
48. Dumanović, J.; Nepovimova, E.; Natić, M.; Kuča, K.; Jaćević, V. The Significance of Reactive Oxygen Species and Antioxidant Defense System in Plants: A Concise Overview. *Front. Plant Sci.* **2021**, *11*, 552969. [CrossRef]
49. Shunji, K.; Masaharu, M.; Masao, K. Studies of the Transient Intermediates in the Photoreduction of Methylene Blue. *Bull. Chem. Soc. Jpn.* **1964**, *37*, 117–124. [CrossRef]
50. Zhang, X.; Qin, J.; Xue, Y.; Yu, P.; Zhang, B.; Wang, L.; Liu, R. Effect of aspect ratio and surface defects on the photocatalytic activity of ZnO nanorods. *Sci. Rep.* **2014**, *4*, 4596. [CrossRef]
51. Shvalagin, V.V.; Stroyuk, A.L.; Kuchmii, S.Y. Photochemical synthesis of ZnO/Ag nanocomposites. *J. Nanoparticle Res.* **2007**, *9*, 427–440. [CrossRef]
52. Zhang, Z.; Yates, J.T., Jr. Band Bending in Semiconductors: Chemical and Physical Consequences at Surfaces and Interfaces. *Chem. Rev.* **2012**, *112*, 5520–5551. [CrossRef]
53. Pala, A.; Kurşun, G. The effect of different nanocatalysts for photocatalytic degradation of Methylene blue. *Environ. Eng.* **2019**, *6*, 79–83. [CrossRef]
54. Nnodim, U.J.; Adogwa, A.A.; Akpan, U.G.; Ani, I.J. Photocatalytic Degradation of Methylene Blue Dye with Green Zinc Oxide Doped with Nitrogen. *J. Clin. Rheumatol. Res.* **2022**, *2*, 59.
55. Chekir, N.; Benhabiles, O.; Tassalit, D.; Laoufi, N.A.; Bentahar, F. Photocatalytic degradation of methylene blue in aqueous suspensions using TiO_2 and ZnO. *Desalination Water Treat.* **2016**, *57*, 6141–6147. [CrossRef]
56. Ranjbari, A.; Kim, J.; Kim, J.H.; Yu, J.; Demeestere, K.; Heynderickx, P.M. Enhancement of commercial ZnO adsorption and photocatalytic degradation capacity of methylene blue by oxygen vacancy modification: Kinetic study. *Catal. Today* **2023**, *413*, 113976. [CrossRef]
57. Priyadharshini, S.S.; Shubha, J.P.; Shivalingappa, J.; Adil, S.F.; Kuniyil, M.; Hatshan, M.R.; Shaik, B.; Kavalli, K. Photocatalytic Degradation of Methylene Blue and Metanil Yellow Dyes Using Green Synthesized Zinc Oxide (ZnO) Nanocrystals. *Crystals* **2022**, *12*, 22. [CrossRef]

Disclaimer/Publisher's Note: The statements, opinions and data contained in all publications are solely those of the individual author(s) and contributor(s) and not of MDPI and/or the editor(s). MDPI and/or the editor(s) disclaim responsibility for any injury to people or property resulting from any ideas, methods, instructions or products referred to in the content.

MDPI
St. Alban-Anlage 66
4052 Basel
Switzerland
www.mdpi.com

Nanomaterials Editorial Office
E-mail: nanomaterials@mdpi.com
www.mdpi.com/journal/nanomaterials

Disclaimer/Publisher's Note: The statements, opinions and data contained in all publications are solely those of the individual author(s) and contributor(s) and not of MDPI and/or the editor(s). MDPI and/or the editor(s) disclaim responsibility for any injury to people or property resulting from any ideas, methods, instructions or products referred to in the content.

www.ingramcontent.com/pod-product-compliance
Lightning Source LLC
LaVergne TN
LVHW070631100526
838202LV00012B/782